T0324139

Derivative Pricing

A Problem-Based Primer

CHAPMAN & HALL/CRC
Financial Mathematics Series

Aims and scope:

The field of financial mathematics forms an ever-expanding slice of the financial sector. This series aims to capture new developments and summarize what is known over the whole spectrum of this field. It will include a broad range of textbooks, reference works and handbooks that are meant to appeal to both academics and practitioners. The inclusion of numerical code and concrete real-world examples is highly encouraged.

Series Editors

M.A.H. Dempster
Centre for Financial Research
Department of Pure Mathematics and Statistics
University of Cambridge

Dilip B. Madan
Robert H. Smith School of Business
University of Maryland

Rama Cont
Department of Mathematics
Imperial College

C++ for Financial Mathematics
John Armstrong

Model-free Hedging
A Martingale Optimal Transport Viewpoint
Pierre Henry-Labordere

Stochastic Finance
A Numeraire Approach
Jan Vecer

Equity-Linked Life Insurance
Partial Hedging Methods
Alexander Melnikov, Amir Nosrati

High-Performance Computing in Finance
Problems, Methods, and Solutions
M. A. H. Dempster, Juho Kanniainen, John Keane, Erik Vynckier

Derivative Pricing
A Problem-Based Primer
Ambrose Lo

For more information about this series please visit: *https://www.crcpress.com/ Chapman-and-HallCRC-Financial-Mathematics-Series/book-series/CHFINANCMTH*

Derivative Pricing
A Problem-Based Primer

Ambrose Lo

CRC Press
Taylor & Francis Group
Boca Raton London New York

CRC Press is an imprint of the
Taylor & Francis Group, an **informa** business

A CHAPMAN & HALL BOOK

CRC Press
Taylor & Francis Group
6000 Broken Sound Parkway NW, Suite 300
Boca Raton, FL 33487-2742

First issued in paperback 2020

© 2018 by Taylor & Francis Group, LLC
CRC Press is an imprint of Taylor & Francis Group, an Informa business

No claim to original U.S. Government works

Version Date: 20180518

ISBN 13: 978-0-367-73421-3 (pbk)
ISBN 13: 978-1-138-03335-1 (hbk)

This book contains information obtained from authentic and highly regarded sources. Reasonable efforts have been made to publish reliable data and information, but the author and publisher cannot assume responsibility for the validity of all materials or the consequences of their use. The authors and publishers have attempted to trace the copyright holders of all material reproduced in this publication and apologize to copyright holders if permission to publish in this form has not been obtained. If any copyright material has not been acknowledged please write and let us know so we may rectify in any future reprint.

Except as permitted under U.S. Copyright Law, no part of this book may be reprinted, reproduced, transmitted, or utilized in any form by any electronic, mechanical, or other means, now known or hereafter invented, including photocopying, microfilming, and recording, or in any information storage or retrieval system, without written permission from the publishers.

For permission to photocopy or use material electronically from this work, please access www.copyright.com (http://www.copyright.com/) or contact the Copyright Clearance Center, Inc. (CCC), 222 Rosewood Drive, Danvers, MA 01923, 978-750-8400. CCC is a not-for-profit organization that provides licenses and registration for a variety of users. For organizations that have been granted a photocopy license by the CCC, a separate system of payment has been arranged.

Trademark Notice: Product or corporate names may be trademarks or registered trademarks, and are used only for identification and explanation without intent to infringe.

Visit the Taylor & Francis Web site at
http://www.taylorandfrancis.com

and the CRC Press Web site at
http://www.crcpress.com

Contents

III Epilogue

9 General Properties of Option Prices

Appendix A   Standard Normal Distribution Table

Appendix B   Solutions to Odd-Numbered End-of-Chapter Problems

Bibliography

Index

List of Figures

List of Tables

Preface

Derivatives, which are financial instruments whose value depends on or is "derived" from (hence the name "derivatives") other more basic underlying variables, have become commonplace in financial markets all over the world. The proliferation of these relatively new financial innovations, options in particular, has underscored the ever-increasing importance of derivative literacy among a wide range of users that span students, practitioners, regulators, and researchers, all of whom are in need of a fundamental understanding of the mechanics, typical uses, and pricing theory of derivatives, though to different extents. Despite the diversity of such users, existing books on the subject have predominantly catered to only a very specific group of users and gone to two extremes. They either adopt a mostly descriptive approach to the intrinsically technical subject of derivatives, with occasional number crunching and slavish applications of pricing formulas taken without proof, or are preoccupied with sophisticated mathematical techniques from such areas as random processes and stochastic calculus, which can be inaccessible to students or practitioners lacking the necessary background and undesirably obscure the underlying conceptual ideas. Neither the "black box" approach nor the "purely mathematical" approach is of much pedagogical value.

Being an outgrowth of my lecture notes for a course entitled *ACTS:4380 Mathematics of Finance II* offered at the University of Iowa for advanced undergraduate and beginning graduate students in actuarial science, this book is a solid attempt to strike a balance between the two aforementioned methods to teach and learn derivatives, and to meet the needs of different types of readers. Adopting a mathematically rigorous yet widely accessible approach that will appeal to a wide variety of audience, the book is conceptually driven and strives to demystify the mechanics of typical derivatives and the fundamental mechanism of derivative pricing methodologies that should be part of the toolkit of every professional these days. This is accomplished by a combination of lucid explanations of the theory and assumptions behind common derivative pricing models, repeated emphasis on a small set of core ideas (e.g., no-arbitrage principle, replication, risk-neutral pricing), and a careful selection of fully worked-out illustrative examples and end-of-chapter problems. Readers of this book will leave with a firm understanding of "what" derivatives are, "how" and, more importantly, "why" derivatives are used and derivative pricing works.

Here is the skeleton of this book, divided into three parts.

- Part I (Chapters 1 to 3) lays the conceptual groundwork of the whole book by setting up the terminology of derivatives commonly encountered in the literature and introducing the definition, mechanics, typical use, and payoff structures of the two primary groups of derivatives, namely, forwards and options, which bestow upon their holders an obligation and a right to trade an underlying asset at a fixed price on a fixed date, respectively. Particular emphasis is placed on how and why a derivative works in a given scenario of interest. In due course, we also present the all-important *no-arbitrage assumption* and the method of *pricing by replication*. In loose terms, the former says that the prices of derivatives should be such that the market does not admit "free lunches," and the latter implements the former using the common-sense idea that if two derivatives possess the same payoff structure at expiration, they must enjoy the same initial price. Underlying the pricing and hedging of derivatives throughout this book, these two vehicles are applied in this part to determine the fair price of a forward, where "fair" is meant in the sense that the resulting price permits no free lunch opportunities.

- Whereas the pricing of forwards is model-independent in that it works for any asset price distribution, the pricing of options depends critically on the probabilistic behavior of the future asset price. In Part II (Chapters 4 to 8), which is the centerpiece of this book, we build upon the background material in Part I and tackle option pricing in two stages—first in the discrete-time binomial tree model (Chapter 4), which is simple, intuitive, and easy to implement, then in the technically more challenging continuous-time Black-Scholes model (Chapters 5 to 8). In this part, the no-arbitrage assumption and the method of replication continue to play a vital role in valuing options and lead to the celebrated *risk-neutral pricing formula*, which asserts that the price of a (European) derivative can be computed as its expected payoff at expiration in a risk-neutral sense, discounted at the risk-free interest rate. The implementation and far-reaching implications of the method of risk-neutral valuation for the pricing and hedging of derivatives are explored in Chapters 6 to 8.

- Finally, we end in Part III (Chapter 9) with a description of some general properties satisfied by option prices when no asset price model is prescribed. Even in this model-free framework setting, there is a rich theory describing the no-arbitrage properties universally satisfied by option prices. Although this part can be read prior to studying Part II, you will find that what you learn from Part II, especially the notion of an exchange option in Chapter 8, will provide you with surprisingly useful insights into the connections between different options.

It deserves mention that this book, as a primer, is indisputably not encyclopedic in scope. The choice of topics is geared towards the derivatives portion (Topics 6 to 10) of the Society of Actuaries' *Investment and Financial Markets* (IFM) Exam,[i] which is typically taken by advanced undergraduate students in actuarial science and allied disciplines. The theory of random processes and stochastic calculus, while conducive to understanding the pricing theory of derivatives in full but often an insurmountable barrier to first-time learners, is not covered in the book, neither are credit and interest rate derivatives (which, without doubt, are important in practice). By concentrating on the most essential conceptual ideas, we realize the huge "payoff" of being able to disseminate these core ideas to readers with minimal mathematical background; it is understandable that individuals interested in using and pricing derivatives nowadays come from a wide variety of background. To be precise, readers are only assumed to have taken a calculus-based probability and statistics course at the level of Hogg, Tanis and Zimmerman (2014) or Hogg, McKean and Craig (2013), where the basic notions of random variables, expectations, variances, are taught, and be able to perform simple discounted cash flow calculations as covered in a theory of interest or corporate finance course. With these modest prerequisites, this book is self-contained, with the necessary mathematical ideas presented progressively as the book unfolds. For readers interested in more advanced aspects of the use and pricing of derivatives, this book will provide them with a springboard for performing further studies in this burgeoning field.

It is widely acknowledged that the best way to learn a subject deeply is to test your understanding with a number of meaningful exercises. With this in mind, this primer lives up to its name and features an abundance of illustrative in-text examples and end-of-chapter problems (to be precise, 177 examples and 209 problems) on different aspects of derivatives. These problems are of a diverse nature and varying levels of difficulty (harder ones

---

[i]Section 4.5, Section 7.3 (the portion on the Black-Scholes equation), Subsection 8.1.2, and Section 9.1 are beyond the scope of the Exam IFM syllabus.

are labeled as [**HARDER!**]); while many emphasize calculating quantities such as payoffs, prices, profits, a primitive skill that most students in a derivatives course need to acquire, at least for exam purposes (in this respect, this book is an ideal exam preparation aid for students who will write Exam IFM), some concern more theoretical aspects of using and pricing derivatives, and consist of true-or-false items or derivations of formulas. All of these problems can be worked out in a pen-and-paper environment with the aid of a scientific calculator and a standard normal distribution function calculator (an example is `https://www.prometric.com/en-us/clients/soa/pages/mfe3f_calculator.aspx`, which is designed for students who will take Exam IFM.). If you do not have Internet access, you may use the less precise standard normal distribution table provided in Appendix A of this book. Readers who attempt these examples and problems seriously will benefit from a much more solid understanding of the relevant topics. To help you check your answers, full solutions to all odd-numbered end-of-chapter problems are provided in Appendix B. A solutions manual with solutions to all problems is available to qualified instructors.

It would be remiss of me not to thank my past ACTS:4380 students for personally class testing earlier versions of the book manuscript and many of the end-of-chapter problems, as well as my esteemed colleague, Professor Elias S.W. Shiu, at the University of Iowa, for sharing with me his old ACTS:4380 notes and questions, from which some of the examples and problems in this book were motivated. I am also grateful to the Society of Actuaries and Casualty Actuarial Society for kindly allowing me to reproduce their past and sample exam questions, of which they own the sole copyright, and which have proved instrumental in illustrating ideas in derivative pricing. Doctoral student Zhaofeng Tang at the University of Iowa merits a special mention for his professional assistance with some of the figures in this book and for meticulously proofreading part of the book manuscript. All errors that remain, typographical or otherwise, are solely mine. To help improve the content of the book, I would deeply appreciate it if you could bring any potential errors you have identified to my attention; my email address is `ambrose-lo@uiowa.edu`. For readers' benefits, an erratum and updates to the book will be maintained on my web page at `https://sites.google.com/site/ambroseloyp/publications/derivative-pricing`.

It is my sincere hope that this book will not only introduce you to the fascinating world of derivatives, but also to instill in you a little of the enthusiasm I have for this subject since my undergraduate studies. Welcome and may the fun begin!

Ambrose Lo, PhD, FSA, CERA
Iowa City, IA
May 2018

Symbols

Symbol Description

S or $S(0)$	time-0 price of an underlying asset	$F_{t,T}$	time-t price of a forward maturing at time T
$S(t)$	time-t price of an underlying asset	$F_{t,T}^{\mathrm{obs}}$	time-t observed price of a forward maturing at time T
$S^a(0)$	time-0 ask price of an underlying asset	$F_{t,T}^{\mathrm{fair}}$	time-t fair price of a forward maturing at time T
$S^b(0)$	time-0 bid price of an underlying asset	V	time-0 price of a generic derivative
$X(t)$	time-t exchange rate	$V^{\max}$	time-0 price of a maximum contingent claim
K	strike price of an option		
K_1	strike price of a gap option	$V^{\min}$	time-0 price of a minimum contingent claim
K_2	payment trigger of a gap option	C	time-0 price of a generic call
r	continuously compounded risk-free interest rate (per annum)	C^E	time-0 price of a generic European call
r^b	continuously compounded borrowing rate (per annum)	C^A	time-0 price of a generic American call
r^l	continuously compounded lending rate (per annum)	$C(K,T)$	time-0 price of a K-strike T-year call
i	effective annual interest rate (per annum)	$C^{\mathrm{gap}}(K_1,K_2)$	time-0 price of a K_1- strike K_2-trigger generic gap call
$\mathrm{PV}_{t,T}$	time-t (present) value of cash flows between time t and time$T$	$C(S(t),K,t,T)$	time-t price of a K-strike call maturing at time T when the time-t stock price is $S(t)$
$\mathrm{FV}_{t,T}$	time-T (future) value of cash flows between time t and time$T$	P	time-0 price of a generic put
		P^E	time-0 price of a generic European put
T	maturity time of a generic derivative	P^A	time-0 price of a generic American put
T_f	maturity time of a futures contract	$P^{\mathrm{gap}}(K_1,K_2)$	time-0 price of a K_1-strike K_2-trigger generic gap put
T_1	maturity time of a compound option	$P(S(t),K,t,T)$	time-t price of a K-strike put maturing at time T when the time-t stock price is $S(t)$
T_2	maturity time of the underlying option of a compound option	BS	Black-Scholes pricing function
		σ	volatility of an underlying asset
$F_{t,T}^P$	time-t price of a prepaid forward maturing at time T	σ_{option}	volatility of an option
		Δ	delta of a generic derivative

Δ_C	delta of a generic call	$\rho(X, Y)$	correlation coefficient between random variables X and Y
Δ_P	delta of a generic put		
Γ	gamma of a generic derivative	$N(\cdot)$	distribution function of the standard normal distribution
Γ_C	gamma of a generic call		
Γ_P	gamma of a generic put	$N'(\cdot)$	density function of the standard normal distribution
Ω	elasticity of a generic derivative		
Ω_C	elasticity of a generic call	$:=$	defined as
Ω_P	elasticity of a generic put	LHS	left-hand side
$\mathbb{E}[X]$	expectation of random variable X	RHS	right-hand side
$\mathrm{Var}(X)$	variance of random variable X	x_+	positive part of real number x
$\mathrm{Cov}(X, Y)$	covariance between random variables X and Y	1_A	indicator function of event A
		ZCB	zero-coupon bond

Part I

Conceptual Foundation on Derivatives

1

An Introduction to Forwards and Options

Chapter overview: This book centers on *derivatives*—not those you have studied in your calculus class, but those instruments whose values (or, in financial parlance, payoffs) depend on or are "derived" from other more basic underlying variables, such as the price of a stock or a commodity, an interest rate, a currency exchange rate, or even non-financial variables like the temperature of a city on a particular day or your semester GPA in college, so long as they can be quantified. These underlying driving variables are called the *underlying asset*, or in short, the *underlying*. To begin our systematic study of derivatives, this preparatory chapter provides a conceptual introduction to the two predominant types of derivatives, namely, forwards and options, and sets up the basic derivatives terminology that will be intensively used throughout this book. These primitive derivatives are basic building blocks of more sophisticated financial instruments that will be examined in later chapters. For each of forwards and options, we analyze its mechanics, typical use, and, most importantly, the structure and derivation of its payoff.

1.1 Forwards

Instead of spoon-feeding you directly with the terms and conditions of different kinds of derivatives, a much better way to get you to better understand their use and mechanics is through a daily-life example most familiar to actuarial students.

Motivating example.

It is now July 1, 3017 (assume that Doomsday has not yet come). You are planning to take Exam X, a notoriously difficult actuarial exam, in November 3017 and will be in severe need of a study manual in August for your exam preparation (assume that 3 months are enough for your study!). According to the Internet (assume that the Internet still exists in the next millennium), the current price of the study manual for Exam X is $250 (per unit). You are worried, however, that the price of the study manual would rise dramatically over the next month, which would seriously jeopardize your financial well-being as a frugal student. Is there a way to protect yourself against adverse increases in the price of the study manual?

Now consider the following "contract" available in the market:

> You will be provided with one copy of the study manual on August 1, 3017 in exchange for $275. The contract needs to be signed today and will bind you legally[i] to pay $275 for the study manual on August 1.

Such a "contract" is the quintessence of a forward contract, the subject of the current section.

---

[i]We ignore counterparty risk, i.e., the risk that one or more parties fail to deliver on the obligations imposed by the derivative.

What is a forward?

In general, a *forward contract*, or simply a *forward*, is an agreement between two parties (also called counterparties) to buy or sell an asset at a certain time in the future at a certain price. The contract is agreed upon and signed today, with one party (the buyer) undertaking to pay the stipulated price in exchange for the asset and the other party (the seller) liable to deliver the asset and receive the same stipulated price at the particular future time specified in the contract. This way, a forward effectively removes the uncertainty that one needs to face with respect to the future price of the underlying asset. Here is a good mnemonic that helps you remember and make sense of a forward:

> A forward allows you to look *forward* in time and lock in the transaction price of an asset you wish to buy or sell in the future.

With respect to terminology, the essential elements in a forward are:

Term	Description	Study Manual Example
Underlying asset	The asset that the forward is based on	Study manual
Expiration date (or maturity date)	The time when the forward has to be settled	August 1, 3017
Forward price	The price that will be paid when the forward is settled	$275
Long or short	If you buy (resp. sell) the forward, you are said to be in a *long* (resp. *short*) forward position. The words long and short can serve as adjectives or verbs, so we may also say that you long (resp. short) the forward.	You are long the forward

The most important ingredient in a forward is the forward price. In this introductory chapter, we take the forward price as given. In Chapter 2, we will study in detail how the forward price can be determined in a way that is fair in a sense to be formalized. Notation-wise, we denote generically by $S(t)$ the time-t price of the underlying asset (the use of "S" is motivated by the fact that the most common underlying asset is a stock; in fact the terms "underlying asset" and "underlying stock" will be used interchangeably in the sequel when no confusion arises), and by $F_{0,T}$ the forward price that is determined at time 0 and will be paid at time T. Unless otherwise stated, in this book we will always measure time in years.

The payoff of a forward contract.

A very useful way to describe a derivative, a forward in particular, is to look at its *payoff*. This concept permeates virtually the entire book and is one of the most important dimensions of a derivative. Loosely speaking, the payoff of a derivative position at a particular time point at or before the expiration date is defined as the value of the position. An equivalent, but more informative and concrete definition that lends itself to practical computations is that the payoff of a derivative at a given point of time equals the amount of money that its holder would have had *if* he/she completely liquidated his/her position, i.e., to sell whatever he/she is holding, and buy back whatever he/she has owed.

Let us apply this line of reasoning to determine the payoff of a T-year long forward (i.e.,

Time	Transaction	Cash Flow
0	You buy a forward.	0
T	You settle the forward by paying the seller the forward price, $F_{0,T}$.	$-F_{0,T}$
	The seller delivers the asset to you.	0
	You sell the asset at the *spot price* (i.e., market price) of the asset, $S(T)$.	$+S(T)$
	Total: $S(T) - F_{0,T}$	

TABLE 1.1
Cash flows associated with a long forward (physical settlement).

the forward expires in T years). Under a long forward, at expiration you pay $F_{0,T}$ for the underlying asset and can cash out by selling the asset at its market price for $S(T)$ (see Table 1.1). On a net basis, you receive $S(T) - F_{0,T}$. This is the payoff of a T-year long forward:

$$\text{Payoff of long forward} = \underbrace{\text{Spot price}}_{\text{receive (income)}} - \underbrace{\text{forward price}}_{\text{pay (cost)}} = S(T) - F_{0,T}. \qquad (1.1.1)$$

The cash flows associated with a short forward are the exact opposite of those of a long forward. Upon receiving the forward price from the buyer, purchasing the underlying asset at the market price for $S(T)$, and delivering the asset thus purchased to the buyer, you as the short forward party receive a payoff of

$$\text{Payoff of short forward} = \text{Forward price} - \text{spot price} = F_{0,T} - S(T), \qquad (1.1.2)$$

which is a mirror image of (1.1.1). You may notice that the sum of (1.1.1) and (1.1.2) is exactly zero. In other words, a forward is a zero-sum game, with what the buyer gains exactly offset by what the seller loses.

Example 1.1.1. (SOA Exam IFM Introductory Derivatives Sample Question 68: Payoff of long/short forward) For a nondividend-paying stock index, the current price is 1100 and the 6-month forward price is 1150. Assume the price of the stock index in 6 months will be 1210.

Which of the following is true regarding forward positions in the stock index?

(A) Long position gains 50

(B) Long position gains 60

(C) Long position gains 110

(D) Short position gains 60

(E) Short position gains 110

Solution. By (1.1.1) and (1.1.2), the long forward will gain $1210 - 1150 = \boxed{60}$ while the short forward will gain $1150 - 1210 = -60$, i.e., a loss of 60. (**Answer: (B)**) □

It deserves mention that a forward imposes upon its counterparties an *obligation* to

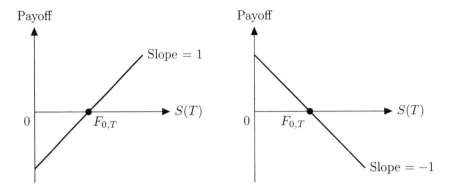

FIGURE 1.1.1
Payoff diagrams of a long forward (left) and a short forward (right).

execute the contract at expiration. Under a long forward position, you are obligated to pay the forward price at the expiration date, regardless of how high or low the then market price is. Even if the ending asset price is less than the forward price, in which case you suffer a loss, you still need to settle the forward. This is a characteristic of a forward that distinguishes it from other derivatives.

Payoff diagrams.

In Part I of this book, we will make intensive use of a visual device called a *payoff diagram*, which displays graphically the payoff of a derivative as a function of the underlying asset price at the point of interest. As we shall see soon, the payoff diagrams of many derivatives possess salient features that say a lot about the properties of the derivatives. In fact, one of the best ways to remember a derivative is arguably to associate it with the geometry of its payoff diagram.

Given (1.1.1) and (1.1.2), the payoff diagrams of a long forward and a short forward are sketched in Figure 1.1.1. They are straight lines cutting the horizontal axis at the forward price $F_{0,T}$ with a slope of 1 and -1, respectively. The linearity arises from a commitment to trade the underlying asset in the future. In contrast, we will see in Section 1.2 that options, the other primary category of derivatives, are characterized by nonlinear payoff functions because of their intrinsic optionality.

Physical vs cash settlement.

In the above discussions, we assumed that at the expiration of the forward, the seller indeed delivers the asset to the buyer for $F_{0,T}$, and the buyer indeed cashes out by selling the asset for $S(T)$, realizing a payoff of $S(T) - F_{0,T}$. This mode of settling a forward involving genuine delivery of the underlying asset is known as *physical* settlement of a forward. On second thought, however, it would make no difference if the seller directly gives the buyer an amount of $S(T) - F_{0,T}$ in cash (this amount can be positive or negative); see Table 1.2. The holder of the long forward will be in the same financial position as when there is genuine physical delivery of the underlying. Settling a forward this way by means of a direct exchange of cash without the corresponding delivery of the underlying asset is known as *cash* (or *financial*) settlement. Both physical settlement and cash settlement apply not only to forwards but also to other derivatives.

The two modes of settlement have their relative merits and demerits. Physical settlement aids our understanding of the mechanics of a derivative and the reason for using that

Time	Transaction	Cash Flow
0	You buy a forward.	0
T	The seller pays you $S(T) - F_{0,T}$.	$S(T) - F_{0,T}$

TABLE 1.2
Cash settlement of a long forward.

derivative in the first place. It is also critical to the fundamental derivation of its payoff formula (the long forward payoff formula $S(T) - F_{0,T}$, while taken by many for granted, can be derived from basics and need not be taken as a definition). However, a physical transaction in practice will incur possibly significant transaction costs and is applicable only to assets for which physical delivery is possible. This excludes more abstract assets such as temperature, your GPA, interest rates, market volatility, all of which cannot be physically traded. Cash settlement is widely applicable and simple to understand, but it does not lend itself to deriving the payoff of a derivative from first principles. Both modes of settlement will be used in the later part of this book, although cash settlement will play the dominant role.

Main motivation for using derivatives: Hedging and speculation.

With the payoff formula determined above, we are in a much better position to understand the typical use of a forward. Back to the study manual example, we again denote by $S(T)$ the unit price of the study manual on August 1. Suppose that the study manual is a necessity to you, so that you must buy the study manual however cheap or expensive it is. Under this assumption, your cash flow on August 1 will be $-S(T)$, which is inherently random. If you couple your position with a long forward on the study manual, then your payoff on August 1 will be constant at $-S(T) + (S(T) - F_{0,T}) = -F_{0,T}$. In effect, what you have to pay to own the study manual is transformed from a random amount of $S(T)$ to the constant forward price $F_{0,T}$. The long forward therefore allows you to *hedge* against the risk associated with the price of the study manual on August 1 by locking in the transaction price.

More generally, *hedging* aims to use the cash flows generated by a derivative to mitigate the (often random) cash flows from a given position. In the case of a forward, the payoff from the long forward counteracts the random cash outflow as a result of buying the underlying at its random future price. With a forward, you are shielded from the price risk of the underlying; the cash flow uncertainty arising from the future transaction has been completely eliminated. Long forwards are commonly employed by business companies to hedge against the future price risk associated with their necessary production inputs and outputs. They are also popular among importers or exporters concerned with the fluctuations of foreign exchange rates in the future.

In contrast to hedging, *speculation* refers to the attempt to profit from anticipated price movements of the underlying asset without an existing exposure in the asset. In the study manual example, if your mom, whom I assume does not need the study manual for her own use (unless she happens to be an actuarial science professor!), is convinced that the price of the study manual will skyrocket (resp. plummet) in August, she may take advantage of her belief by buying (resp. selling) a forward. Should the price of the study manual go up (resp. go down) as expected, she can reap a huge payoff from her long (resp. short) forward position. Of course, she may also suffer a huge loss from the forward if the price of the study manual does not move in the direction she predicts.

1.2   Options

1.2.1   Call Options

Motivation.

Recall from Section 1.1 that if you enter into a forward, you are required under any circumstances to settle the forward at the expiration date, even if doing so results in a negative payoff. In the study manual example, if the price of the study manual on August 1, 3017 is only $260, it seems stupid (of course, with the benefit of hindsight!) to buy the manual at the forward price of $275—you will pay $15 more! You may wonder:

> Is there a derivative that entitles you the *option* to buy an asset if and only if doing so is to your interest, i.e., you have the right to walk away from the deal if you wish to?

Derivatives that give you an option to buy or sell an asset are naturally called *options* (pun intended!). There are two main types of options, namely, call options and put options. They are intended to provide one-sided protection against unfavorable movements in the price of the underlying asset.

Definition and terminology of a call option.

A *call* option (or a *call* in short) gives its holder the *right*, but not the obligation, to buy the underlying asset at a prespecified price. Here is a cheap mnemonic:

> By means of a call, you have the option to "call" the asset from someone and own it.

In addition to "underlying asset" and "long or short," which enjoy the same definition as a forward, the following terms are very useful in specifying an option in general and a call in particular.

1. *Exercise:* Exercising an option refers to the act of making use of the option to trade the underlying asset. If you exercise a call, you pay a certain price (see the next point) in return for the underlying asset.

2. *Strike price:* The *strike price*, also known as *exercise price* or simply the *strike*, is the price specified in the contract at which the option holder can exercise the option. In the case of a call option, the strike price is what the holder can choose to pay for the underlying asset. We shall denote the strike price of an option generically by K.

3. *Expiration:* The expiration date is the time when the option holder must decide whether to exercise the option. As in Section 1.1, the generic symbol for the time to expiration is T.

4. *Exercise style:* Exercise style is a characteristic unique to options but not forwards. It governs when an option can be exercised.

   - If the option can only be exercised at the expiration date, it is called a *European*-style or simply *European* option.

   - If the option can be exercised anytime prior to or on the expiration date, it is called an *American*-style or, in short, *American* option.

- A *Bermudan*-style option allows its holder to exercise it during only specified periods, but not throughout the entire life of the option.

Here is a mnemonic:

$$\boxed{E}\text{uropean} \;=\; \boxed{E}\text{xpiration}$$
$$\boxed{A}\text{merican} \;=\; \boxed{A}\text{nytime}$$
$$\boxed{B}\text{ermudan} \;=\; \boxed{B}\text{between}$$

Unless otherwise stated, options that are studied in the remainder of this book are European, whose analysis is mathematically more tractable. It is easy to conceive that American options are more valuable than otherwise identical European and Bermudan options because of the higher degree of freedom in relation to the time of exercise. We will see in Chapter 9 that in some cases, it turns out that the opportunity to early exercise an option is not exploited, so that European and otherwise identical American and Bermudan options do share the same price.

Payoff of a call option.

To determine the payoff of a long European call at the expiration time T, consider two cases:

Case 1. If $S(T) > K$, you will exercise the call to use the strike price K to buy the asset and sell it in the market immediately for the market price $S(T)$, earning $S(T) - K$. Here, we are assuming that the call is settled physically.

Case 2. If $S(T) \le K$, it is irrational to exercise the call to buy the asset for K because you are better off buying the asset directly in the market at the cheaper price of $S(T)$. In this case, you will opt out of the call, which becomes worthless, and your payoff is zero.

To sum up, the payoff of holding a European[ii] call is

$$\text{Long call payoff} \;=\; \begin{cases} S(T) - K, & \text{if } S(T) > K \\ 0, & \text{if } S(T) \le K \end{cases}$$

$$=\; \boxed{(S(T) - K)_+},$$

where $(\cdot)_+$ is the positive part function defined by $x_+ := \max(x, 0)$ for any $x \in \mathbb{R}$. For example, $6_+ = 6$, $11.3_+ = 11.3$, $(-4)_+ = 0$, and $(-9.21)_+ = 0$. The positive part function signifies the optionality inherent in options and is characteristic of option payoffs.

The payoff diagram of a long call is graphed in Figure 1.2.1. Unlike the payoff of a long forward, that of a long call is zero if the spot price is less than the strike price, and becomes a straight line emanating from the strike price and pointing to the right with a slope of 1 otherwise. The turning point at the strike price is an indication of the optionality of the call—the right is exercised when and only when $S(T) \ge K$. In essence, the payoff of a long call position is obtained from that of a long forward position by zeroing the negative portion of the latter and keeping the positive part intact.

[ii]The time-t payoff of a T-year K-strike American call is $(S(t) - K)_+$.

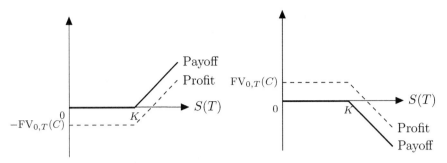

FIGURE 1.2.1
Payoff and profit diagrams of a long call (left) and a short call (right).

Time	Transaction	Cash Flow
0	You buy a call and pay the call premium C.	$-C$
T	You can decide whether to exercise the call, depending on the spot price.	$(S(T) - K)_+$

TABLE 1.3
Cash flows associated with a long European call position.

Option premium.

By entering into a call option, you have the option to benefit from a rise in the price of asset without the need for bearing any downside risk. In other words, your payoff at expiration must be non-negative regardless of the spot price. For the call to be fair to the seller, you must pay him/her an amount called the *option premium*, which we designate as C, at the inception of the option as a form of compensation—the upside protection offered by the call is not free! The premium is also called the *price* of the call option. Table 1.3 shows what and when you pay or receive if you buy a European call.

Parenthetically, the determination of the fair price of an option is a highly nontrivial task, much more technically complicated than that of the fair price of a forward, and is the subject of Part II of this book.

Profit of a call option.

The payoff of a derivative captures its value at a particular instant and ignores the cash flows at other times, particularly the initial cost to set up the derivative. A more global perspective on a derivative is through its *profit*, defined as the payoff of the derivative less the *future value* of cash flows at previous time points.

In the case of a long call, the profit is obtained by subtracting the future value of the option premium from the option payoff, i.e.,

$$\boxed{\text{Long call profit} = (S(T) - K)_+ - \text{FV}_{0,T}(C),} \qquad (1.2.1)$$

where $\text{FV}_{0,T}(\cdot)$ denotes the future value from time 0 to time T. If r represents the continuously compounded risk-free interest rate,[iii] then $\text{FV}_{0,T}(C) = Ce^{rT}$; if i is the effective annual interest rate, then $\text{FV}_{0,T}(C) = C(1 + i)^T$.

We note in passing that because a forward entails no initial investment by definition, its profit coincides with its payoff given in (1.1.1) and (1.1.2).

[iii]Throughout this book, interest rates are measured per annum.

Example 1.2.1. (Given the profit, find the spot price) You buy a 50-strike 6-month call option on a stock at a price of 5. The continuously compounded risk-free interest rate is 5%.

At the end of 6 months, the profit from the long call option is 4.

Calculate the price of the stock at the end of 6 months.

Ambrose's comments:

Problems directly asking for the payoff or profit of a derivative given other inputs via using formulas like (1.1.1) and (1.2.1) are somewhat too simple. A more realistic and non-trivial problem may turn things around and ask that you determine the unknown stock price at expiration to achieve a certain profit.

Solution. In terms of the 6-month stock price $S(0.5)$, the profit from the long call is $(S(0.5) - 50)_+ - 5e^{0.05(0.5)}$. Because the positive part function takes different expressions depending on the sign of the argument, we solve the equation $(S(0.5) - 50)_+ - 5e^{0.05(0.5)} = 4$ for $S(0.5)$ in two cases:

Case 1. If $S(0.5) < 50$, then the equation is $-5e^{0.05(0.5)} = 4$, which has no solution.

Case 2. If $S(0.5) \geq 50$, then the equation becomes $(S(0.5) - 50) - 5e^{0.05(0.5)} = 4$, which gives $S(0.5) = 59.1266$.

The only possible solution is $S(0.5) = \boxed{59.1266}$. $\square$

Example 1.2.2. (SOA Exam IFM Introductory Derivatives Sample Question 11: Comparing the profits of three calls) Stock XYZ has the following characteristics:

- The current price is 40.

- The price of a 35-strike 1-year European call option is 9.12.

- The price of a 40-strike 1-year European call option is 6.22.

- The price of a 45-strike 1-year European call option is 4.08.

The annual effective risk-free interest rate is 8%.

Let S be the price of the stock one year from now.

All call positions being compared are long.

Determine the range for S such that the 45-strike call produces a higher profit than the 40-strike call, but a lower profit than the 35-strike call.

(A) $S < 38.13$

(B) $38.13 < S < 40.44$

(C) $40.44 < S < 42.31$

(D) $S > 42.31$

(E) The range is empty

We present two solutions, one algebraic and one geometric.

Solution 1 (Algebraic). Denote by Pr_K the profit of a K-strike 1-year European call. We first express each Pr_K in terms of S:

$$\mathrm{Pr}_{35} = (S - 35)_+ - 9.12(1.08) = (S - 35)_+ - 9.8496 \qquad (1.2.2)$$
$$\mathrm{Pr}_{40} = (S - 40)_+ - 6.22(1.08) = (S - 40)_+ - 6.7176 \qquad (1.2.3)$$
$$\mathrm{Pr}_{45} = (S - 45)_+ - 4.08(1.08) = (S - 45)_+ - 4.4064 \qquad (1.2.4)$$

To find the range for S such that $\mathrm{Pr}_{40} < \mathrm{Pr}_{45} < \mathrm{Pr}_{35}$, we consider each of the following cases:

Case 1. If $S < 35$, then all three positive part functions in (1.2.2), (1.2.3), and (1.2.4) vanish, so we have $\mathrm{Pr}_{35} < \mathrm{Pr}_{40} < \mathrm{Pr}_{45}$, which is not the order we seek.

Case 2. If $35 \leq S < 40$, then $\mathrm{Pr}_{35} = S - 44.8496$, $\mathrm{Pr}_{40} = -6.7176$, and $\mathrm{Pr}_{45} = -4.4064$. To get the order of profits we want, we solve $-6.7176 < -4.4064 < S - 44.8496$, resulting in $S > 40.4432$, which contradicts the assumption $35 \leq S < 40$.

Case 3. If $40 \leq S < 45$, then $\mathrm{Pr}_{35} = S - 44.8496$, $\mathrm{Pr}_{40} = S - 46.7176$, and $\mathrm{Pr}_{45} = -4.4064$. To get the desired order of profits, we solve

$$S - 46.7176 < -4.4064 < S - 44.8496,$$

which gives $40.4432 < S < 42.3112$. Note that this range for S satisfies the hypothesis $40 \leq S < 45$.

Case 4. If $S \geq 45$, then $\mathrm{Pr}_{35} = S - 44.8496$, $\mathrm{Pr}_{40} = S - 46.7176$, and $\mathrm{Pr}_{45} = S - 49.4064$. To get the order of profits we want, we solve

$$S - 46.7176 < S - 49.4064 < S - 44.8496,$$

in which the first inequality has no solution. Therefore, in this range of S, it is impossible that $\mathrm{Pr}_{40} < \mathrm{Pr}_{45} < \mathrm{Pr}_{35}$.

In conclusion, the required range for S is $S \in \boxed{(40.44, 42.31)}$. $\square$

Solution 2 (Geometric). We first compute the future value of the 3 call premiums:

Strike	FV of Call Premium
35	$9.12 \times 1.08 = 9.8496$
40	$6.22 \times 1.08 = 6.7176$
45	$4.08 \times 1.08 = 4.4064$

Now we sketch the three call profit functions in Figure 1.2.2 (the diagram is not and need not be drawn to scale). Observe that the three profit functions are ordered as $\mathrm{Pr}_{40} < \mathrm{Pr}_{45} < \mathrm{Pr}_{35}$ when and only when S lies in the interval AB. To find the horizontal coordinates of the two points, A and B, recall that the slope of the profit function of a call must be 1 when the asset price is beyond the strike price. With this in mind, we can calculate the horizontal coordinates of A and B as $35 + [-4.41 - (-9.85)] = 40.44$ and $40 + [-4.41 - (-6.72)] = 42.31$. The required answer is $\boxed{S \in (40.44, 42.31)}$.

(Answer: (C)) $\square$

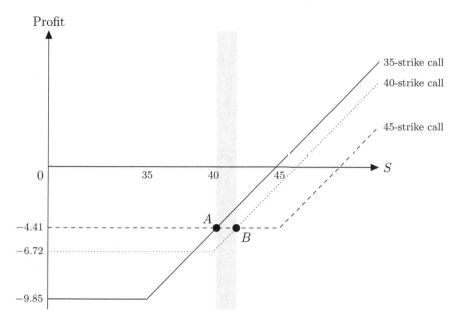

FIGURE 1.2.2
The profit functions of the three calls in Example 1.2.2.

Remark. (i) The SOA provides the current stock price (= 40), which is not necessary for solving the problem.

(ii) See whether you prefer a "geometric" or an "algebraic" solution.

Written call option.

What happens to the seller (a.k.a. *writer*) of a call option? He/she receives the option premium from the outset and has the *obligation*, not the right, to deliver the underlying asset if the buyer chooses to exercise the call. The payoff and profit of the writer are exactly the opposite of those of the buyer:

$$\text{Short call payoff} \quad = \quad -(S(T) - K)_+,$$
$$\text{Short call profit} \quad = \quad \text{FV}_{0,T}(C) - (S(T) - K)_+.$$

Note that the payoff of the call writer is always non-positive. This explains why he/she must insist on receiving compensation in the form of the call option premium at time 0.

Example 1.2.3. (Call vs forward) Investor A wrote a 104-strike 1-year call option whose price is 2. Investor B entered into a 1-year long forward with a forward price of 105.

The continuously compounded risk-free interest rate is 5%.

It turns out that Investor A and Investor B earned the same profit.

Calculate the 1-year stock price.

Solution. Equating the profits of the short call and long forward, we solve the equation

$$\underbrace{2e^{0.05} - (S(1) - 104)_+}_{\text{short call}} = \underbrace{S(1) - 105}_{\text{long forward}}$$

for $S(1)$. Depending on whether $S(1) < 104$ or $S(1) \geq 104$, we have:

Case 1.   If $S(1) < 104$, then the equation above is $2e^{0.05} = S(1) - 105$, which implies $S(1) = 107.10$, a contradiction to the hypothesis $S(1) < 104$!

Case 2.   If $S(1) \geq 104$, then the equation becomes $2e^{0.05} - (S(1) - 104) = S(1) - 105$, resulting in $S(1) = \boxed{105.5513}$, which complies with the hypothesis $S(1) \geq 104$. $\square$

Example 1.2.4. (SOA Exam IFM Introductory Derivatives Sample Question 42: Profit of a written covered call) An investor purchases a nondividend-paying stock and writes a t-year, European call option for this stock, with call premium C. The stock price at time of purchase and strike price are both K.

Assume that there are no transaction costs.

The risk-free annual force of interest is a constant r. Let S represent the stock price at time t.

$S > K$.

Determine an algebraic expression for the investor's profit at expiration.

(A) Ce^{rt}

(B) $C(1 + rt) - S + K$

(C) $Ce^{rt} - S + K$

(D) $Ce^{rt} + K(1 - e^{rt})$

(E) $C(1 + r)^t + K[1 - (1 + r)^t]$

Solution. The time-0 investment is $S(0) - C = K - C$ and the time-t payoff is $S - (S - K)_+ = S - (S - K) = K$ because $S > K$. The profit at expiration is $K - (K - C)e^{rt} = \boxed{Ce^{rt} + K(1 - e^{rt})}$. **(Answer: (D))** $\square$

1.2.2   Put Options

Definition of a put option.

It is now August 1, 3017, and you have got hold of the study manual of Exam X via the long forward or the long call, whichever way you prefer. You are not yet done with Exam X, but you are pretty sure that you will pass. Why still keep the abominable study manual and not sell it to the market after the exam, say in December 3017? A put option that expires in December 3017 can come to your rescue.

Whereas call options discussed in the preceding subsection allow their holders to buy a particular asset in the future, a *put* option (or a *put* in short) gives its holder the right to *sell* a certain asset for a certain price at a certain date or in a time period. Here is a cheap mnemonic:

> By means of a put option, you have the option to "put" the underlying asset to someone.

All contractual terms that apply to a call (e.g., strike price, expiration date, exercise style, premium, etc.) easily carry over to a put. In particular, the strike price K of a put is the price for which the put holder may elect to *sell* the underlying asset. At the other end of the transaction, the *seller* of a put is a potential *buyer* of the asset. Should the put holder decide to exercise the put, the seller is *obligated* to buy the asset against his/her own will and suffer a negative payoff.

Payoff and profit of a put option.

In parallel with the call payoff, we consider two cases to determine the payoff of a European put at the expiration time T:

Case 1.   If $S(T) \leq K$, you will buy the asset for $S(T)$ and exercise the option to sell the asset for a higher price of K, earning $K - S(T)$.

Case 2.   If $S(T) > K$, it is inadvisable to exercise the put because of selling the asset (K) at less than what it is worth in the market ($S(T)$). You will then forfeit the put, in which case your payoff is zero.

Combining both cases, we see that the payoff of holding a European put option is

$$\text{Long put payoff} \quad = \quad \begin{cases} K - S(T), & \text{if } S(T) \leq K \\ 0, & \text{if } S(T) > K \end{cases}$$

$$= \quad \boxed{(K - S(T))_+}.$$

The profit of the long put is simply the excess of its payoff over the future value of the put premium:

$$\text{Long put profit} = (K - S(T))_+ - \text{FV}_{0,T}(P).$$

The payoff and profit diagrams of a long put and short put are given in Figure 1.2.3.

Example 1.2.5. (SOA Exam IFM Introductory Derivatives Sample Question 48: Warm-up payoff manipulations) For a certain stock, Investor A purchases a 45-strike call option while Investor B purchases a 135-strike put option. Both options are European with the same expiration date. Assume that there are no transaction costs.

If the final stock price at expiration is S, Investor A's payoff will be 12.

Calculate Investor B's payoff at expiration, if the final stock price is S.

(A) 0

(B) 12

(C) 36

(D) 57

(E) 78

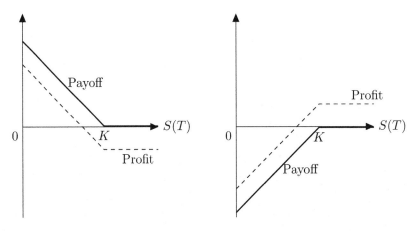

FIGURE 1.2.3
Payoff and profit diagrams of a long put (left) and a short put (right).

Solution. Considering Investor A, we have $(S - 45)_+ = 12$, or $S = 57$. For the same final stock price, the payoff of Investor B at expiration is $(135 - S)_+ = \boxed{78}$. **(Answer: (E))**   □

Example 1.2.6. (SOA Exam IFM Introductory Derivatives Sample Question 33: Payoff of American, European and Bermudan options) Several years ago, John bought three separate 6-month options on the same stock.

- Option I was an American-style put with strike price 20.

- Option II was a Bermudan-style call with strike price 25, where exercise was allowed at any time following an initial 3-month period of call protection.

- Option III was a European-style put with strike price 30.

When the options were bought, the stock price was 20.
    When the options expired, the stock price was 26.
    The table below gives the maximum and minimum stock prices during the 6-month period:

Time Period	1st 3 months of Option Term	2nd 3 months of Option Term
Maximum Stock Price	24	28
Minimum Stock Price	18	22

John exercised each option at the optimal time.
    Rank the three options, from highest to lowest payoff.

(A) I > II > III

(B) I > III > II

(C) II > I > III

(D) III > I > II

(E) III > II > I

Solution. The payoff of each option is determined and tabulated below:

Option	Remark	Payoff
I	• As an American option, it can be exercised at any time during the 6-month period. • As a put, its payoff is greatest when the stock price is smallest.	$20 - 18 = 2$
II	• As a Bermuda option, it can be exercised at any time during the second 3-month period. • As a call, its payoff is greatest when the stock price is greatest.	$28 - 25 = 3$
III	• As a European option, it can be exercised only at maturity. • We can directly apply the usual put payoff formula $(K - S(T))_+$.	$(30 - 26)_+ = 4$

It follows that the payoffs of the three options in descending order are $\boxed{\text{III} > \text{II} > \text{I}}$. (**Answer: (E)**)

Example 1.2.7. (SOA Exam IFM Introductory Derivatives Sample Question 35: Simple profit calculation for puts) A customer buys a 50-strike put on an index when the market price of the index is also 50. The premium for the put is 5. Assume that the option contract is for an underlying 100 units of the index.

Calculate the customer's profit if the index declines to 45 at expiration.

(A) -1000

(B) -500

(C) 0

(D) 500

(E) 1000

Ambrose's comments:

Be careful! The put option is on 100 units (not just one) of the index.

Solution. The customer's profit if the index declines to 45 at expiration is $100[(50 - 45)_+ - 5] = \boxed{0}$. (**Answer: (C)**)   □

Remark. This example makes no mention of the interest rate, so we have no choice but to assume it to be zero.

Example 1.2.8. (SOA Exam IFM Introductory Derivatives Sample Question 12: Long put, short put, same profit) Consider a European put option on a stock index without dividends, with 6 months to expiration and a strike price of 1,000. Suppose that the annual nominal risk-free rate is 4% convertible semiannually, and that the put costs 74.20 today.

  Calculate the price that the index must be in 6 months so that being long in the put would produce the same profit as being short in the put.

(A) 922.83

(B) 924.32

(C) 1,000.00

(D) 1,075.68

(E) 1,077.17

Ambrose's comments:

This example shares a similar spirit as Example 1.2.1, with the quantity of interest being the asset price at expiration resulting in a certain profit.

Solution. The profit of the long put is $(1,000 - S(0.5))_+ - 74.2(1.02)$, while that of the short put is $74.2(1.02) - (1,000 - S(0.5))_+$. Equating these two gives $(1,000 - S(0.5))_+ = 75.684$, which in turn implies that $S(0.5) = \boxed{924.32}$. (**Answer: (B)**)   □

Remark. If being long and being short in the same derivative share the same profit, both positions must have a zero profit.

Example 1.2.9. (SOA Exam IFM Introductory Derivatives Sample Question 62: Expected profit of a put) The price of an asset will either rise by 25% or fall by 40% in 1 year, with equal probability. A European put option on this asset matures after 1 year.

  Assume the following:

• Price of the asset today: 100

• Strike price of the put option: 130

- Put option premium: 7

- Annual effective risk free rate: 3%

Calculate the expected profit of the put option.

(A) 12.79

(B) 15.89

(C) 22.69

(D) 27.79

(E) 30.29

Solution. The 1-year payoff of the put option will be either $(130 - 100 \times 1.25)_+ = 5$ or $(130 - 100 \times 0.6)_+ = 70$ with equal probability. The expected profit is $(5 + 70)/2 - 7(1.03) = \boxed{30.29}$. (**Answer: (E)**) □

1.3 Classification of Derivatives

Thus far, we have introduced altogether six different positions: forwards, calls, and puts, each of which can be long or short. In this section, we present several ways to compare and contrast different derivatives. These easy comparisons also provide opportunities for us to review the material covered in Sections 1.1 and 1.2.

Universal comparison.

Derivatives (not only forwards and options) can be classified according to the following general criteria:

Criterion 1. *Long or short with respect to the underlying asset:* By definition, a position is *long* (resp. *short*) with respect to the underlying asset if it benefits from increases (resp. decreases) in the price of the underlying asset, or, mathematically speaking, has a payoff function which is increasing (resp. decreasing) in the asset price. For example, a long forward is long with respect to the underlying asset, but a long put is short.

Note that the usage of the words "long" and "short" here is different from that in Sections 1.1 and 1.2, where "long" and "short" are synonyms for "buy" and "sell." You should be cautioned that your position with respect to the underlying asset can be different from your position in the derivative. For example, a short put is, by definition, short in the put, but actually long in the asset. Indeed, a way to distinguish whether a position is long or short with respect to the underlying asset is to see if it represents a right or an obligation to *buy* or *sell the asset.* A short put carries an obligation to buy the asset if the put holder exercises the option, so is fundamentally a long position.

Criterion 2. *Maximum profit and maximum loss:* Of particular interest is the maximum and minimum profit of a derivative. Unlimited maximum profit (e.g., long forward, long call) is certainly desirable, but unlimited loss (e.g., short forward, short call) is a great cause for concern—it is possible that you go bankrupt!

Example 1.3.1. (SOA Exam IFM Introductory Derivatives Sample Question 26: Unlimited loss) Determine which, if any, of the following positions has or have an unlimited loss potential from adverse price movements in the underlying asset, regardless of the initial premium received.

    I.   Short 1 forward contract
    II.   Short 1 call option
    III. Short 1 put option

(A) None

(B) I and II only

(C) I and III only

(D) II and III only

(E) The correct answer is not given by (A), (B), (C), or (D)

Solution. Only a short forward and a short call have an unlimited loss potential because their payoff and profit functions exhibit an indefinite decreasing trend from a certain point onward. **(Answer: (B))** $\square$

Example 1.3.2. (SOA Exam IFM Introductory Derivatives Sample Question 49: Maximum loss of a long put) The market price of Stock A is 50. A customer buys a 50-strike put contract on Stock A for 500. The put contract is for 100 shares of A.

    Calculate the customer's maximum possible loss.

(A) 0

(B) 5

(C) 50

(D) 500

(E) 5000

Solution. Because the customer is long a put option, his/her maximum loss is attained when the price of Stock A at maturity is 50 or beyond. In that case, the payoff of the put is zero and the customer will have lost the entire initial investment of $\boxed{500}$. **(Answer: (D))** $\square$

Remark. As in Example 1.2.7, we tacitly assume that the interest rate, which is not given, is zero.

Criterion 3. *Asset price contingency:* Derivatives can also be compared with respect to the conditions that trigger the settlement of the derivative. For forwards,

Position	Long/short w.r.t. Asset	Max. Profit	Min. Profit	Asset Price Contingency
Long forward	Long	$+\infty$	$-F_{0,T}$	Always
Short forward	Short	$F_{0,T}$	$-\infty$	Always
Long call	Long	$+\infty$	$-\mathrm{FV}_{0,T}(C)$	$S(T) > K$
Short call	Short	$\mathrm{FV}_{0,T}(C)$	$-\infty$	$S(T) > K$
Long put	Short	$K - \mathrm{FV}_{0,T}(P)$	$-\mathrm{FV}_{0,T}(P)$	$S(T) < K$
Short put	Long	$\mathrm{FV}_{0,T}(P)$	$\mathrm{FV}_{0,T}(P) - K$	$S(T) < K$

TABLE 1.5
Different criteria to compare derivatives.

settlement is an obligation and always takes place; for options, it is at the discretion of the buyer, depending on the relative magnitude of the spot price at expiration and the strike price. Note that whether a derivative is settled does not depend on whether it is long or short.

These three criteria are applied to compare the six positions involving forwards and options; see Table 1.5.

Comparison between options and forwards.

There are two critical differences between options and forwards.

1. Forward holders have the *obligation* to buy/sell the asset no matter how favorable/ unfavorable the market is. In contrast, option holders have the *right* to buy/sell the underlying asset, depending on the market situation. In short, forwards carry commitment while options endow their holders with discretion.

2. Purchasing options require an upfront investment in the form of the option premium, while it is costless to enter into a forward.

Comparison across options: Moneyness.

Moneyness describes whether the payoff of an option would be positive or negative *had* the option been exercised immediately. This concept applies to both European, American, and Bermudan options even though European and Bermudan options may not be exercised before expiration.

- *In-the-money:* An *in-the-money* option is one with a strictly positive payoff if exercised immediately. A call is in-the-money if the current asset price is greater than the strike price, while a put is in-the-money if the current asset price is less than the strike price.

- *Out-of-the-money:* An *out-of-the-money* option is one with a strictly negative payoff if exercised immediately. The conditions for being out-of-the-money are the opposite of being in-the-money.

- *At-the-money:* An option is said to be *at-the-money* if it is both in-the-money and out-of-the-money. In other words, the payoff if exercised immediately is zero, which happens if and only if the current asset price is equal to the strike price.

The concept of moneyness will be useful when one wants to control the amount of insurance to purchase; see Chapter 3 for more details.

Example 1.3.3. (SOA Exam IFM Introductory Derivatives Sample Question 44: Simple deductions for moneyness – I) You are given the following information about two options, A and B:

(i) Option A is a one-year European put with exercise price 45.

(ii) Option B is a one-year American call with exercise price 55.

(iii) Both options are based on the same underlying asset, a stock that pays no dividends.

(iv) Both options go into effect at the same time and expire at $t = 1$.

You are also given the following information about the stock price:

(i) The initial stock price is 50.

(ii) The stock price at expiration is also 50.

(iii) The minimum stock price (from $t = 0$ to $t = 1$) is 46.

(iv) The maximum stock price (from $t = 0$ to $t = 1$) is 58.

Determine which of the following statements is true.

(A) Both options A and B are "at-the-money" at expiration.

(B) Both options A and B are "in-the-money" at expiration.

(C) Both options A and B are "out-of-the-money" throughout each option's term.

(D) Only option A is ever "in-the-money" at some time during its term.

(E) Only option B is ever "in-the-money" at some time during its term.

Solution. At expiration, the price is 50 and both options are out-of-the-money, eliminating Answers (A) and (B). With a strike price of 45 and a minimum stock price of 46, option A with payoff $(45 - S(t))_+ = 0$ for all $t \in [0, 1]$ is never in-the-money, eliminating Answer (D). With a strike price of 55, option B will be in-the-money at the time the stock price is 58 with a payoff of $58 - 55 = 3$, eliminating Answer (C) and verifying $\boxed{\textbf{Answer (E)}}$. $\square$

Example 1.3.4. (SOA Exam IFM Introductory Derivatives Sample Question 61: Simple deductions for moneyness – II) An investor purchased Option A and Option B for a certain stock today, with strike prices 70 and 80, respectively. Both options are European one-year put options.

Determine which statement is true about the moneyness of these options, based on a particular stock price.

(A) If Option A is in-the-money, then Option B is in-the-money.

(B) If Option A is at-the-money, then Option B is out-of-the-money.

(C) If Option A is in-the-money, then Option B is out-of-the-money.

(D) If Option A is out-of-the-money, then Option B is in-the-money.

(E) If Option A is out-of-the-money, then Option B is out-of-the-money.

Solution. Fix any time $t \in [0,1]$ and let Payoff_A and Payoff_B be the time-t payoffs of Option A and Option B, respectively. Notice the order $\text{Payoff}_B \geq \text{Payoff}_A$.

If Option A is in-the-money at time t, i.e., $\text{Payoff}_A > 0$, then necessarily $\text{Payoff}_B > 0$, i.e., Option B is also in-the-money. (**Answer: (A)**) □

Example 1.3.5. (SOA Exam IFM Introductory Derivatives Sample Question 66: Simple deductions for moneyness − III) The current price of a stock is 80. Both call and put options on this stock are available for purchase at a strike price of 65.

Determine which of the following statements about these options is true.

(A) Both the call and put options are at-the-money.

(B) Both the call and put options are in-the-money.

(C) Both the call and put options are out-of-the-money.

(D) The call option is in-the-money, but the put option is out-of-the-money.

(E) The call option is out-of-the-money, but the put option is in-the-money.

Solution. The current stock price, 80, is higher than the strike price, 65. Since the call option would have a positive payoff of 15 if exercised immediately, it is in-the-money. On the other hand, the put option would have a negative payoff of −15 if exercised immediately, so it is out-of-the-money. (**Answer: (D)**) □

1.4   Problems

Problem 1.4.1. (Payoff/profit of a forward for different ending stock prices)
Aaron has purchased a forward contract on a stock. You are given:

(i) If the stock price at expiration is S, his payoff would be $-\$5$.

(ii) If the stock price at expiration is $1.1S$, his payoff would be $\$1$.

Calculate Aaron's profit on the long forward if the stock price at expiration is $1.2S$.

Problem 1.4.2. (Long put vs short forward) You are given:

(i) The current price of a 100-strike 9-month European put option is 12.

(ii) A 9-month forward has a forward price of 105.

(iii) The continuously compounded risk-free interest rate is 3%.

Calculate the stock price after 9 months such that the long put option and the *short* forward contract have the same profit.

Problem 1.4.3. (Call vs put) Jack buys a 50-strike 6-month European call option on stock ABC at a price of 8. Rose buys a 50-strike 6-month European put option on the same stock at a price of 6.

The continuously compounded risk-free interest rate is 4%.

6 months later, Jack suffers a loss while Rose realizes a profit, with Rose's profit being twice as large as Jack's loss.

Calculate the price of stock ABC at the end of 6 months.

Problem 1.4.4. (Maximum and minimum profits of a short put) Bob writes a two-year 100-strike European put with a premium of $10. The continuously compounded risk-free interest rate is 4%.

Calculate the difference between Bob's maximum profit and his minimum profit.

Problem 1.4.5. (Put version of Example 1.2.1: Comparing the profits of three puts) You are given the following premiums of one-year European put options on stock ABC for various strike prices:

Strike	Put Premium
35	0.44
40	1.99
45	5.08

The effective annual risk-free interest rate is 8%.

Let $S(1)$ be the price of the stock one year from now.

Determine the range for $S(1)$ such that the 35-strike short put produces a higher profit than the 45-strike short put, but a lower profit than the 40-strike short put.

(Note: All put positions being compared are short.)

Problem 1.4.6. (European, American, and Bermuda options) Once upon a time, Leo entered into three separate positions involving 2-year options on the same stock.

- Option I was a *short* American-style call with strike price 30.

- Option II was a long Bermuda-style put with strike price 28, where exercise was allowed at any time following an initial 1-year period of put protection.

- Option III was a long European-style put with strike price 20.

At inception, the stock price was 27.
When the options expired, the stock price was 30.
The table below gives the maximum and minimum stock price during the 2-year period:

Time Period	1st year of Option Term	2nd year of Option Term
Maximum Stock Price	28	32
Minimum Stock Price	25	24

Each option was exercised by its holder at the optimal time.
Calculate the sum of the payoffs of the three options.

Problem 1.4.7. (Similarities between a long call and a short put) Determine which of the following statements about a long European call option and a short European put option on the same underlying asset is/are correct.

I. Both are long with respect to the underlying asset.

II. Both involve a possible purchase of the underlying asset in the future.

III. Both give you the right but not the obligation to buy the underlying asset at the strike price in the future.

(A) None

(B) I and II only

(C) I and III only

(D) II and III only

(E) The correct answer is not given by (A), (B), (C), or (D)

Problem 1.4.8. (Simple deduction for moneyness) An investor purchased Call X and Call Y for a certain stock today, with strike prices 50 and 60, respectively. Both options are European options with the same time to expiration.

Determine which of the following statements is true about the moneyness of these options, based on a particular stock price.

(A) If Call X is in-the-money, then Call Y is in-the-money.

(B) If Call X is at-the-money, then Call Y is in-the-money.

(C) If Call X is in-the-money, then Call Y is out-of-the-money.

(D) If Call X is out-of-the-money, then Call Y is in-the-money.

(E) If Call X is out-of-the-money, then Call Y is out-of-the-money.

2

Forwards and Futures

Chapter overview: Building upon the conceptual foundation laid in Chapter 1, this chapter explores in greater depth forward contracts on financial instruments, particularly stocks and stock indexes. Section 2.1 presents four different ways to own a stock at a fixed future time point, leading naturally to the introduction of the notions of prepaid forward and forward contracts. The discussion is expanded in Sections 2.2 and 2.3, where we study in detail how these contracts are priced and hedged for stocks that pay different modes of dividends under the all-important no-arbitrage assumption. In the course of our derivations, we present a synthetic construction of a prepaid forward and a forward which not only allows us to replicate the payoff of a forward by means of the underlying asset and the risk-free asset, but also furnishes a recipe to effect an arbitrage strategy if the observed price in the market deviates from the fair price. Finally, Section 2.4 concludes the chapter with a brief introduction of futures contracts, which are a variant of forwards, and their mechanics.

2.1 Alternative Ways to Buy a Stock

Four ways to own a stock.

There are intricately more ways to buy and own a stock than one might have imagined, if the payment and physical receipt of the stock are separated as two different activities. Table 2.1 lists altogether four different ways that result in the ownership of one unit of a stock at a fixed expiration time T.

1. *Outright purchase:* The simplest method is to pay for the stock at the outset and own it immediately.

2. *Fully leveraged purchase:* One can also receive the stock now and fund the required investment of $S(0)$ by making a loan to be repaid at time T. Given a continuously compounded risk-free interest rate of r, the required payment at time T is $S(0)e^{rT}$.

3. *Prepaid forward contract:* If you pay for the stock now and receive it at time T, then you are said to have entered into a *prepaid forward* contract. We denote the payment made at time 0 under a prepaid forward contract, known as the *prepaid forward price*, by $F_{0,T}^P$, where the symbol F suggests "forward" and the superscript P signifies "prepaid."

   The difference between an outright purchase and a prepaid forward is that with the latter, you own the stock at time T, rather than at time 0.

4. *Forward contract:* As discussed in Chapter 1, we can also own the stock at time T via a forward contract by agreeing today to pay the prespecified forward price at time T and owning the stock. We denote, as in Chapter 1, the forward price by $F_{0,T}$.

Strategy	Time of Payment	Time of Receipt of Asset	Amount of Payment
Outright purchase	0	0	$S(0)$
Fully leveraged purchase	T	0	$S(0)e^{rT}$
Prepaid forward contract	0	T	$F_{0,T}^{P}$
Forward contract	T	T	$F_{0,T}$

TABLE 2.1
Four different ways to own one share of stock at time T.

The focus of this chapter is on the third and fourth arrangements. In fact, the principal objective of this chapter is to find the *fair* values of $F_{0,T}^{P}$ and $F_{0,T}$—"fair" is meant in the sense that the price will not give rise to "free lunch" opportunities (to be defined below) in the market. The technique used to derive the fair prices is known as the *no-arbitrage argument*, the mastery of which is equally important as being able to calculate $F_{0,T}^{P}$ and $F_{0,T}$ proficiently.

Example 2.1.1. (SOA Exam IFM Introductory Derivatives Sample Question 7: Name of arrangement) A nondividend-paying stock currently sells for 100. One year from now the stock sells for 110. The continuously compounded risk-free interest rate is 6%. A trader purchases the stock in the following manner:

- The trader pays 100 today

- The trader takes possession of the stock in one year

Determine which of the following describes this arrangement.

(A) Outright purchase

(B) Fully leveraged purchase

(C) Prepaid forward contract

(D) Forward contract

(E) This arrangement is not possible due to arbitrage opportunities

Solution. Because the trader pays now and possesses the stock in one year, such an arrangement is a $\boxed{\text{prepaid forward contract}}$. **(Answer: (C))**   □

No-arbitrage principle.

The most important conceptual tool in this chapter is the *no-arbitrage principle*. At its core is the simple but reasonable idea that the prices of derivatives in a market should be set in a way that prohibits *arbitrage opportunities*, informally known as "free lunch," which allow investors to earn profits for sure and are something too good to be true in reality. Those prices are referred to as *fair prices* because they ensure that the market does not allow any completely "unfair" phenomenon. Symbolically:

> Market (Observed) price = Fair price $\quad\Leftrightarrow\quad$ No arbitrage opportunity

The no-arbitrage assumption, which will be maintained throughout this book, is very natural but critical. It imposes the minimum form of regularity governing the prices of derivatives in the market and provides the theoretical basis for pricing and hedging all the derivatives in this book. If it does not hold, then it is possible to earn risk-free profits by buying underpriced derivatives and selling overpriced derivatives—the famous adage of "buy low, sell high."

Terminology-wise, a set of financial transactions designed to exploit an arbitrage opportunity is called an *arbitrage strategy*. It is a portfolio that carries only cash inflows, random or non-random, now or in the future, but not cash outflows. Holding such a portfolio will therefore always entitle us to a positive profit. Meanwhile, an individual engaging in an arbitrage strategy is termed an *arbitrageur*. In the later part of this chapter, we shall learn, when the market price deviates from the fair price, how to construct an arbitrage strategy, i.e., we will learn how to become a successful arbitrageur!

2.2 Prepaid Forwards

We separate the discussion of prepaid forwards according to whether the underlying stock pays dividends, and in what manner.

2.2.1 Nondividend-paying Stocks

We start by deriving the price of a prepaid forward on a nondividend-paying stock using two different methods. The more complicated case of dividend-paying stocks will be treated in the next subsection.

Method 1 (Pricing by "common sense").

Let's first discuss how to find the fair prepaid forward price using a common-sense approach. Intuitively, in the absence of dividends, it makes no difference between:

(a) Outright purchase (owning the stock starting from time 0 and holding it until time T)

(b) Prepaid forward (owning the stock starting from time T)

The cost of Method (a) is the current stock price $S(0)$, and that of Method (b) is the prepaid forward price $F_{0,T}^P$. These two methods both give rise to a payoff of $S(T)$ at time T[i] and entail no other cash flows between time 0 and time T other than the initial investment because of the absence of stock dividends. Due to the same financial nature of Methods (a) and (b), they must have the same time-0 cost, that is,

> $F_{0,T}^P = S(0).$ (for nondividend-paying stocks)

---

[i]We ignore non-pecuniary benefits associated with owning a stock, such as voting and control rights. These benefits are hardly quantifiable.

Method 2 (Pricing by replication).

Pricing by replication is a more general and systematic method that can not only be easily extended to more complicated scenarios when common sense fails, but also be employed to take advantage of mispricing in the market. It is illustrated below for proving $F_{0,T}^P = S(0)$.

The following timeline diagram documents the cash flows associated with a long prepaid forward:

$$\begin{array}{ccccc} & -F_{0,T}^P \text{ (target)} & & +S(T) & \\ \hline \text{Time} & 0 & \cdots & T & \\ & \text{You pay the} & & \text{The prepaid forward matures;} & \\ & \text{prepaid forward price} & & \text{you receive one unit of the asset} & \end{array}$$

As its name suggests, the replication method entails trading other securities in the market, forming the *replicating portfolio*, to "replicate" the cash flows of the target derivative of interest, which is the prepaid forward in our case. What securities do we have at our disposal? In our simple market, we can trade the underlying stock and borrow or lend money. To reproduce the cash inflow of $S(T)$ at time T, the simplest way is to buy one unit of the stock at time 0 and sell it at time T. Doing so results in the following cash flows:

$$\begin{array}{ccccc} & -S(0) & & +S(T) & \\ \hline \text{Time} & 0 & \cdots & T & \\ & \text{You pay the} & & \text{You sell the stock and} & \\ & \text{time-0 stock price} \quad \leftarrow & \text{Hold the stock} \quad \rightarrow & \text{receive the time-}T \text{ price} & \end{array}$$

Note that there are no cash flows between time 0 and time T because the stock is assumed to pay no dividends. This allows us to restrict our attention to only time 0 and time T.

As soon as the target derivative (the long prepaid forward in our case), has been replicated, the method of replication says that the fair price of our target is simply equal to the cost of creating the replicating portfolio (to buy one unit of the stock at time 0 in our case), the price of which is easily determined by market prices. After all, the long prepaid forward and the long stock have the same payoff at time T and should naturally cost the same at time 0. Our conclusion is that the fair value of the prepaid forward price should coincide with the time-0 stock price:

$$\boxed{F_{0,T}^P = S(0). \quad \text{(for nondividend-paying stocks)}}$$

In making this conclusion, we are using the following perfectly reasonable idea, which will be repeatedly used in this book:

Two positions having the same *payoff* (at a reference future time) command the same *price* (at time 0).

What if this perfectly reasonable idea turns out to be wrong in the market? That is, what if we observe a prepaid forward price different from the current stock price? In this case, arbitrage opportunities exist and the method of replication has provided us with the recipe to take advantage of the mispricing. Consider:

Transaction	Cash Flows	
	Time 0	**Time T**
Buy 1 unit of stock @ $S(0)$	$-S(0)$	$+S(T)$
Sell prepaid forward @ $F_{0,T}^P$	$+F_{0,T}^P$	$-S(T)$
Total	$F_{0,T}^P - S(0)$	0

TABLE 2.2
Trading strategies to effect arbitrage when $F_{0,T}^P > S(0)$.

Case 1. If $F_{0,T}^P > S(0)$, we can follow the "buy low, sell high" strategy by buying one unit of the stock ("low") and selling the prepaid forward ("high"), immediately realizing a positive cash inflow of $F_{0,T}^P - S(0)$ (> 0). Buying the stock at time 0 guarantees that at time T we have at hand the stock, which we are obligated to deliver to the buyer of the prepaid forward. Note that the total cash flow at time T is exactly zero, but we already earn a positive cash inflow of $F_{0,T}^P - S(0)$ at time 0. This creates a risk-free profit and thus an arbitrage strategy (see Table 2.2 for the transactions and the associated cash flows).

Case 2. If $F_{0,T}^P < S(0)$, then reversing the transactions in Table 2.2 (i.e., buying the prepaid forward and short selling 1 unit of the stock) shows that an arbitrage is possible.

Combining the two cases, we infer that the only fair value of $F_{0,T}^P$ is $S(0)$.

In summary, the method of pricing by replication is a very useful tool to implement the no-arbitrage principle. It underlies the pricing of essentially all derivatives (not just forwards and options) and is one of the most fundamental ideas in this book. You should make every effort to master it fully!

2.2.2 Dividend-paying Stocks

General prepaid forward price formula.

In the presence of stock dividends, an outright purchase and a prepaid forward do make a difference. The reason is that if you own the stock starting from time 0 and hold it until time T, you receive the dividends payable from time 0 to time T as well, but the owner of the T-year prepaid forward does not. Because of the loss of dividends, intuition suggests and formal no-arbitrage arguments can be used to confirm that the holder of a long prepaid forward should be compensated (relative to an outright purchase) by having to pay *less* than the stock holder by the *price of the stock dividends*[ii] payable over the life of the prepaid forward, i.e.,

$$F_{0,T}^P = S(0) - \text{Price of dividends}.$$

We distinguish between two types of dividend payments.

Case 1: Discrete, non-random dividends.

Suppose it is expected that at each *known* time t_i, the stock will make a dividend payment with a *known* amount of $D(t_i)$, where $0 < t_1 < t_2 < \cdots < t_n \leq T$ and T is the maturity time of a T-year prepaid forward in question. If there is a dividend payable at time T, we assume that it is paid immediately before the prepaid forward matures. To replicate the long prepaid

---

[ii]We prefer not to use the term "present value of dividends" because stochastic dividends, discounted for time value for money, remain stochastic and thus cannot serve as a price.

forward by buying one unit of the stock, we will end up with the additional cash inflows of $D(t_i)$ at times $t_1, \ldots, t_n$ due to the discrete dividends that the holder of the stock is entitled to. To completely replicate the long prepaid forward, these interim cash inflows need to be eliminated. This can be accomplished by making a loan of $\mathrm{PV}_{0,t_i}(D(t_i)) = D(t_i)e^{-rt_i}$ to be repaid at time t_i. Here shows the complete replicating portfolio and the associated cash flows:

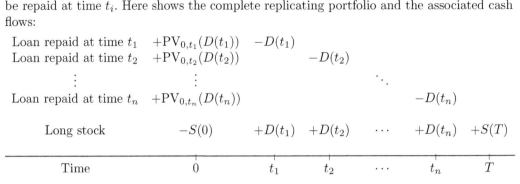

The fair value of the prepaid forward price then equals the cost of setting up the above replicating portfolio, which means that

$$F_{0,T}^P = S(0) - \sum_{i=1}^{n} \mathrm{PV}_{0,t_i}(D(t_i)) = S(0) - \sum_{i=1}^{n} D(t_i)e^{-rt_i}. \tag{2.2.1}$$

Again, in the event that the observed value of $F_{0,T}^P$ differs from (2.2.1), the replicating portfolio will supply the necessary ingredients for effecting an arbitrage strategy, as we will see in the next subsection.

Example 2.2.1. (SOA Exam IFM Introductory Derivatives Sample Question 71: Calculation of $F_{0,T}^P$ with discrete dividends) A certain stock costs 40 today and will pay an annual dividend of 6 for the next 4 years. An investor wishes to purchase a 4-year prepaid forward contract for this stock. The first dividend will be paid one year from today and the last dividend will be paid just prior to delivery of the stock. Assume an annual effective interest rate of 5%.

Calculate the price of the prepaid forward contract.

(A) 12.85

(B) 13.16

(C) 17.29

(D) 18.72

(E) 21.28

Solution. By (2.2.1), the price of the 4-year prepaid forward contract is

$$\begin{aligned} F_{0,4}^P &= S(0) - \mathrm{PV}_{0,4}(\mathrm{Div}) \\ &= 40 - 6(1/1.05 + 1/1.05^2 + 1/1.05^3 + 1/1.05^4) \\ &= \boxed{18.72}. \quad \textbf{(Answer: (D))} \end{aligned}$$

$\square$

Case 2: Continuous proportional dividends.

For stock indexes comprising a number of stocks, it is common to model the dividends as being paid continuously at a rate proportional to the level of the stock index. This means that there is a non-negative constant δ, called the *dividend yield*, such that for each unit of the stock index, the amount of (stochastic) dividends paid between time t and $t + \mathrm{d}t$ for any infinitesimally small $\mathrm{d}t$ is $S(t)\delta\,\mathrm{d}t$, where $S(t)$ is the time-t stock index price. Throughout this book, we assume that the dividends received are not paid out in cash, but are reinvested immediately in the stock, resulting in more and more shares as time goes by.

To determine the increase in the number of shares, let $N(t)$ be the number of shares of the stock we hold at time t under the reinvestment policy. We give a calculus-based proof for the expression of $N(t)$. Between time t and time $t + \mathrm{d}t$, the amount of dividend payment is $S(t)\delta\,\mathrm{d}t$ per share, so the total amount of dividends we receive is $N(t)S(t)\delta\,\mathrm{d}t$. Reinvesting this amount in the stock allows us to buy $N(t)S(t)\delta\,\mathrm{d}t/S(t) = N(t)\delta\,\mathrm{d}t$ more shares. In other words, the change in the number of shares is given by

$$\mathrm{d}N(t) = N(t + \mathrm{d}t) - N(t) = N(t)\delta\,\mathrm{d}t,$$

which means that

$$\frac{\mathrm{d}N(t)}{\mathrm{d}t} = \delta N(t).$$

The solution to this separable ordinary differential equation in $N(t)$ with initial shares $N(0)$ is given by[iii]

$$N(t) = N(0)\mathrm{e}^{\delta t}.$$

In particular, 1 share (i.e., $N(0) = 1$) at time 0 will grow to $\mathrm{e}^{\delta T}$ shares at time T. By proportion, to obtain 1 share at time T and to replicate the payoff of $S(T)$, it suffices to buy $\mathrm{e}^{-\delta T}$ shares at time 0 and reinvest all dividends in the stock between time 0 and time T, which is what the replicating portfolio entails. It follows that the fair prepaid forward price in the presence of continuous dividends is the cost of buying $\mathrm{e}^{-\delta T}$ shares at time 0, or

$$\boxed{F_{0,T}^{P} = S(0)\mathrm{e}^{-\delta T}.} \tag{2.2.2}$$

Adjusting the initial position to offset the effect of dividend income so that exactly one unit of the underlying stock is received at expiration is called *tailing* the position.

You may draw an analogy between the continuous dividend yield acting on the number of shares and the force of interest acting on the amount of money in a risk-free account as you have seen in your theory of interest class. If the continuous dividend yield is δ, then 1 share at time 0 will accumulate to $\mathrm{e}^{\delta T}$ shares at time T. Likewise, if the force of interest is δ, then \$1 invested in the risk-free money account at time 0 will grow to \$$\mathrm{e}^{\delta T}$ at time T.

Let's review what we have learned from Cases 1 and 2 via a challenging example that combines discrete and continuous dividends.

Example 2.2.2. [HARDER!] (Discrete plus continuous dividends) For $t \geq 0$, let $S(t)$ be the time-t price of Stock ABC. You are given:

(i) $S(0) = 100$

[iii]If $\delta = \delta(t)$ is a deterministic function of time t, then it can be shown that the number of shares at time t is given by

$$N(t) = N(0)\exp\left(\int_{0}^{t}\delta(s)\,\mathrm{d}s\right).$$

(ii) At time 0.5, a cash dividend of $10 per share will be paid.

(iii) From time 0.75 to time 1, dividends are paid continuously at a rate proportional to its price. The dividend yield is 10%.

(iv) The continuously compounded risk-free interest rate is 8%.

Calculate the price of a one-year prepaid forward contract on stock ABC.

Ambrose's comments:

This is a relatively hard problem combining discrete and continuous proportional dividends. It nicely illustrates the futility of slavishly memorizing prepaid forward price formulas without a solid understanding of the underlying dividend-paying mechanics. To help you answer this question, think about how you can end up with exactly 1 unit of stock at time 1.

Solution. To receive exactly one unit of stock ABC at time 1, we should start with only $e^{-\delta(1-0.75)} = e^{-0.025}$ shares of stock ABC because of the reinvestment of the continuous proportional dividends between time 0.75 and time 1. With $e^{-0.025}$ shares at time 0 and at time 0.5, we will receive a cash dividend of $\$10e^{-0.025}$ at time 0.5:

Method	Cash Flows		
	$t = 0$	$t = 0.5$	$t = 1$
Outright purchase	$-S(0)e^{-0.025}$	$10e^{-0.025}$	$S(T)$
Prepaid forward	$-F_{0,1}^P$	0	$S(T)$

To fully imitate the one-year prepaid forward, the replicating portfolio consists of:

1. Buying $e^{-0.025}$ number of shares of stock ABC at time 0

2. Making a loan of $\mathrm{PV}_{0,0.5}(10e^{-0.025})$ at time 0 and repaying it at time 0.5.

The one-year prepaid forward price equals the cost of setting up the replicating portfolio, which in turn is

$$
\begin{aligned}
F_{0,1}^P &= S(0)e^{-0.025} - 10e^{-0.025} \times e^{-0.5r} \\
&= 100e^{-0.25(0.1)} - 10e^{-0.25(0.1)} \times e^{-0.08(0.5)} \\
&= \boxed{88.16}.
\end{aligned}
$$

□

Epilogue: Significance of prepaid forward prices.

The notion of prepaid forward prices is not a very popular and widely used one in the literature.[iv] However, it is a very useful concept which permeates the whole book. Its significance is (at least) twofold:

1. As we will see in the next section, prepaid forward prices and forward prices are inti-

---

[iv] As far as the author is aware, prepaid forward prices are introduced in McDonald (2013) (see Chapter 5 therein).

mately connected. We can understand the pricing of prepaid forwards if we understand the pricing of forwards, and the other way round.

2. Prepaid forward prices are also objects of independent interest. They provide a unifying treatment of the current price (or value) of a general cash inflow, *random or non-random*, in the future. As we have seen earlier, the T-year prepaid forward price of a stock, $F_{0,T}^P$, captures what you have to pay today in order to receive exactly one unit of the stock at a future time T, with a random payoff of $S(T)$. To receive a constant cash inflow of $\$K$ at time T, we need only pay $\$\mathrm{PV}_{0,T}(K) = \$Ke^{-rT}$ today. We can symbolize this with the prepaid forward price notation by writing

$$F_{0,T}^P(K) = Ke^{-rT}. \tag{2.2.3}$$

When confusion arises, we write $F_{0,T}^P(S)$ to signify the prepaid forward price on the stock to distinguish it from the present value of cash.

2.3 Forwards

2.3.1 Forward Prices

Most elementary textbooks on financial derivatives treat forward contracts directly, but now that we are armed with the formulas of the fair prices of prepaid forwards, the corresponding formulas of forwards can be obtained with considerable ease.

Going from the prepaid forward price to the forward price.

Think in this way:

What is the difference between a prepaid forward and a forward?

As we have seen in Section 2.1, whether you buy a prepaid forward or a forward, you will always end up receiving one unit of the stock at time T. You do not receive the interim dividends, if any, with both contracts. The only difference is the time of payment:

- With a prepaid forward, you pay the prepaid forward price $F_{0,T}^P$ at time 0.

- With a forward, you pay the forward price $F_{0,T}$ at time T.

Since both prices are constants, $F_{0,T}$ is simply the time-T future value of $F_{0,T}^P$ accumulated at the risk-free interest rate, i.e.,

$$\boxed{F_{0,T} = \mathrm{FV}_{0,T}(F_{0,T}^P) \quad \text{or} \quad F_{0,T}^P = \mathrm{PV}_{0,T}(F_{0,T}).} \tag{2.3.1}$$

Explicit forward price formulas.

Accumulating the time-0 prepaid forward price formulas at the continuously compounded risk-free interest rate r from time 0 to time T, we readily have the following forward price formulas:

$$F_{0,T} = \begin{cases} S(0)e^{rT} - \underbrace{\sum_{i=1}^n D(t_i)e^{r(T-t_i)}}_{\mathrm{FV}_{0,T}(\mathrm{Div})}, & \text{for discrete dividends} \\ S(0)e^{(r-\delta)T}, & \text{for continuous proportional dividends} \end{cases} \tag{2.3.2}$$

Example 2.3.1. (SOA Exam IFM Introductory Derivatives Sample Question 29: Ranking of different methods) The dividend yield on a stock and the interest rate used to discount the stock's cash flows are both continuously compounded. The dividend yield is less than the interest rate, but both are positive.

The following table shows four methods to buy the stock and the total payment needed for each method. The payment amounts are as of the time of payment and have not been discounted to the present date.

METHOD	TOTAL PAYMENT
Outright purchase	A
Fully leveraged purchase	B
Prepaid forward contract	C
Forward contract	D

Determine which of the following is the correct ranking, from smallest to largest, for the amount of payment needed to acquire the stock.

(A) C < A < D < B

(B) A < C < D < B

(C) D < C < A < B

(D) C < A < B < D

(E) A < C < B < D

Solution. The expressions for A, B, C, and D are

$$
\begin{aligned}
A &= S(0), \\
B &= S(0)\mathrm{e}^{rT}, \\
C &= F^P_{0,T} = S(0)\mathrm{e}^{-\delta T}, \\
D &= F_{0,T} = S(0)\mathrm{e}^{(r-\delta)T}.
\end{aligned}
$$

As $\mathrm{e}^{-\delta T} < 1 < \mathrm{e}^{(r-\delta)T} < \mathrm{e}^{rT}$, the correct ranking is $\boxed{\text{C} < \text{A} < \text{D} < \text{B}}$. **(Answer: (A))** $\qquad\square$

Example 2.3.2. (SOA Exam IFM Introductory Derivatives Sample Question 37: Discrete dividends − I) A one-year forward contract on a stock has a price of $75. The stock is expected to pay a dividend of $1.50 at two future times, six months from now and one year from now, and the annual effective risk-free interest rate is 6%.

Calculate the current stock price.

(A) 70.75

(B) 73.63

(C) 75.81

(D) 77.87

(E) 78.04

Solution. By the first formula in (2.3.2), we solve the equation

$$S(0)(1.06) - \underbrace{1.5(1.06^{1/2} + 1)}_{\text{FV}_{0,1}(Div)} = F_{0,1} = 75$$

yields $S(0) = \boxed{73.63}$. (**Answer: (B)**) □

Example 2.3.3. (SOA Exam IFM Introductory Derivatives Sample Question 51: Discrete dividends − II) You are given the following information about Stock XYZ:

(i) The current price of the stock is 35 per share.

(ii) The expected continuously compounded rate of return is 8%.

(iii) The stock pays semi-annual dividends of 0.32 per share, with the next dividend to be paid two months from now.

The continuously compounded risk-free interest rate is 4%.
    Calculate the current one-year forward price for stock XYZ.

(A) 34.37

(B) 35.77

(C) 36.43

(D) 37.23

(E) 37.92

Solution. By the first formula in (2.3.2) again, the one-year forward price is

$$F_{0,1} = 35e^{0.04} - 0.32(e^{0.04(10/12)} + e^{0.04(4/12)}) = \boxed{35.77}. \quad (\textbf{Answer: (B)})$$

□

Remark. The continuously compounded rate of return on the stock is not needed.

Remark: Relationship between the forward price and expected future stock price.

You may be tempted to think that forward prices are a good estimator of the expected stock prices in the future. Such an impression may stem from the very definition that the forward price is what you agree to pay in the future in place of the random future stock price for owning the stock. Intuition suggests that the forward price, as a replacement of the expected future stock price, should estimate the latter reasonably well. It turns out, however, that the forward price is a systematic *underestimator* of the expected future stock price.
    For concreteness, let α be the continuously compounded expected rate of return on the

stock. By definition, α is the continuously compounded rate at which the initial stock price accumulates to the expected stock price, i.e., α satisfies $\mathbb{E}[S(T)] = S(0)e^{\alpha T}$. Because the stock is risky in the sense that its future price movements are random, investors (to be precise, risk-averse investors) naturally demand that $\alpha > r$, so that the stock is worse than the risk-free asset because of the presence of price variability but outperforms the risk-free asset with respect to the expected rate of return. The difference $\alpha - r$ is known as the *risk premium*. It follows that

$$\mathbb{E}[S(T)] = S(0)e^{\alpha T} > S(0)e^{rT} > S(0)e^{rT} - \mathrm{FV}_{0,T}(\mathrm{Div}) = F_{0,T}$$

for any positive T, meaning that the expected stock price at expiration is always *higher* than the forward price, no matter whether the stock pays dividends, and in what manner.

Example 2.3.4. (SOA Exam IFM Introductory Derivatives Sample Question 6: Bounds on forward price/expected stock price – I) The following relates to one share of XYZ stock:

- The current price is 100.

- The forward price for delivery in one year is 105.

- P is the expected price in one year.

Determine which of the following statements about P is TRUE.

(A) $P < 100$

(B) $P = 100$

(C) $100 < P < 105$

(D) $P = 105$

(E) $P > 105$

Solution. The expected 1-year stock price should be higher than the 1-year forward price. (**Answer: (E)**) $\square$

Example 2.3.5. (SOA Exam IFM Introductory Derivatives Sample Question 38: Bounds on forward price/expected stock price – II) The current price of a medical company's stock is 75. The expected value of the stock price in three years is 90 per share. The stock pays no dividends.
    You are also given:

(i) The risk-free interest rate is positive.

(ii) There are no transaction costs.

(iii) Investors require compensation for risk.

The price of a three-year forward on a share of this stock is X, and at this price an investor is willing to enter into the forward.
    Determine what can be concluded about X.

Transaction	Cash Flows	
	Time 0	Time T
Buy $e^{-\delta T}$ shares of stock	$-S(0)e^{-\delta T}$	$+S(T)$
Borrow $S(0)e^{-\delta T}$	$+S(0)e^{-\delta T}$	$-F_{0,T} = -S(0)e^{(r-\delta)T}$
Total	0	$S(T) - F_{0,T}$

TABLE 2.3
Demonstration of (2.3.3) in the case of continuous proportional dividends.

(A) $X < 75$

(B) $X = 75$

(C) $75 < X < 90$

(D) $X = 90$

(E) $90 < X$

Solution. Because the stock pays no dividends, $X = F_{0,3} = S(0)e^{3r} > S(0) = 75$. Moreover, $X \leq \mathbb{E}[S(3)] = 90$. **(Answer: (C))** □

2.3.2 Cash-and-Carry Arbitrage

It bears mention that the forward price formulas given in (2.3.2) represent the fair price of a forward. Should the price of the forward in the market be different from the fair price, it is possible to develop an arbitrage strategy to reap risk-free profits. The key vehicle is a synthetic construction of a forward by the underlying stock coupled with borrowing or lending. In what follows, we assume that the stock pays continuous proportional dividends with a dividend yield of δ.

Synthetic forwards.

Because a forward is identical to a prepaid forward except that the payment is made at expiration instead of at the outset, a forward can be replicated by the replicating portfolio of the prepaid forward together with a loan that erases the initial investment (recall that a genuine forward carries no initial investment) and defers the payment to the time of expiration. We know from Subsection 2.2.2 that it suffices to buy $e^{-\delta T}$ shares of the stock to replicate the prepaid forward, which costs $S(0)e^{-\delta T}$ at time 0, so at the same time we also take out a loan of $S(0)e^{-\delta T}$ and repay it at time T with interest.

Table 2.3 confirms that these two actions successfully reproduce the cash flows of a long forward. The *synthetic construction* can be written symbolically as

$$\text{(Genuine) Forward} = \underbrace{\text{Stock} + \text{Borrowing}}_{\text{Synthetic forward}}, \qquad (2.3.3)$$

where "=" indicates having the same cash flows at every point of time.

Transaction	Cash Flows	
	Time 0	**Time T**
Buy $e^{-\delta T}$ shares of stock	$-S(0)e^{-\delta T}$	$+S(T)$
Borrow $S(0)e^{-\delta T}$	$+S(0)e^{-\delta T}$	$-S(0)e^{(r-\delta)T}$
Short forward	0	$F_{0,T}^{\text{obs}} - S(T)$
Total	0	$F_{0,T}^{\text{obs}} - S(0)e^{(r-\delta)T} > 0$

TABLE 2.4
Transactions and cash flows for a cash-and-carry arbitrage.

Example 2.3.6. (SOA Exam IFM Introductory Derivatives Sample Question 56: Rearranging the synthetic forward equation) Determine which of the following positions has the same cash flows as a short stock position.

(A) Long forward and long zero-coupon bond

(B) Long forward and short forward

(C) Long forward and short zero-coupon bond

(D) Long zero-coupon bond and short forward

(E) Short forward and short zero-coupon bond

Solution. Rearranging (2.3.3) yields

$$-\text{Stock} = -\text{Forward} + \text{Borrowing},$$

which corresponds to (E). **(Answer: (E))**   □

(Reverse) Cash-and-carry arbitrage.

Given the synthetic forward construction, we now put ourselves in the shoes of an arbitrageur and demonstrate how risk-free profits can be earned when the observed forward price $F_{0,T}^{\text{obs}}$ is not equal to the fair price $F_{0,T}^{\text{fair}} = S(0)e^{(r-\delta)T}$.

Case 1.   Suppose that $F_{0,T}^{\text{obs}} > F_{0,T}^{\text{fair}} = S(0)e^{(r-\delta)T}$ (i.e., the observed forward price is too high, or the forward is *overpriced*). We follow the "buy low, sell high" rule, which means selling the overpriced forward in the market and buying the synthetic forward in accordance with (2.3.3). Table 2.4 shows the transactions, called *cash-and-carry*, and the corresponding cash flows. Note that the cash inflow at time T involves quantities that are already known at time 0.

Here is a mnemonic about why we call the transactions in Table 2.4 "cash-and-carry":

Borrow the "cash" required and

"carry" the stock from time 0 to time T.

Example 2.3.7. (SOA Exam IFM Introductory Derivatives Sample Question 21: Cash-and-carry arbitrage in action!) A market maker in stock index forward contracts observes a 6-month forward price of 112 on the index. The index spot price is 110 and the continuously compounded dividend yield on the index is 2%.

The continuously compounded risk-free interest rate is 5%.

Describe actions the market maker could take to exploit an arbitrage opportunity and calculate the resulting profit (per index unit).

(A) Buy observed forward, sell synthetic forward, Profit = 0.34

(B) Buy observed forward, sell synthetic forward, Profit = 0.78

(C) Buy observed forward, sell synthetic forward, Profit = 1.35

(D) Sell observed forward, buy synthetic forward, Profit = 0.78

(E) Sell observed forward, buy synthetic forward, Profit = 0.34

Solution. The fair 6-month price of the forward contract is

$$F_{0,0.5}^{\text{fair}} = S(0)e^{(r-\delta)T} = 110e^{(0.05-0.02)(0.5)} = 111.6624,$$

which is 0.34 less than the observed price. Thus we can engage in cash-and-carry arbitrage by selling the observed forward at 112 and buying a synthetic forward (i.e., buying $e^{-0.01}$ units of the index and borrowing $110e^{-0.01}$) at 111.6624, realizing a profit of $112 - 111.6624 = \boxed{0.34}$. **(Answer: (E))** □

Remark. The arbitrage profit, realized at expiration, is equal to the absolute value of the difference between the observed forward price and the fair forward price.

Case 2. Suppose that $F_{0,T}^{\text{obs}} < F_{0,T}^{\text{fair}} = S(0)e^{(r-\delta)T}$ (i.e., the observed forward price is too low, or the forward is *underpriced*). In this case, we simply take the opposite strategy of Case 1, namely:

1. Buy the observed forward in the market.
2. Sell the synthetic forward, i.e., short sell the stock and lend the proceeds.

These transactions constitute a *reverse cash-and-carry arbitrage* whose components and associated cash flows are shown in Table 2.5.

Example 2.3.8. (SOA Exam IFM Introductory Derivatives Sample Question 73) The current price of a nondividend-paying stock is 100. The annual effective risk-free interest rate is 4%, and there are no transaction costs.

The stock's two-year forward price is mispriced at 108, so to exploit this mispricing, an investor can short a share of the stock for 100 and simultaneously take a long position in a two-year forward contract. The investor can then invest the 100 at the risk-free rate, and finally buy back the share of stock at the forward price after two years.

Determine which term best describes this strategy.

(A) Hedging

Transaction	Cash Flows	
	Time 0	Time T
Short $e^{-\delta T}$ shares of stock	$+S(0)e^{-\delta T}$	$-S(T)$
Lend $S(0)e^{-\delta T}$	$-S(0)e^{-\delta T}$	$+S(0)e^{(r-\delta)T}$
Long forward	0	$S(T) - F_{0,T}^{\text{obs}}$
Total	0	$S(0)e^{(r-\delta)T} - F_{0,T}^{\text{obs}} > 0$

TABLE 2.5
Transactions and cash flows for a reverse cash-and-carry arbitrage.

(B) Immunization

(C) Arbitrage

(D) Paylater

(E) Diversification

Solution. The fair 2-year forward price is $F_{0,2}^{\text{fair}} = 100(1.04)^2 = 108.16$, which is higher than the observed forward price of 108. The forward in the market is underpriced, and a reverse cash-and-carry $\boxed{\text{arbitrage}}$ should be undertaken. **(Answer: (C))**   □

Example 2.3.9. (Reverse cash-and-carry arbitrage with discrete dividends)
The current stock price is 80. A dividend of 2 will be paid 6 months from now. The continuously compounded risk-free interest rate is 6%.

If you observe a 1-year forward price of 82, describe actions you could take as an arbitrageur, and calculate the resulting arbitrage profit (per stock unit).

Ambrose's comments:

This example is a "reverse" and "discrete" counterpart of Example 2.3.7. To avoid repetition, the stock is changed to pay discrete dividends.

Solution. The fair 1-year forward price is $F_{0,1}^{\text{fair}} = 80e^{0.06} - 2e^{0.03} = 82.8860$, which is higher than the observed price. To effect a reverse cash-and-carry arbitrage, we buy the observed forward, short sell the stock, and borrow and lend as follows:

Transaction	Cash Flows		
	Time 0	Time $1/2$	Time 1
Short 1 share of stock	$+80$	-2	$-S(1$
Lend $S(0)$ at time 0	-80	0	$+80e^{0.06}$
Borrow 2 at time 0.5	0	$+2$	$-2e^{0.03}$
Long forward	0	0	$S(1) - 82$
Total	0	0	$\boxed{0.8860}$

Note the dividend of 2, which we have to *pay* in 6 months because of selling the stock.

□

Sidebar: Short selling a stock.

In Example 2.3.9 above, we sell a share of the stock without actually owning it in the first place. This possibly strange act is known as *short selling*, which refers to selling an asset that you do not own. Although short selling is mathematically the opposite of an ordinary purchase, practically it is not as simple as it may seem. The precise procedure is as follows:

Step 1.   Borrow the asset from a third party (e.g., your broker).

Step 2.   Immediately sell the asset to the market, thereby creating a *short* position, and receive the proceeds.

Step 3.   Buy back the same asset some time later and return it to the lender. The short position in the asset is said to be *closed out*.

Short selling is therefore essentially the same as borrowing money and paying back a random amount which is determined by the future asset price.

Some of the primary motivates for engaging in a short sale are:

1. *Speculation:* A short sale makes money only if the asset price goes down, or you sell high first and buy low later. Therefore, short-selling can be regarded as speculating that the asset price will drop in the future.

2. *Hedging:* A short sale can be undertaken to offset asset price risk. For example, a long forward coupled with a short sale will eliminate any uncertainty arising from future asset price movements.

3. *Arbitrage:* When you engage in a reverse cash-and-carry arbitrage, part of your transactions is to short sell the stock, as in Example 2.3.9.

Note that in the presence of dividends, a short seller must pay the broker any dividends payable by the asset that has been shorted. The broker will transfer such dividends to the lender from whom the asset has been borrowed.

Example 2.3.10. (Illustration of cash flows under a short sale) On January 1, the price of a stock is $120 per share. An investor short sells 500 shares and closes out the position by buying them back on March 1 when the price per share is $100. A dividend of $1 *per share* is paid on February 1.
     The cash flow of the investor is as follows:

Time	Transaction	Cash Flow
Jan 1	Borrow 500 shares and sell them at $120	$60,000
Feb 1	Pay dividend	−$500
Mar 1	Buy 500 shares at $100 and repay short sale	−$50,000

     Note that these cash flows are exactly the opposite of an investor who is long 500 shares.

2.3.3   Digression: Market Frictions

Thus far, we have been discussing the trading of derivatives in an idealistic, friction-free market. There are no taxes, transaction costs, bid/ask spreads, and no disparity between borrowing and lending interest rates. In this brief subsection, we take into account these practical factors and explore how they impact on fair forward prices.

Bid-ask spread	For any given asset, the price at which you can buy is called the *ask price*, $S^a(0)$, while the price at which you can sell is called the *bid price*, $S^b(0)$. At any moment of time, the ask price must be higher than or equal to the bid price, or else you can buy the asset at the ask price and immediately sell the asset at the bid price, realizing a risk-free profit. The difference between the two prices is called the *bid-ask spread*, which is a source of income to market makers.
Disparity between borrowing and lending rates	In general, the borrowing rate r^b and lending rate r^l need not be identical. To avoid arbitrage, it must hold that $r^b \geq r^l$ (why?).
Transaction costs	Buying or selling derivatives involves transaction costs, which can be a fixed amount or a variable amount proportional to the scale of transaction.

Example 2.3.11. (How much transaction cost do you incur?) You observe two prices $65.1 and $65.2 quoted in the market for the stock of Company ABC.

The brokerage commission includes:

(i) 0.3% of the transaction amount

(ii) A fixed cost of $50 per transaction.

Calculate how much you gain/loss if you purchase 200 shares and then sell them all immediately.

Solution. Among the two observed prices, the bid price must be lower than or equal to the ask price, so the bid price is $65.1 and the ask price is $65.2.

To complete the desired transaction, you first spend $200 \times 65.2 \times (1 + 0.3\%) + 50 = \$13{,}129.12$ for purchasing 200 shares at the ask price, then receive $200 \times 65.1 \times (1 - 0.3\%) - 50 = \$12{,}930.94$ upon selling the 200 shares at the bid price. The amount lost is $13{,}129.12 - 12{,}930.94 = \boxed{\$198.18}$.   □

The following example shows how market frictions affect no-arbitrage forward prices.

Example 2.3.12. [HARDER!] (SOA Exam IFM Introductory Derivatives Sample Question 52: No-arbitrage forward price) The ask price for a share of ABC company is 100.50 and the bid price is 100. Suppose an investor can borrow at an annual effective rate of 3.05% and lend (i.e., save) at an annual effective rate of 3%. Assume there are no transaction costs and no dividends.

Determine which of the following strategies does not create an arbitrage opportunity.

(A) Short sell one share, and enter into a long one-year forward contract on one share with a forward price of 102.50.

(B) Short sell one share, and enter into a long one-year forward contract on one share with a forward price of 102.75.

(C) Short sell one share, and enter into a long one-year forward contract on one share with a forward price of 103.00.

(D) Purchase one share with borrowed money, and enter into a short one-year forward contract on one share with a forward price of 103.60.

(E) Purchase one share with borrowed money, and enter into a short one-year forward contract on one share with a forward price of 103.75.

Solution. • *Prelude:* In the presence of market frictions, the fair forward price will no longer be a unique value. Instead, it will be relaxed to an *interval* whose length is governed by the extent to which the market frictions exist. It can be shown by no-arbitrage arguments that market frictions enter the no-arbitrage forward price interval in such a way to make it as wide as possible, e.g., if there exist a bid-ask spread and different interest rates for borrowing and lending, then the bid price will go with the lending rate to form the lower bound of the interval, and the ask price will go with the borrowing rate to form the upper bound:

$$F_{0,T}^{\text{fair}} \in [S^b(0)e^{(r^l-\delta)T}, S^a(0)e^{(r^b-\delta)T}].$$

• *Back to this example:* Given the above result, the no-arbitrage interval for the 1-year forward price can be (and will be) shown to be

$$[100(1.03), 100.50(1.0305)] = [103, 103.56525].$$

Observe that the forward prices in (A) and (B) are strictly less than the lower bound of 103 whereas those in (D) and (E) are strictly greater than the upper bound of 103.56525. For the purpose of illustration, we consider (A), (B), and (C), where $F_{0,1}^{\text{obs}} \leq 103$, and show that there is an arbitrage opportunity as long as $F_{0,1}^{\text{obs}} < 103$.

Because the observed forward is underpriced, we pursue a reverse cash-and-carry arbitrage strategy by buying the forward and selling the synthetic forward by short selling one share (recall that there are no dividends) and lending the proceeds. The following table shows the resulting cash flows:

	Cash Flows	
Transaction	Time 0	Time 1
Short one share of stock	$S^b(0) = 100$	$-S(1)$
Lend $S^b(0)$	$-S^b(0) = -100$	$100(1.03) = 103$
Long forward	0	$S(1) - F_{0,1}^{\text{obs}}$
Total	0	$103 - F_{0,1}^{\text{obs}} \geq 0$

Note that in the "Time 0" column, it is the *bid* price that is used (we are selling the stock to the market) and in the "Time 1 column" the *lending* rate is used to accumulate the proceeds of 100. Now observe that the cash flow at time 1 is strictly positive so long as $F_{0,1}^{\text{obs}} < 103$ and an arbitrage strategy is created. If $F_{0,1}^{\text{obs}} = 103$, then the cash flows at both time 0 and time 1 are zero. Nothing is paid and received. (**Answer: (C)**)

$\square$

Remark. Using the above strategy, we can conclude that $F_{0,1}^{\text{obs}}$ should be at least 103 to preclude arbitrage opportunities. The upper bound of 103.56525 can be established by a cash-and-carry arbitrage strategy.

2.4   Futures

We conclude this chapter with a brief discussion on futures contracts and look at how they are implemented in practice.

2.4.1   Differences between Futures and Forwards

Futures vs forwards.

Futures contracts (make sure that you write 'futures," not 'future!") are similar to forward contracts in the sense that both of them involve a commitment to buy or sell the underlying asset at a prespecified price on a particular date. However, there are important contractual differences between them with respect to the time of settlement, liquidity, uniformity, the issue of credit risk, and the existence of price limits (see Table 2.7). These will become clearer as we learn the mechanics of futures in the next subsection.

Example 2.4.1. (SOA Exam IFM Introductory Derivatives Sample Question 30: Forward vs futures – I) Determine which of the following is NOT a distinguishing characteristic of futures contracts, relative to forward contracts.

(A) Contracts are settled daily, and marked-to-market.

(B) Contracts are more liquid, as one can offset an obligation by taking the opposite position.

(C) Contracts are more customized to suit the buyer's needs.

(D) Contracts are structured to minimize the effects of credit risk.

(E) Contracts have price limits, beyond which trading may be temporarily halted.

Solution. It is forwards that are more customized and futures that are more standardized. (**Answer: (C)**) $\square$

	Forwards	Futures
1.	Settled at expiration.	Settled at the end of every mark-to-market period (see Subsection 2.4.2).
2.	Traded over the counter, forwards are relatively illiquid.	Being exchange-traded, futures are liquid. It is easy to offset obligations by entering into opposite positions.
3.	Customized to suit the needs of buyer and seller.	Standardized with respect to expiration date, size, underlying asset, etc.
4.	Credit risk remains—the buyer/seller may fail to fulfill his/her obligations.	Credit risk is minimized as a result of marking to market.
5.	No price limits.	Price limits are imposed, triggering temporary halts in trading if the futures price moves dramatically.

TABLE 2.7
Differences between forwards and futures.

Example 2.4.2. (SOA Exam IFM Introductory Derivatives Sample Question 69: Forward vs futures – II) Determine which of the following statements about futures and forward contracts is false.

(A) Frequent marking-to-market and settlement of a futures contract can lead to pricing differences between a futures contract and an otherwise identical forward contract.

(B) Over-the-counter forward contracts can be customized to suit the buyer or seller, whereas futures contracts are standardized.

(C) Users of forward contracts are more able to minimize credit risk than are users of futures contracts.

(D) Forward contracts can be used to synthetically switch a portfolio invested in stocks into bonds.

(E) The holder of a long futures contract must place a fraction of the cost with an intermediary and provide assurances on the remaining purchase price.

Solution. Futures contracts are more useful than forwards for minimizing credit risk. This is due to the typically daily settlement of futures contracts. (**Answer: (C)**) □

2.4.2   Marking to Market

How does marking to market work?

As an illustration of how a futures contract works, consider the S&P 500 futures contract, which has the following specifications:

Underlying	S&P 500 index
Notional value (size)	$250× S&P 500 futures price (i.e., multiplier = 250)
Months	March, June, September, December
Settlement	Cash-settled
Mark-to-market frequency	Daily

We also assume:

Margin	10% (of the notional value of the futures contract)
Risk-free rate	6%, compounded continuously per annum
Current futures price	1,100

With a slight abuse of notation, we denote the Day-t price of the futures maturing at Day n for $t \leq n$ by $F_{t,n}$.

We now discuss various components that play a role in implementing a futures contract.

- *Margin:* Both buyers and sellers of a futures contract have to make a deposit, known as the *margin*, into an interest-bearing performance bond. This protects both parties against the failure to meet obligations. With the margin being 10% of the notional value of the contract, the *initial margin balance* is

$$
\begin{aligned}
B_0 &= \boxed{\text{Margin} \times \text{Multiplier} \times F_{0,n}} \quad\quad (2.4.1)\\
&= 10\% \times 250 \times 1,100 \\
&= 27,500.
\end{aligned}
$$

Denote by B_t the balance in the margin account at the end of Day t, for $t = 1, 2, \ldots, n$.

- *Mark-to-market proceeds:* Under an S&P 500 futures contract, adjustments will be made daily to account for the changes in futures price. At the end of day t, the margin account receives a payment, known as a *mark-to-market proceed*, which reflects the change in value of the futures contract and is given mathematically, for a long futures position,[v] by

$$
\boxed{D_t = \text{Multiplier} \times \underbrace{(F_{t,n} - F_{t-1,n})}_{\text{change in futures price}},} \quad\quad (2.4.2)
$$

which can be positive (the margin account gets credited) or negative (the margin account gets debited).

To make sense of (2.4.2), we can think in this way:

▷ Consider $t = 1$ (i.e., Day 1). If the futures contracts *were* settled at the end of Day 1, you would be obligated to pay $F_{0,n}$. In contrast, a futures that is entered on Day 1 would require a payment of $F_{1,n}$. You therefore *save* $F_{1,n} - F_{0,n}$ (which can be positive or negative) per unit of the index by owning the futures purchased on Day 0. Multiplying this quantity by the multiplier specified in the futures contract yields the mark-to-market proceed on Day 1.

▷ On Day 2, the amount of "saving" is

$$
F_{2,n} - F_{0,n} = (F_{2,n} - F_{1,n}) + \underbrace{(F_{1,n} - F_{0,n})}_{\text{Accounted for by } D_1}.
$$

---

[v]The mark-to-market proceed for a *short* futures position is

$$
D_t = \text{Multiplier} \times (F_{t-1,n} - F_{t,n}).
$$

Day	Futures Price	Mark-to-Market Proceeds	Margin Account Balance
0	1,100	(NA)	27,500
1	1,050	$-12,500$	15,004.52
2	1,120	17,500	32,506.99
3	1,180	15,000	47,512.33
4	1,200	5000	52,520.14

TABLE 2.8
Mark-to-market proceeds and margin account balance over 4 days from a long position in
S&P 500 futures contract.

Since the second term has been included in D_1, only the first term enters the
mark-to-market proceed on Day 2.

▷ Inductively, we can regard each D_t as the *incremental* "saving" or as the *incremental value* brought by the futures on Day t.

- *Recursive calculations of margin account balances:* The balance in the margin account
  at the end of Day t can be determined recursively as

$$B_t = B_{t-1} \underbrace{e^{r/365}}_{\substack{\text{accumulate previous} \\ \text{balance w/ interest}}} + D_t, \quad t = 1, 2, \ldots, n,$$

with the starting value B_0 given in (2.4.1).

As an example, suppose that the futures price drops from $1,100$ to $1,050$ on Day 1.
Then the mark-to-market proceed on Day 1 is

$$250 \times (1,050 - 1,100) = -12,500.$$

Since the initial balance has also earned interest for one day, the final balance at the
end of Day 1 is

$$27,500e^{0.06/365} - 12,500 = 15,004.52.$$

- *Maintenance margin and margin call:* Since the mark-to-market proceeds can be negative, it can happen that the margin account balance will drop substantially below the
  initial margin. For this reason, there is usually a minimum level of the margin account
  balance, called the *maintenance margin* (often set at 70%-80% of the initial margin),
  below which the investor will receive a *margin call*, requiring him/her to make an additional deposit to bring the balance back to the initial margin level.

- *Overall profit:* The profit of the futures on Day t, should you choose to leave, is obtained
  by subtracting the future value of the original margin deposit from the final margin
  balance:

$$\text{Profit}_t = B_t - B_0 e^{r \times n/365}.$$

Table 2.8 displays the evolution of the margin account balance over 4 days with hypothetical futures prices. The profit of the futures position is

$$52,520.14 - 27,500e^{0.06 \times 4/365} = 25,002.05.$$

In contrast, the profit of a 4-day forward is

$$250(1,200 - 1,100) = 25,000,$$

which is slightly lower.

We end with two examples concerned with determining the mark-to-market proceeds and margin account balances for a small number of mark-to-market periods, e.g., one or two, which is amenable to pen-and-paper calculations.

Example 2.4.3. (SOA Exam IFM Introductory Derivatives Sample Question 45: Given the mark-to-market proceed) An investor enters a long position in a futures contract on an index (F) with a notional value of $200 \times F$, expiring in one year. The index pays a continuously compounded dividend yield of 4%, and the continuously compounded risk-free interest rate is 2%.

At the time of purchase, the index price is 1100. Three months later, the investor has sustained a loss of 100. Assume the margin account earns an interest rate of 0%.

Let S be the price of the index at the end of month three.

Calculate S.

(A) 1078

(B) 1085

(C) 1094

(D) 1105

(E) 1110

Solution. Since the margin account does not earn interest, the mark-to-market frequency plays no role. The time-0 and time-0.25 futures prices, assumed to be equal to forward prices, are, respectively,

$$F_{0,1} = 1,100e^{(0.02-0.04)(1)} = 1,078.2185$$

and

$$F_{0.25,1} = Se^{(0.02-0.04)(0.75)} = 0.985112S.$$

Solving $200(F_{0.25,1} - F_{0,1}) = -100$ results in $S = \boxed{1,094.01}$. **(Answer: (C))** □

Remark. It can be shown that when the risk-free interest rate is constant, futures prices coincide with forward prices (see, e.g., page 143 of McDonald (2013) for discussions).

Example 2.4.4. (SOA Exam IFM Introductory Derivatives Sample Question 32: Range of values of index price) Judy decides to take a short position in 20 contracts of S&P 500 futures. Each contract is for the delivery of 250 units of the index at a price of 1500 per unit, exactly one month from now. The initial margin is 5% of the notional value, and the maintenance margin is 90% of the initial margin. Judy earns a continuously compounded risk-free interest rate of 4% on her margin balance. The position is marked-to-market on a daily basis.

On the day of the first marking-to-market, the value of the index drops to 1498. On the day of the second marking-to-market, the value of the index is X and Judy is not required to add anything to the margin account.

Calculate the largest possible value of X.

(A) 1490.50

(B) 1492.50

(C) 1500.50

(D) 1505.50

(E) 1507.50

Solution. The initial margin is $5\% \times 20 \times 250 \times 1,500 = 375,000$. The maintenance margin is $0.9(375,000) = 337,500$

On the day of the first marking-to-market, the margin account balance is

$$375,000e^{0.04/365} + 20(250)(1,500 - 1,498) = 385,041.0981.$$

On the day of the second marking-to-market, the margin account balance becomes

$$385,041.0981e^{0.04/365} + 20(250)(1,498 - X).$$

For this to be greater than $337,500$, we have $X < \boxed{1,507.52}$. (**Answer: (E)**) □

Remark. Strictly speaking, the mark-to-market payments should be based on the changes in the *futures* price, not the index price after one day. However, the one-day futures price is not given in the question. Also refer to the (correct) solution of the previous example.

2.5   Problems

Prepaid forwards

Problem 2.5.1. (Direct application of prepaid forward price formula) The current price of a stock is 50. It pays a dividend of $1 every 3 months, with the first dividend to be paid 3 months from today and the last dividend to be paid in 1 year.

The continuously compounded risk-free interest rate is 6%.

Calculate the price of a prepaid forward contract that expires 1 year from today, immediately after the last dividend.

Problem 2.5.2. (Given the forward price, deduce the interest rate – I) The current price of a stock is 60. A dividend of 2 will be paid 6 months from now.

The one-year forward price is 61.80.

Calculate the continuously compounded risk-free annual rate of interest.

Problem 2.5.3. (Given the forward price, deduce the interest rate – II) The current price of stock XYZ is $80. A one-year forward contract on stock XYZ has a price of $84. Stock XYZ is expected to pay a dividend of $2 six months from now and a dividend of $3 one year from now, immediately before the one-year forward contract expires.

Calculate the effective annual risk-free interest rate, assuming that it is non-negative.

Problem 2.5.4. (Price of continuous random dividends) The current price of a stock is 100. The stock pays dividends continuously at a rate proportional to its price. The dividend yields is 3%.

The continuously compounded risk-free interest rate is 7%.

Calculate the *price* of the stream of dividends to be paid in the next 5 years.

(Note: Because the dividends are stochastic, their present value is also stochastic and hence cannot be their price.)

Forwards

Problem 2.5.5. (SOA Exam IFM Introductory Derivatives Sample Question 20: Geometrically increasing discrete dividends) The current price of a stock is 200, and the continuously compounded risk-free interest rate is 4%. A dividend will be paid every quarter for the next 3 years, with the first dividend occurring 3 months from now. The amount of the first dividend is 1.50, but each subsequent dividend will be 1% higher than the one previously paid.

Calculate the fair price of a 3-year forward contract on this stock.

(A) 200

(B) 205

(C) 210

(D) 215

(E) 220

Problem 2.5.6. (Piecewise constant dividend yield) It is now January 1, 3018. You are given:

(i) The current price of the stock is 1,000.

(ii) The stock pays dividends continuously at a rate proportional to its price. The dividend yield changes throughout the year. In March, June, September, and December, the dividend yield is 3%. In other months, the dividend yield is 2%.

(iii) The continuously compounded risk-free interest rate is 9%.

Calculate the price of a 1-year forward contract.

(Hint: How many shares should you buy at time 0 to end up with exactly one share in one year?)

Problem 2.5.7. (Arbitraging a mispriced forward) The current price of stock XYZ is 120. Stock XYZ pays dividends continuously at a rate proportional to its price. The dividend yield is 4%. The continuously compounded risk-free interest rate is 6%.

You observe a 1-year forward price of 121 on stock XYZ.

Describe, giving as many details as possible, actions you could take at time 0 to exploit an arbitrage opportunity and calculate the resulting profit (per unit of stock XYZ at expiration).

Problem 2.5.8. (Fair dividend yield) You are given:

(i) The current price of a stock is 1,000.

(ii) The stock pays dividends continuously at a rate proportional to its price.

(iii) The continuously compounded risk-free interest rate is 5%.

(iv) A 6-month forward price of 1,020 is observed in the market.

Describe actions you could take to exploit an arbitrage opportunity and calculate the resulting profit (per stock unit) in each of the following cases:

(a) The dividend yield of the stock is 0.5%

(b) The dividend yield of the stock is 2%

Problem 2.5.9. (Short selling stocks with continuous proportional dividends and transaction costs) You short sold 100 shares of stock X on November 1, 3016 and closed your position on November 1, 3018.

You are given:

(i) Stock X pays dividends continuously at a rate proportional to its price. The dividend yield is 4%.

(ii) The continuously compounded risk-free interest rate is 5%.

(iii) The commission rate (for purchase or sale) is 2% of the transaction amount.

(iv) The following bid and ask prices of stock X (per share) observed at various time points:

Date	Bid Price	Ask Price
November 1, 3016	95.0	95.5
November 1, 3017	98.0	98.5
November 1, 3018	100.0	100.5

Calculate your profit measured as of November 1, 3018.

Problem 2.5.10. (SOA Exam IFM Introductory Derivatives Sample Question 70: Bounds on forward price/expected stock price) Investors in a certain stock demand to be compensated for risk. The current stock price is 100.

The stock pays dividends at a rate proportional to its price. The dividend yield is 2%. The continuously compounded risk-free interest rate is 5%.

Assume there are no transaction costs.

Let X represent the expected value of the stock price 2 years from today. Assume it is known that X is a whole number.

Determine which of the following statements is true about X.

(A) The only possible value of X is 105.

(B) The largest possible value of X is 106.

(C) The smallest possible value of X is 107.

(D) The largest possible value of X is 110.

(E) The smallest possible value of X is 111.

Problem 2.5.11. (Comparison of various forward-related quantities) Consider a stock which pays dividends continuously at a rate proportional to its price. The dividend yield is less than the interest rate, but both are positive and continuously compounded.

Rank the following quantities in ascending order (i.e., from lowest to highest):

$$
\begin{array}{rcl}
(A) & = & \text{Current stock price} \\
(B) & = & \text{One-year forward price} \\
(C) & = & \text{Two-year forward price} \\
(D) & = & \text{Two-year prepaid forward price} \\
(E) & = & \text{Expected stock price at the end of two years}
\end{array}
$$

Problem 2.5.12. (No-arbitrage interval with discrete dividends) You are given the following information:

(i) The current bid price and ask price of stock X are 50 and 51, respectively.

(ii) A dividend of 3 will be paid 6 months from now.

(iii) The continuously compounded risk-free interest rate is 6%.

(iv) The one-year forward price on stock X is 55.

(v) The only transaction costs are:

- A $1.5 transaction fee, paid at time 0, for buying or selling each unit of stock X.

- A $1 transaction fee, paid at time 0, for buying or selling a forward contract on stock X.

Describe how arbitrage profits *in one year* can be locked in using actions taken *at time 0 only*. Calculate the resulting profit (per stock unit).

Problem 2.5.13. (No-arbitrage interval with continuous dividends) You are given the following information:

(i) The current bid price and ask price of stock Y are 40 and 41, respectively.

(ii) Stock Y pays dividends continuously at a rate proportional to its price. The dividend yield is 3%.

(iii) The continuously compounded lending and borrowing rates are 6% and 7%, respectively.

(iv) The only transaction costs are:

- A $1 transaction fee, paid at time 0, for buying or selling each unit of stock B.
- A $2 transaction fee, paid at expiration, for settling a forward contract on stock B.

(v) The 3-year forward price on stock B is 38.

Describe actions you could take at time 0 to exploit an arbitrage opportunity (if any). Calculate the resulting profit (per stock unit).

Futures

Problem 2.5.14. (Margin account balance calculation given futures prices) You are given the following historical futures prices of the P&K 689 Index observed at different time points for various maturities:

Observation Date	Futures Price		
	1-month	2-month	3-month
March 1, 2018	624	626	629
April 1, 2018	712	715	718
May 1, 2018	640	643	645

On March 1, 2018, Cyrus decides to take a long position in ten 3-month P&K 689 index futures. Each contract permits the delivery of 250 units of the index. The initial margin is 10% of the notional value, and the maintenance margin is 75% of the initial margin. Cyrus earns a continuously compounded risk-free interest rate of 5% on his margin balance. The position is marked-to-market on a monthly basis.

Calculate the balance in Cyrus's margin account on the day of the second marking-to-market (i.e., on May 1, 2018).

(Hint: Only some of the futures prices are useful.)

Problem 2.5.15. (Margin call) The P&Q futures is currently trading at 1,629. The P&Q index pays dividends continuously at a rate proportional to its price. The dividend yield is 2%.

Today Peter enters into eight 3-month P&Q long futures contracts. Each contract permits the delivery of 250 units of the index. The initial margin is 10% of the notional value, and the maintenance margin is 70% of the initial margin. Peter earns the continuously compounded risk-free interest rate of 6% on his margin balance. The position is marked-to-market on a monthly basis.

Determine the range of values for the P&Q index price 1 month from now which results in a margin call.

3

Option Strategies

Chapter overview: In this chapter, we turn our attention to options and option strategies. We shall combine different option positions and obtain in this way a wide array of new derivatives which allow investors to better manage risk or speculate on asset price movements. In Section 3.1, we illustrate the use of a long put and a long call to hedge against the risks inherent in a long asset and a short asset, respectively. These results bear particular relevance to a business entity which sells its products or purchases its production inputs regularly at random future prices as part of its operating cycle. In Section 3.2, we demonstrate how calls and puts can be coupled to artificially create forwards. In the course of doing so, we obtain an all-important relationship known as put-call parity governing the prices of otherwise identical European call and put options. In Sections 3.3 and 3.4, we explore a number of new option strategies that involve the combinations of options of different types and/or strike prices and can be appropriately adopted to satisfy various risk management needs. The put-call parity developed in Section 3.2 allows us to understand the properties of and relations between the costs of some of these strategies.

Note that much of this conceptual chapter can be understood algebraically (i.e., by combining the payoff formulas of different positions), graphically (i.e., by looking at the payoff diagram of different positions), or by verbal reasoning (i.e., by thinking about how different positions work). Although all of these three approaches will be discussed, it is highly suggested that you strive to master the latter two approaches because of their pedagogical value— they are the key to understanding portfolios of options. As a key learning objective of this chapter, you should be acquainted with the shape and characteristics of the payoff diagram of a multitude of option strategies. You will find that the geometry of the payoff diagrams of these strategies can go a long way towards helping you remember their definitions.

3.1 Basic Insurance Strategies

This section introduces how options can be combined with a long or short position in the underlying asset to protect yourself against some adverse market scenarios. In each case, if we are long (resp. short) with respect to the underlying asset, then we take a counteracting position in an option which is short (resp. long) with respect to the asset.

3.1.1 Insuring a Long Position: Floors

Motivation.

At time 0, suppose that you invest in a certain asset, which for concreteness we assume is a share of a hypothetical stock called ABC, and plan to sell it in T years. If the price of stock ABC drops substantially after T years, then, against your original will, you can only sell

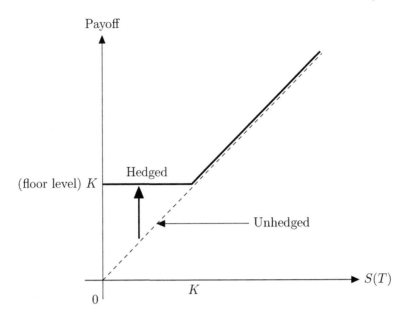

FIGURE 3.1.1
The payoff diagrams of a long asset (unhedged, dashed) and a long asset coupled with a long K-strike put (hedged, bold).

the stock for much less than the initial stock price. To hedge against this "downside risk," a put option on stock ABC can be of use.

Payoff and profit.

If you buy a K-strike T-year put option on one share of stock ABC, then your total time-T payoff is

$$\text{Payoff} = \underbrace{S(T)}_{\text{long asset}} + \underbrace{(K - S(T))_+}_{\text{long put}} = \begin{cases} K, & \text{if } S(T) < K, \\ S(T), & \text{if } S(T) \geq K, \end{cases}$$

or more compactly,

$$\text{Payoff} = \max(S(T), K),$$

which is bounded from below by the strike price K (see Figure 3.1.1 for the payoff diagram, where the payoff of the unhedged long asset between $S(T) = 0$ and $S(T) = K$ is pushed upward due to the long K-strike put). In other words, the downside risk is completely eliminated. For this reason, the put,[i] in the presence of a long stock position, is also called a *floor*—it places a "floor" on the sale price of the stock ABC (see the floor level of K in Figure 3.1.1).

A floor is a classical example of how options can serve as insurance. Intuitively, the put creates value in this context by allowing you to sell a share of stock ABC for at least K. If the time-T stock price of stock ABC is higher than K, then you can simply scrap the put and sell the stock at the higher time-T stock price. In the event that the time-T price of stock ABC falls below K, you can exercise the put option and sell the stock for K, which is the guaranteed minimum sale price. Of course, the downside protection that the floor

---

[i]A portfolio consisting of a long asset plus a long put is sometimes referred to as a *protective put*.

creates comes with a price: you need to pay the put option premium upfront (in addition to the time-0 price of stock ABC), which carries an interest cost.

Example 3.1.1. (SOA Exam IFM Introductory Derivatives Sample Question 75: Hedging an implicit long position) Determine which of the following risk management techniques can hedge the financial risk of an oil producer arising from the price of the oil that it sells.

I. Short forward position on the price of oil

II. Long put option on the price of oil

III. Long call option on the price of oil

(A) I only

(B) II only

(C) III only

(D) I, II, and III

(E) The correct answer is not given by (A), (B), (C), or (D)

Solution. Because the oil producer is to sell oil in the future, he/she will benefit from increases in the price of the oil. More precisely, his/her payoff of selling *each unit* of oil is the future (random) oil price, $S(T)$. As a result, he/she is *long* with respect to oil. The oil producer is therefore in need of positions which can help him hedge against the downside risk he/she faces arising from oil price. Here $\boxed{\text{I and II}}$ can serve this purpose.

For I, entering into a short forward position means that the oil producer agrees to sell its oil for a predetermined price in contrast to a random price at a fixed time in the future, which protects the producer from decreases in oil price.

For II, buying a put option allows the producer to sell oil for a minimum price, the strike price, which protects the producer from drops in oil price below the strike price. This sets up a floor. **(Answer: (E))**   □

Remark. (i) Watch out! The producer is not short with respect to oil, although he/she is to *sell* oil.

(ii) Note that III protects the buyer of oil, who is vulnerable to increases in oil price, not the seller.

Example 3.1.2. (SOA Exam FM Derivatives Markets Sample Question 19: Expected profit of a floor) A producer of gold has expenses of 800 per ounce of gold produced. Assume that the cost of all other production-related expenses is negligible and that the producer will be able to sell all gold produced at the market price. In one year, the market price of gold will be one of three possible prices, corresponding to the following probability table:

Gold Price in 1-year	Probability
750 per ounce	0.2
850 per ounce	0.5
950 per ounce	0.3

The producer hedges the price of gold by buying a 1-year put option with an exercise price of 900 per ounce. The option costs 100 per ounce now, and the continuously compounded risk-free interest rate is 6%.

Calculate the expected 1-year profit per ounce of gold produced.

(A) 0.00

(B) 3.17

(C) 6.33

(D) 8.82

(E) 11.74

Ambrose's comments:

This FM sample question was deleted from the set of MFE/IFM Introductory Derivatives sample questions. However, it can be done using only the knowledge of a floor.

Solution. Let's determine the 1-year profit for each of the three 1-year gold prices:

1-year Gold Price	Profit per Ounce of Gold
750 per ounce	$750 - 800 + (900 - 750)_+ - 100e^{0.06} = -6.1837$
850 per ounce	$850 - 800 + (900 - 850)_+ - 100e^{0.06} = -6.1837$
950 per ounce	$950 - 800 + (900 - 950)_+ - 100e^{0.06} = 43.8163$

The expected 1-year profit is these three possible profits weighed by their respective probabilities, or $(0.2 + 0.5)(-6.1837) + (0.3)(43.8163) = \boxed{8.82}$.
(**Answer: (D)**)

$\square$

Remark. (i) The profit when the gold price per ounce is 750 equals that when the gold price per ounce is 850 because of the floor of 900.

(ii) The expected 1-year profit without hedging is 60, which is much higher than 8.82, but in case the 1-year gold price turns out to be 750, your profit can be as low as -50.

An observation on Figure 3.1.1.

If you observe Figure 3.1.1 carefully, you will notice that the hedged position created by a long asset and a long put resembles a long call in shape. The payoff function is initially horizontal, then becomes upward sloping to the right of the strike price, just like a long call. The similarity between the hedged long asset and a long call (with the same strike) does make intuitive sense. When we own an asset and hedge it with a long put, we place a "floor" on the downside while still being able to benefit from the upside. Likewise, a long call, by definition, has a limited loss on the downside and allows its holder to profit from the upside. The difference between the payoff function of the hedged long position and that of a long call is that the former is an upward translation of the latter by K. Although the hedged long position and long call do not share the same payoff, towards the end of this section we will see that they indeed have the same profit.

3.1.2 Insuring a Short Position: Caps

Motivation.

A cap is used in the "mirror scenario" corresponding to a floor. Here you short sell one share of stock ABC at time 0 and plan to buy it back after T years in the hope that the stock price will plummet. In the adverse situation that the T-year stock price skyrockets, you will be exposed to a huge, indeed, infinite loss. Your short position can be insured by a long call on stock ABC to protect against repurchasing the stock at an exorbitant price.

Payoff and profit.

The payoff of a short asset position coupled with a long call is given by

$$\text{Payoff} = \underbrace{-S(T)}_{\text{short asset}} + \underbrace{(S(T) - K)_+}_{\text{long call}} = \begin{cases} -S(T), & \text{if } S(T) < K, \\ -K, & \text{if } S(T) \geq K, \end{cases}$$

or more compactly,

$$\text{Payoff} = -\min(S(T), K),$$

which is capped at $-K$ (see Figure 3.1.2 for the payoff diagram, where the payoff of the unhedged short asset is pushed upward to the right of the strike price K due to the long K-strike call). Note that the upside risk is eliminated: if the stock price after T years rises above K, we can still use $\$K$ instead of the higher price of $\$S(T)$ to close the short stock position. For this reason, the call option is also called a *cap* because it effectively "caps" the minimum payoff of the short asset position at $-K$. As in the case of a floor, the upside protection comes with a price: we need to pay the call option premium at time 0.

Example 3.1.3. (SOA Exam IFM Introductory Derivatives Sample Question 13: Profit on a cap) A trader shorts one share of a stock index for 50 and buys a 60-strike European call option on that stock that expires in 2 years for 10. Assume the annual effective risk-free interest rate is 3%.

   The stock index increases to 75 after 2 years.

   Calculate the profit on your combined position, and determine an alternative name for this combined position.

	Profit	Name
(A)	−22.64	Floor
(B)	−17.56	Floor
(C)	−22.64	Cap
(D)	−17.56	Cap
(E)	−22.64	"Written" Covered Call

Solution. A position consisting of a short asset position and a long call is, by definition, a cap. This eliminates Answers (A), (B), and (E). It remains to determine the profit on the cap, which is the payoff at time 2 less the future value of the initial investment. The payoff at time 2 is

$$\underbrace{-S(2)}_{\text{short stock}} + \underbrace{(S(2) - 60)_+}_{\text{long call}} = -75 + (75 - 60)_+ = -60,$$

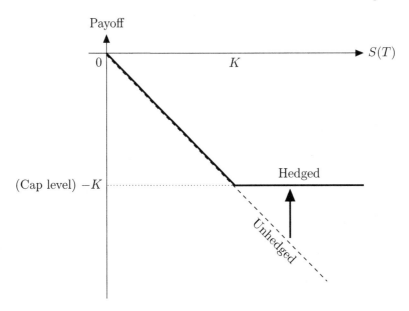

FIGURE 3.1.2
The payoff diagrams of a short asset (unhedged, dashed) and a short asset coupled with a long K-strike call (hedged, bold).

---

which is the cap level, and the investment made at time 0 is the call price less the initial stock price, or $10 - 50 = -40$, i.e., \$40 is received at time 0. The 2-year profit is $-60 - (-40)(1.03)^2 = \boxed{-17.564}$. **(Answer: (B))** $\square$

Remark. The concept of a written covered call will be covered in the next subsection.

Example 3.1.4. (SOA Exam IFM Introductory Derivatives Sample Question 74: Hedging an implicit short position) Consider an airline company that faces risk concerning the price of jet fuel.

Select the hedging strategy that best protects the company against an increase in the price of jet fuel.

(A) Buying calls on jet fuel

(B) Buying collars on jet fuel

(C) Buying puts on jet fuel

(D) Selling puts on jet fuel

(E) Selling calls on jet fuel

Ambrose's comments:

This sample question is the opposite of Example 3.1.1 (IFM Introductory Derivatives Sample Question 75).

Solution. The airline company pays for jet fuel as an input and is therefore susceptible to increases in the price of jet fuel. It is short with respect to jet fuel. Answers (B), (C), and (E) are all short with respect to jet fuel, so that leaves only Answers (A) and (D). While both (A) and (D) are long with respect to jet fuel, (D) will leave the right tail risk associated with jet fuel unhedged, but (A), which gives rise to a cap, can make the airline invulnerable to extreme increases in jet fuel price. **(Answer: (A))** □

3.1.3 Selling Insurance

We now turn to how options can be *sold* for risk management purposes. The primary motive for selling options is to earn the option premium upfront, with the trade-off being that your payoff and profit will be lower than otherwise when the terminal asset price falls within a certain region.

Short covered calls.

We refer to writing a call in conjunction with a long position in the underlying asset as *writing a covered call*, where the word "covered" emanates from the fact that the short call, which alone can have an infinite loss, is covered by the corresponding long asset position. In contrast, a *naked call* is written when we simply sell a call without taking an offsetting long position in the underlying asset. This can be a highly dangerous activity because there is no limit to your maximum loss—as dangerous as being "naked!"

The payoff as a result of writing a covered call is

$$\underbrace{S(T)}_{\text{long asset}} \underbrace{-(S(T) - K)_+}_{\text{short call}} = \begin{cases} S(T), & \text{if } S(T) \le K \\ K, & \text{if } S(T) > K \end{cases}$$

$$= \min(S(T), K),$$

which is undesirably bounded from above by the strike price K. When $S(T)$ is higher than K, we have the obligation to sell the asset for K, meaning that we lose the potential upside gain on the underlying asset. To compensate for this loss of potential upside gain, we receive the call premium at time 0 and earn interest on it.

Example 3.1.5. (SOA Exam IFM Introductory Derivatives Sample Question 46: Simple true-or-false statements) Determine which of the following statements about options is true.

(A) Naked writing is the practice of buying options without taking an offsetting position in the underlying asset.

(B) A covered call involves taking a long position in an asset together with a written call on the same asset.

(C) An American style option can only be exercised during specified periods, but not for the entire life of the option.

(D) A Bermudan style option allows the buyer the right to exercise at any time during the life of the option.

(E) An in-the-money option is one which would have a positive profit if exercised immediately.

Solution. Statement (A) is false because naked "writing" involves selling, not buying, options. Statements (C) and (D) would be true if "American" and "Bermudan" were swapped. Statement (E) is also false because being in-the-money means that there is a positive payoff, not necessarily a positive profit. Only Statement (B) is correct. (**Answer: (B)**)   □

Example 3.1.6. (SOA Exam IFM Introductory Derivatives Sample Question 47: Composition of a written covered call) An investor has written a covered call. Determine which of the following represents the investor's position.

(A) Short the call and short the stock

(B) Short the call and long the stock

(C) Short the call and no position on the stock

(D) Long the call and short the stock

(E) Long the call and long the stock

Ambrose's comments:

When we speak of a written covered call, it implicitly means that a long asset position is in existence, or else the covered call is not well defined. If there is an answer choice saying "Short the call" only, don't choose it!

Solution. By definition, writing a covered call requires shorting the call option along with simultaneous ownership in the underlying asset. (**Answer: (B)**)   □

Short covered puts.

Analogous to writing a covered call, we call a short put that comes hand in hand with a short position in the underlying asset a *written covered put*. The payoff of writing a covered put is

$$\underbrace{-S(T)}_{\text{short asset}} \underbrace{-(K-S(T))_+}_{\text{short put}} = \begin{cases} -K, & \text{if } S(T) \leq K \\ -S(T), & \text{if } S(T) > K \end{cases}$$

$$= -\max(S(T), K).$$

Although the terminal payoff is always negative, you receive the put premium initially.

In practice, a written covered put is rarely used because it leaves the unlimited upside loss potential of the original short asset position unchanged. It can be as dangerous as writing a naked call!

3.1.4 A Simple but Useful Observation: Parallel Payoffs, Identical Profit

Before closing this section, we find it beneficial to bring forward a simple observation that will be immensely useful throughout this chapter. In simple terms, this observation says that two positions, say A and B, that enjoy parallel payoff functions (i.e., one payoff function is always higher than the other payoff function by a constant amount) must possess an identical profit function. Mathematically: (the subscripts A and B denote the payoff and profit functions of the respective positions)

Parallel payoffs, identical profit

If $\text{Payoff}_A = \text{Payoff}_B + c$ for some real constant c, then $\text{Profit}_A = \text{Profit}_B$.

The validity of this assertion can be argued as follows. Because the two payoff functions are parallel, their profit functions, being the payoff functions translated downward by the future value of the initial investment, are parallel as well. If $\text{Profit}_A \neq \text{Profit}_B$, then one profit function will be everywhere higher than the other profit function. Without loss of generality, we assume that $\text{Profit}_A > \text{Profit}_B$. Since Position A is always more attractive than Position B in the sense of having a higher profit, we take a long position in A and a short position in B. The overall profit is given by $\text{Profit}_A - \text{Profit}_B$, which is strictly positive by hypothesis. We have thus constructed an arbitrage strategy. By the no-arbitrage principle, $\text{Profit}_A \neq \text{Profit}_B$ cannot be true, and we must have the equality $\text{Profit}_A = \text{Profit}_B$.

A useful special case of the above result is that when the two payoff functions are identical (i.e., $c = 0$), so are the two profit functions.

Example 3.1.7. (Identical profit) Assume the same underlying asset, same time to expiration, and same strike price for all concerned options.

Which of the following must have the same profit as a written covered call?

(A) Short call

(B) Long call

(C) Short put

(D) Long put

(E) Written covered put

Solution. The payoff diagram of a written covered call is shown below.

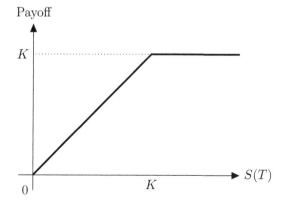

From this figure, one observes that the payoff function of a written covered call has the same shape as that of a short put. It follows from the "parallel payoffs, identical profit" phenomenon that a written covered call has the same profit function as a short put. **(Answer: (C))**   □

3.2   Put-call Parity

In Section 3.1, we saw that call and put options, when combined with positions in the underlying asset, are related in some ways. In this section, we make this relationship, called put-call parity, mathematically precise. We first introduce the concept of a synthetic forward constructed from options (thus distinct from that of Section 2.3 constructed using the underlying asset and borrowing or lending). This is crucial to deriving and understanding put-call parity.

3.2.1   Synthetic Forwards

What is a synthetic forward?

Perhaps to your astonishment, it is possible to combine options, which are meant to provide "options" in the first place, to create an *obligation* to buy or sell an asset, thereby imitating a forward. To see this, consider the following portfolio of options, which we call Portfolio $(*)$ for simplicity:

> **Portfolio $(*)$**
>
> Buy a European call option and sell a European put option, with both of them having the same underlying asset, strike price K, and time to expiration T.

What happens at the end of T years?

Case 1.   If $S(T) \geq K$, you will exercise the call option to buy the asset at a price of K, while the holder of the put option you sold will not.

Case 2.   If $S(T) < K$, then the holder of the put option you sold will exercise the put, forcing you to use K to *buy* the asset. Meanwhile, although holding a call option, you will be better off not to exercise it.

Combining both cases, we see that no matter what the asset price at expiration is, you always end up buying the underlying asset for K. The purchase price is thus guaranteed. With the asset at hand, the payoff of Portfolio $(*)$ at expiration is $S(T) - K$, which is the same as the payoff of a long forward contract with forward price K (see Figure 3.2.1 for a pictorial illustration of the payoff diagram of Portfolio $(*)$, where the turning points in the payoffs of the two options are smoothed into a straight line). In other words, we could use a long call plus a short put to mimic a long forward. We call Portfolio $(*)$ a *synthetic long forward*, where the word "synthetic" is meant in the sense that the long forward is created from other derivatives, to distinguish it from a *genuine* forward.

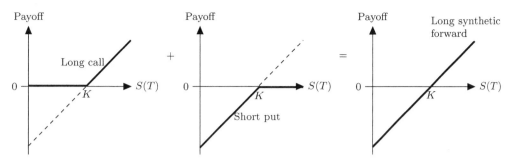

FIGURE 3.2.1
The payoff diagram of a long synthetic forward constructed by K-strike long call and short put options.

The equivalence between the synthetic long forward constructed by K-strike options and a long forward with a forward price of K can also been seen algebraically. The time-T payoff of Portfolio $(*)$ is

$$\text{Payoff of Portfolio } (*) = \underbrace{(S(T) - K)_+}_{\text{long call}} \underbrace{-(K - S(T))_+}_{\text{short put}}.$$

Due to the identity $x_+ - (-x)_+ = x$ for any real number x, the preceding payoff can be simplified into

$$\text{Payoff of Portfolio } (*) = S(T) - K,$$

without the use of the positive part function. This linear payoff function is exactly the payoff of a long forward with a forward price of K.

Example 3.2.1. (Profit on a synthetic forward) You enter into a long synthetic forward on a stock using 1-year European options. The price of a 50-strike call option is 6.8 and the price of a 50-strike put option is 4.2.
  The continuously compounded risk-free interest rate is 3%.
  The stock sells for 54 after 1 year.
  Determine your profit from the long synthetic forward.

Solution. The long synthetic forward consists of buying the 50-strike call and selling the 50-strike put. The initial investment required is $C(50) - P(50) = 6.8 - 4.2 = 2.6$. After 1 year, the payoff of the synthetic forward is $S(1) - 50 = 54 - 50 = 4$. The 1-year profit is $4 - 2.6e^{0.03} = \boxed{1.3208}$. $\square$

Differences between "synthetic" and "genuine" forwards.

There are two major differences between a synthetic forward and a genuine forward.

1. A genuine forward contract, by definition, has a zero premium; we neither pay nor receive money when we enter into a forward position at time 0. In contrast, the investment of a synthetic long (resp. short) forward at time 0 is $C(K, T) - P(K, T)$ (resp. $P(K, T) - C(K, T)$), which is generally non-zero.

   Here and in the remainder of this chapter, we denote the premium of a K-strike T-year call (resp. put) option by $C(K,T)$ (resp. $P(K,T)$). When no confusion arises, we may omit some or all of the arguments in $C(\cdot,\cdot)$ and $P(\cdot,\cdot)$.

2. With a synthetic forward, you pay the common strike price K for the asset, while you pay the fair forward price $F_{0,T}$ with a genuine forward. There can be many choices for K due to the availability of options with different strike prices in the market, whereas there is only one fair value for $F_{0,T}$.

3.2.2 The Put-call Parity Equation

Familiar idea: Same payoff, same price.

As we have seen in the previous subsection, a long synthetic forward, consisting of a long call and an otherwise identical short put, possesses the same payoff at expiration as a genuine forward with a forward price of K, which, in view of its payoff formula $S(T) - K$, is equivalent to a long asset coupled with a short zero-coupon bond with a face value of K. Symbolically and mathematically, we write

$$
\begin{array}{ccccccc}
\text{Long} & & \text{Short} & & \text{Long} & & \text{Short ZCB} \\
\text{call} & + & \text{put} & = & \text{asset} & + & \text{with face value } K
\end{array}
\quad \text{(in terms of \underline{payoff})}
$$

$$
(S(T) - K)_+ \;\; - \;\; (K - S(T))_+ \;\; = \;\; S(T) \;\; - \;\; K
$$

By the no-arbitrage principle, the two sides of the preceding equation should give rise to the same price (or cost) at time 0:

$$
\begin{array}{ccccccc}
\text{Long} & & \text{Short} & & \text{Long} & & \text{Short ZCB} \\
\text{call} & + & \text{put} & = & \text{asset} & + & \text{with face value } K
\end{array}
\quad \text{(in terms of \underline{price})}
$$

$$
C(K,T) \;\; - \;\; P(K,T) \;\; = \;\; F_{0,T}^P \;\; - \;\; \mathrm{PV}_{0,T}(K)
$$

Note that the price of receiving *exactly one unit* of the asset at time T is generally not the current stock price $S(0)$, but the T-year prepaid forward price $F_{0,T}^P$ as discussed in Section 2.2. The price equality above,

$$\boxed{C(K,T) - P(K,T) = F_{0,T}^P - \mathrm{PV}_{0,T}(K) = \mathrm{PV}_{0,T}(F_{0,T} - K),} \qquad (3.2.1)$$

is known as *put-call parity*, which provides a mathematical equation governing the prices of otherwise identical European call and put options, the (prepaid) forward price for underlying asset, and the common strike price. It is one of the most fundamental results in the theory of options and will be used intensively in the remainder of this book.

How to better remember put-call parity?

If you have trouble writing the put-call parity equation correctly, struggling whether the right-hand side should be $F_{0,T}^P - \mathrm{PV}_{0,T}(K)$ or $\mathrm{PV}_{0,T}(K) - F_{0,T}^P$, you may keep the following in mind:

- The left-hand side of put-call parity gives the (time-0) price of the long synthetic forward when viewing it from the perspective of *options* (a long call plus a short put). This is the very definition of a synthetic forward.

- The right-hand side treats the long synthetic forward as a long *forward* with a forward price of K and expresses its (time-0) price as the fair price for receiving $S(T) - K$ at time T.

The put-call parity is therefore a mathematical manifestation of the dual identity of a synthetic forward—it originates from a combination of *options*, but can also serve as a *forward*.

Furthermore, (3.2.1) can be recast in terms of the prepaid forward notation as the symmetric and appealing form

$$C(K,T) - P(K,T) = F_{0,T}^P(S) - F_{0,T}^P(K), \tag{3.2.2}$$

which expresses the difference between the call price and the put price as the difference between the prepaid forward price of the stock (which you own by exercising the call) and the prepaid forward price of the strike (which you own by exercising the put).

Example 3.2.2. (SOA Exam IFM Introductory Derivatives Sample Question 65: Which one is put-call parity?) Assume that a single stock is the underlying asset for a forward contract, a K-strike call option, and a K-strike put option.
   Assume also that all three derivatives are evaluated at the same point in time. Which of the following formulas represents put-call parity?

(A) Call Premium – Put Premium = Present Value (Forward Price – K)

(B) Call Premium – Put Premium = Present Value (Forward Price)

(C) Put Premium – Call Premium = 0

(D) Put Premium – Call Premium = Present Value (Forward Price – K)

(E) Put Premium – Call Premium = Present Value (Forward Price)

Ambrose's comments:

It is surprising that a sample exam question can be as easy as recalling the form of put-call parity!

Solution. (**Answer: (A)**) No need for explanations!   □

Consequences of put-call parity.

Inspecting (3.2.1), we can make the following useful observations:

$$\underbrace{F_{0,T} > K}_{\substack{\text{pay less relative} \\ \text{to a genuine forward}}} \qquad \Leftrightarrow \qquad \underbrace{C(K,T) - P(K,T) > 0}_{\substack{\text{need to pay extra} \\ \text{at time 0}}}$$

and

$$\underbrace{F_{0,T} < K}_{\substack{\text{pay more relative} \\ \text{to a genuine forward}}} \qquad \Leftrightarrow \qquad \underbrace{C(K,T) - P(K,T) < 0}_{\substack{\text{get compensated} \\ \text{at time 0}}}.$$

These two equivalences shed light on the economic considerations underlying the use of a synthetic forward versus a genuine forward.

Case 1.   If the forward price $F_{0,T}$ is higher than the strike price K, then it follows from put-call parity that $C(K,T) - P(K,T) > 0$. This makes sense because the holder of the long synthetic forward has the benefit of using a price lower than the fair forward price to buy the underlying asset at time T, and this benefit requires *paying* an extra amount equal to $C(K,T) - P(K,T)$ at time 0 when setting up the long synthetic forward.

Case 2. If the forward price $F_{0,T}$ is lower than the strike price K, then put-call parity implies that $C(K,T) - P(K,T) < 0$, which means that you actually *receive* a positive amount equal to $P(K,T) - C(K,T)$ when creating a synthetic forward at time 0. This amount serves as a way of compensation for buying the underlying asset at a price higher than the fair forward price at time T.

In particular, the cost required to set up the synthetic forward is zero if and only if the common strike price of the options equals the forward price:

$$F_{0,T} = K \quad \Leftrightarrow \quad C(K,T) = P(K,T).$$

In this case, the synthetic forward is nothing but a genuine forward, which requires zero investment.

Example 3.2.3. (SOA Exam IFM Introductory Derivatives Sample Question 5: Cost of a synthetic long forward) The PS index has the following characteristics:

- One share of the PS index currently sells for 1,000.

- The PS index does not pay dividends.

Sam wants to lock in the ability to buy this index in one year for a price of 1,025. He can do this by buying or selling European put and call options with a strike price of 1,025.
   The annual effective risk-free interest rate is 5%.
   Determine which of the following gives the hedging strategy that will achieve Sam's objective and also gives the cost today of establishing this position.

(A) Buy the put and sell the call, receive 23.81

(B) Buy the put and sell the call, spend 23.81

(C) Buy the put and sell the call, no cost

(D) Buy the call and sell the put, receive 23.81

(E) Buy the call and sell the put, spend 23.81

Solution. Because Sam wants to lock in the ability to *buy* the index for 1,025 in one year, he should enter into a long position in a synthetic forward. This means buying a 1,025-strike call and selling a 1,025-strike put, leaving only Answers (D) and (E).
   The cost today of establishing Sam's position, by virtue of put-call parity, is
$F_{0,1}^P(S) - F_{0,1}^P(K) = S(0) - F_{0,1}^P(K) = 1,000 - 1,025/1.05 = \boxed{23.81}$. **(Answer: (E))** □

Example 3.2.4. (SOA Exam IFM Introductory Derivatives Sample Question 53: Direct application of put-call parity) For each ton of a certain type of rice commodity, the four-year forward price is 300. A four-year 400-strike European call option costs 110.

The annual risk-free force of interest is a constant 6.5%.

Calculate the cost of a four-year 400-strike European put option for this rice commodity.

(A) 10.00

(B) 32.89

(C) 118.42

(D) 187.11

(E) 210.00

Solution. By put-call parity,

$$\underbrace{C(400, 4)}_{110} - P(400, 4) = \mathrm{PV}_{0,4}(F_{0,4} - 400) = e^{-0.065(4)}(300 - 400),$$

which gives $P(400, 4) = \boxed{187.11}$. (**Answer: (D)**) □

Example 3.2.5. (SOA Exam IFM Introductory Derivatives Sample Question 41: Which is most costly?) XYZ stock pays no dividends and its current price is 100.

Assume the put, the call and the forward on XYZ stock are available and are priced so there are no arbitrage opportunities. Also, assume there are no transaction costs.

The annual effective risk-free interest rate is 1%.

Determine which of the following strategies currently has the highest net premium.

(A) Long a six-month 100-strike put and short a six-month 100-strike call

(B) Long a six-month forward on the stock

(C) Long a six-month 101-strike put and short a six-month 101-strike call

(D) Short a six-month forward on the stock

(E) Long a six-month 105-strike put and short a six-month 105-strike call

Solution. Note that buying or selling a (genuine) forward as in (B) and (D) entails no cost at time 0, while each of (A), (C), and (E) sets up a synthetic forward and requires a non-zero net investment (unless the strike price happens to be the fair forward price). By put-call parity, the cost of each of (A), (C), and (E) takes the form $P(K) - C(K) = \mathrm{PV}_{0,0.5}(K - F_{0,0.5}) = K/1.01^{0.5} - 100$, which increases with the strike price K. When $K = 105$ (i.e., the highest strike price), $\mathrm{PV}_{0,0.5}(K - F_{0,0.5}) = 105/1.01 - 100 > 0$. In other words, the net premium for (E) is the highest among the five choices. (**Answer: (E)**) □

Example 3.2.6. (SOA Exam IFM Introductory Derivatives Sample Question 72: Synthetic short forward) CornGrower is going to sell corn in one year. In order to lock in a fixed selling price, CornGrower buys a put option and sells a call option on each bushel, each with the same strike price and the same one-year expiration date.

    The current price of corn is 3.59 per bushel, and the net premium that CornGrower pays now to lock in the future price is 0.10 per bushel.

    The continuously compounded risk-free interest rate is 4%.

    Calculate the fixed selling price per bushel one year from now.

(A) 3.49

(B) 3.63

(C) 3.69

(D) 3.74

(E) 3.84

Solution. To lock in a fixed selling price, CornGrower enters into a *short* synthetic forward by buying a put option on oil and selling an otherwise identical call option. We are given that the cost of setting up this short synthetic forward is 0.10. This means that $P - C = 0.1$. By put-call parity (assuming that corn is nondividend-paying),

$$-0.1 = C - P = S(0) - Ke^{-rT} = 3.59 - Ke^{-0.04},$$

which gives $K = \boxed{3.8406}$. (**Answer: (E)**)   □

Remark. If you incorrectly take $C - P = +0.1$, you will end up with Answer (B). The answer choices for this question were very carefully set!

Example 3.2.7. (SOA Exam IFM Introductory Derivatives Sample Question 14: Given $C(K_1) - C(K_2)$, find $P(K_2) - P(K_1)$) The current price of a nondividend-paying stock is 40 and the continuously compounded risk-free interest rate is 8%. You are given that the price of a 35-strike call option is 3.35 higher than the price of a 40-strike call option, where both options expire in 3 months.

    Calculate the amount by which the price of an otherwise equivalent 40-strike put option exceeds the price of an otherwise equivalent 35-strike put option.

(A) 1.55

(B) 1.65

(C) 1.75

(D) 3.25

(E) 3.35

Ambrose's comments:

A single application of put-call parity may seem too easy (do you agree?). For this reason, some problems may provide information about options with two different strike prices. In this case, we have to apply put-call parity twice, one for the first strike, and one for the second strike. Sometimes, solving a 2×2 linear system of equations is required.

Solution. Two applications of put-call parity show that

$$\begin{cases} C(35) - P(35) = \text{PV}_{0,0.25}(F_{0,1/4} - 35) \\ C(40) - P(40) = \text{PV}_{0,0.25}(F_{0,1/4} - 40) \end{cases}.$$

Subtracting the second equation from the first one, we get

$$\underbrace{[C(35) - C(40)]}_{3.35} + [P(40) - P(35)] = \text{PV}_{0,0.25}(5)$$

$$P(40) - P(35) = \boxed{1.55}. \quad (\textbf{Answer: (A)})$$

□

Example 3.2.8. (SOA Exam IFM Introductory Derivatives Sample Question 40: Put-call parity for two assets) An investor is analyzing the costs of two-year, European options for aluminum and zinc at a particular strike price.

For each ton of aluminum, the two-year forward price is 1400, a call option costs 700, and a put option costs 550.

For each ton of zinc, the two-year forward price is 1600 and a put option costs 550.

The annual effective risk-free interest rate is 6%.

Calculate the cost of a call option per ton of zinc.

(A) 522

(B) 800

(C) 878

(D) 900

(E) 1231

Solution. Applying put-call parity to options on aluminum and zinc yields

$$\begin{cases} 700 - 550 = \dfrac{1}{1.06^2}(1,400 - K) \\ C - 550 = \dfrac{1}{1.06^2}(1,600 - K) \end{cases}.$$

Subtracting the first equation from the second one gives $C = \boxed{878}$. (**Answer: (C)**) □

Parity arbitrage.

Put-call parity is a mathematical equation that relates the *fair* prices of European call and put options having the same strike price, maturity date and underlying asset. If it is violated, then it is possible to design an arbitrage strategy, called a *parity arbitrage* in this context, to earn risk-free profits. The next example illustrates how this can be done.

Example 3.2.9. (Parity arbitrage) You are given:

(i) The price of a nondividend-paying stock is $31.

(ii) The continuously compounded risk-free interest rate is 10%.

(iii) The price of a 3-month 30-strike European call option is $3.

(iv) The price of a 3-month 30-strike European put option is $2.25. Construct a trading strategy that will generate risk-free arbitrage profits at time 0.

Solution. It is easy to see that the call and put prices violate put-call parity:

- LHS:
$$C(30, 0.25) - P(30, 0.25) = 3 - 2.25 = 0.75,$$

- RHS:
$$F_{0,T}^P(S) - F_{0,T}^P(K) = S(0) - Ke^{-rT} = 31 - 30e^{-(0.1)(0.25)} = 1.7407.$$

To exploit arbitrage profits, we "buy the LHS" ("low") and "sell the RHS" ("high") by engaging in the following transactions:

	Cash Flows	
Transaction	Time 0	Time 0.25
Buy a 3-month 30-strike call	−3	$(S(0.25) - 30)_+$
Sell a 3-month 30-strike put	+2.25	$(30 - S(0.25))_+$
Short sell one share of the stock	+31	$-S(0.25)$
Lend $30e^{-(0.1)(0.25)} = 29.2593$	−29.2593	30
Total	0.9907	0

□

3.3   Spreads and Collars

This section and the next continue the spirit of Sections 3.1 and 3.2, and present several common strategies involving two or more options of possibly different types and strike prices. We shall examine the composition and payoff structure of and the motivation underlying each option strategy.

3.3.1 Spreads

A *spread* is a position typically consisting of only call or only put options (with the exception of the box spread), some purchased and some written, all having the same underlying asset, time to maturity, exercise style, but different strike prices. We present four most common spreads in this subsection.

Spread #1: Bull spreads.

Motive.

We have learnt several strategies to exploit the belief that an asset will appreciate, including: (1) A long forward; (2) A long call; (3) A short put. A *bull spread* is a strategy which enables you to take advantage of a rise in the asset price, but at a lower financing cost than a long call. As a trade-off, you are forced to forgo some of the upside payoff when the terminal asset price indeed rises substantially.

Composition.

A bull spread can be set up by buying a low-strike call and selling an otherwise identical high-strike call. We denote the two strike prices by K_1 and K_2, where $K_1 < K_2$, and refer to the resulting position as a K_1-K_2 *(call) bull spread*. Its payoff is

$$\underbrace{(S(T) - K_1)_+}_{\text{Long } K_1\text{-strike call}} \quad \underbrace{-(S(T) - K_2)_+}_{\text{Short } K_2\text{-strike call}} = \begin{cases} 0, & \text{if } S(T) < K_1, \\ S(T) - K_1, & \text{if } K_1 \leq S(T) < K_2, \\ K_2 - K_1, & \text{if } K_2 \leq S(T), \end{cases}$$

and is plotted in Figure 3.3.1. Here are some tips on sketching the payoff diagram more easily:

> In drawing the payoff diagram of the bull spread, it will be more convenient for you to work *from left to right* and adjust the slope of the payoff function as we pass through each strike price. This is because the payoff function of a call option is always zero to the left of the strike price. Therefore, the payoff function of the call bull spread must be zero to the left of the lower strike price K_1, then direct upward with a slope of $+1$ to the right of K_1 owing to the long K_1-strike call. Beyond K_2, the slope of the payoff function is adjusted downward by 1 as a result of the short K_2-strike call, becoming zero. This means that the payoff function levels off thereafter.

Coupling the long K_1-strike call with a short K_2-strike call has two effects. On the positive side, a K_1-K_2 long bull spread requires only $C(K_1) - C(K_2)$ as the initial investment whereas a K_1-strike long call costs $C(K_1)$. The K_2-strike call premium serves as a source of extra income that counteracts the original investment of $C(K_1)$. On the negative side, the maximum payoff of the bull spread is capped at $K_2 - K_1$, however high the terminal asset price is. When $S(T) \geq K_2$, every extra unit of payoff we gain from the long K_1-strike call is exactly offset by the additional loss on the short K_2-strike call, resulting in a flat payoff function.

Mnemonics.

Observe that:

(a) The payoff is a non-decreasing function of the terminal stock price. This explains why the strategy is called "bull," or long with respect to the underlying asset.

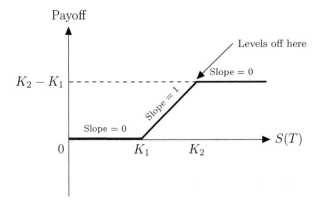

FIGURE 3.3.1
Payoff diagram of a call K_1-K_2 bull spread.

(b) The payoff increases only over a "spread" (i.e., the interval $[K_1, K_2]$), outside which the payoff becomes level.

These two observations may help you remember why a "bull spread" is called so.

Construction by put options.

Interestingly, a K_1-K_2 bull spread can also be created by buying a K_1-strike put and selling an otherwise identical K_2-strike put. Example 3.3.1 below explores the similarities and differences between a *call* bull spread and a *put* bull spread.

Example 3.3.1. (Call bull spread vs put bull spread) Student A constructs a K_1-K_2 bull spread using call options, while Student B constructs a K_1-K_2 bull spread using put options. All options are European with the same underlying asset and time to expiration.

Determine which of the following statements about these two bull spreads is/are correct.

  I. Students A and B both have a short position with respect to the underlying asset.

 II. Students A and B have the same payoff at expiration.

III. Students A and B have the same profit.

(A) I only

(B) II only

(C) III only

(D) I and III only

(E) The correct answer is not given by (A), (B), (C), or (D)

Solution. Central to the comparison between the call bull spread and the put bull spread is the payoff diagram of the latter. Because put options are used and the payoff of each is zero to the right of the strike price, we find it convenient to work from right to left. To the right of K_2, none of the put options pay off, so the payoff of the put bull spread is zero. To the left of K_2, the K_2-strike put starts to pay off. Because we are short the K_2-strike put, the payoff function traverses *downward* until it hits the lower strike price K_1 and becomes flat as a result of the long K_1-strike put.

The payoff diagram of the put bull spread is shown below.

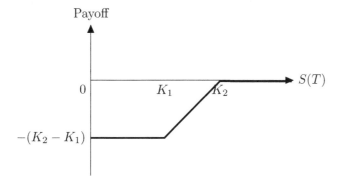

Given the payoff diagram of the put bull spread, we are now ready to determine the truth value of each statement.

I. False. Both bull spreads have a higher payoff as the terminal price of the underlying asset increases and thus are long with respect to the underlying asset.

II. False. The payoff of the call bull spread is always higher than that of the put bull spread by $K_2 - K_1$. The former is always non-negative while the latter is always non-positive.

III. True. Because the two payoff functions are parallel, they must also share the same profit function because of the "parallel payoffs, identical profit" observation in Subsection 3.1.4.

(**Answer: (C)**) □

Example 3.3.2. (SOA Exam IFM Introductory Derivatives Sample Question 50: Break-even price for a put bull spread) An investor bought a 70-strike European put option on an index with six months to expiration. The premium for this option was 1.

The investor also wrote an 80-strike European put option on the same index with six months to expiration. The premium for this option was 8.

The six-month interest rate is 0%.

Calculate the index price at expiration that will allow the investor to break even.

(A) 63

(B) 73

(C) 77

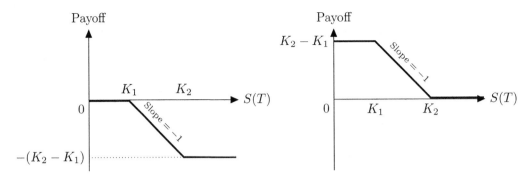

FIGURE 3.3.2
Payoff diagram of K_1-K_2 bear spreads constructed by calls (left) and by puts (right).

(D) 80

(E) 87

Solution. Buying a low-strike put and selling a high-strike put, the investor has set up a put bull spread. The initial investment is $1 - 8 = -7$, or the investor receives 7 at the outset. Because the interest rate is zero, to break even the investor has to lose 7 (i.e., the payoff has to be -7) from the put bull spread. From the payoff diagram, this is attained when $S(0.5) = 80 - 7 = \boxed{73}$. (**Answer: (B)**)   □

Spread #2: Bear Spreads.

As its name suggests, a K_1-K_2 *bear spread* is the "bearish" counterpart of a bull spread[ii] and can be constructed by selling a low-strike option and buying a high-strike option of the same type. Like a bull spread, there are two possible modes of construction:

(a) Selling a K_1-strike call and buying an otherwise identical K_2-strike call.

(b) Selling a K_1-strike put and buying an otherwise identical K_2-strike put.

The payoff diagrams are given in Figure 3.3.2. As an exercise, try to sketch these two diagrams using the slope adjustment tips given on page 75.

Example 3.3.3. (Investment of a put bear spread given call prices) You are given:

(i) The current price of a stock is 70.

(ii) The continuously compounded risk-free interest rate is 5%.

(iii) The price of a 70-strike 1-year European call option is 5.

(iv) The price of a 75-strike 1-year European call option is 3.

[ii]Note that a bull spread from the perspective of the purchaser is a bear spread from the perspective of the writer.

Calculate the amount of investment required to create a 1-year 70-75 European put bear spread.

Ambrose's comments:

This question nicely combines put-call parity discussed in Section 3.2 with bear spread.

Solution. The 70-75 put bear spread is set up by a short 70-strike put and a long 75-strike put. The investment required is $P(75) - P(70)$. By put-call parity, we have

$$\begin{cases} C(70) - P(70) = F_{0,1}^P - \text{PV}_{0,1}(70) \\ C(75) - P(75) = F_{0,1}^P - \text{PV}_{0,1}(75) \end{cases}.$$

Subtracting the second equation from the first yields

$$[P(75) - P(70)] + \underbrace{[C(70) - C(75)]}_{5-3} = \text{PV}_{0,1}(5) = 5e^{-0.05(1)},$$

which implies $P(75) - P(70) = \boxed{2.7561}$. $\square$

Spread #3: Box Spreads.

A *box spread* is an option strategy whose payoff and profit do not vary with the terminal price of the underlying asset. It is a synthetic risk-free investment created using options.

Construction 1: By synthetic forwards.

The standard way to construct a box spread is to combine a synthetic long forward at one forward price K_1 and a synthetic short forward at another forward price K_2. This effectively means that at expiration you pay K_1 and receive K_2.
    To create a box spread this way, you:

Buy a K_1-strike call, sell a K_1-strike put *(synthetic long forward)*, and

sell a K_2-strike call, buy a K_2-strike put *(synthetic short forward)*.

The initial investment is

$$C(K_1) - P(K_1) - C(K_2) + P(K_2),$$

which, by put-call parity, equals $\text{PV}_{0,T}(K_2 - K_1)$. The overall payoff at expiration is

$$\text{Payoff} = \underbrace{(S(T) - K_1)}_{\text{Long } K_1\text{-strike synthetic forward}} + \underbrace{(K_2 - S(T))}_{\text{Short } K_2\text{-strike synthetic forward}} = K_2 - K_1,$$

which is a constant. Therefore, a box spread has no asset price risk and is financially equivalent to lending when $K_1 < K_2$ (in which case the initial investment and the payoff are both fixed and positive) and borrowing when $K_1 > K_2$ (in which case the initial investment and the payoff are both fixed and negative).

Example 3.3.4. (SOA Exam IFM Introductory Derivatives Sample Question 17: Which one is a short position?) The current price for a stock index is 1,000. The following premiums exist for various options to buy or sell the stock index six months from now:

Strike Price	Call Premium	Put Premium
950	120.41	51.78
1,000	93.81	74.20
1,050	71.80	101.21

Strategy I is to buy the 1,050-strike call and to sell the 950-strike call.
Strategy II is to buy the 1,050-strike put and to sell the 950-strike put.
Strategy III is to buy the 950-strike call, sell the 1,000-strike call, sell the 950-strike put, and buy the 1,000-strike put.
Assume that the price of the stock index in 6 months will be between 950 and 1,050. Determine which, if any, of the three strategies will have greater payoffs in six months for lower prices of the stock index than for relatively higher prices.

(A) None

(B) I and II only

(C) I and III only

(D) II and III only

(E) The correct answer is not given by (A), (B), (C), or (D)

Solution. We need to judge which strategy is short with respect to the underlying asset.

- Strategy I is a call bear spread and bear spreads perform better when the price of the underlying asset is relatively low

- Strategy II is also a bear spread – it is a put bear spread.

- Strategy III is a box spread, which has no price risk. The payoff is constant at $1,000 - 950 = 50$, regardless of the price of the underlying asset in 6 months.

Only Strategies I and II are short. (**Answer: (B)**)   □

Remark. The option premiums given in the question seem completely redundant...

Construction 2: By bull and bear spreads.

The positions taken in the K_1-strike and K_2-strike options can be summarized in the following table or "box" (this is the reason why "box" spreads are called so):

Strike price	Call	Put
K_1	Long	Short
K_2	Short	Long

Reading the table horizontally gives Construction 1. Traversing the table vertically and assuming that $K_1 < K_2$, we obtain a second way of constructing a box spread:

A long K_1-K_2 call bull spread and a long K_1-K_2 put bear spread.

Example 3.3.5. (SOA Exam IFM Introductory Derivatives Sample Question 55: To create a box spread) Box spreads are used to guarantee a fixed cash flow in the future. Thus, they are purely a means of borrowing or lending money, and have no stock price risk.

Consider a box spread based on two distinct strike prices (K, L) that is used to lend money, so that there is a positive cost to this transaction up front, but a guaranteed positive payoff at expiration.

Determine which of the following sets of transactions is equivalent to this type of box spread.

(A) A long position in a (K, L) bull spread using calls and a long position in a (K, L) bear spread using puts.

(B) A long position in a (K, L) bull spread using calls and a short position in a (K, L) bear spread using puts.

(C) A long position in a (K, L) bull spread using calls and a long position in a (K, L) bull spread using puts.

(D) A short position in a (K, L) bull spread using calls and a short position in a (K, L) bear spread using puts.

(E) A short position in a (K, L) bull spread using calls and a short position in a (K, L) bull spread using puts.

Solution. To result in a positive initial cost and a guaranteed positive payoff at expiration, we should take a long position in a synthetic forward (long call and short put) and a short position in a synthetic forward at a higher forward price (short call and long put). If $K < L$, this requires that a K-L call bull spread be long and a K-L put bear spread be long. **(Answer: (A))** □

Application of box spreads: Arbitrage mispricing.

If the end result of a box spread is pure borrowing or lending, why do people take all the fuss to use a box spread in the first place? Buying or selling so many options may incur huge transaction costs. Because a box spread is a replication of a risk-free investment, an important practical application of box spread is to exploit the mispricing of options arising from the discrepancy between the interest rate implicit in a box spread and the interest rate observed in the market. The following example illustrates how this can be done given a set of real data.

Example 3.3.6. (Arbitrage by box spread) The following table shows 1-year European call and put option premiums at two strike prices:

Strike Price	Call Premium	Put Premium
40	2.58	1.79
45	1.17	5.08

The continuously compounded risk-free interest rate is 6%.

Describe actions you could take to construct an arbitrage strategy that results in profits at the end of one year using the above options and/or zero-coupon bonds only.

Solution. We begin by determining the risk-free interest rate implied by a box spread and see if it is higher or lower than the interest rate in the market. The initial investment of the box spread is $2.58 + 5.08 - 1.17 - 1.79 = 4.7$ while its 1-year payoff is $45 - 40 = 5$. The implicit interest rate satisfies $e^r = 5/4.7$, so $r = 6.19\%$, which is higher than the observed r.

To construct an arbitrage strategy, we buy the more attractive position, which is the box spread, and sell the less attractive position, which is the risk-free investment in the market. The following table summarizes the positions to be taken in different derivatives:

Derivative	Position
40-strike call	Buy
40-strike put	Sell
45-strike call	Sell
45-strike put	Buy
1-year zero-coupon bond	Sell one with face value $\$4.7e^{0.06}$ (equivalently, borrow \$4.7)

By taking this set of transactions, the initial investment is zero, and the 1-year payoff is $45 - 40 - 4.7e^{0.06} = \boxed{0.009368}$. □

Spread #4: Ratio Spreads.

Composition.

A *ratio spread* is constructed by buying and selling unequal numbers of options of the same type (call or put) and time to maturity but at different strikes, e.g., buying m calls at one strike and selling n calls at another strike with $n \neq m$.

Motives.

The reason for trading a ratio spread depends on which type of options is used, and the relative values of m and n. By varying the type and number of options, a wide variety of payoff structures can be accommodated, making it difficult to give a unified motive behind a ratio spread. Also, it is possible to adjust the ratio $m : n$ so that the initial investment of a ratio spread is zero.

Example 3.3.7. (SOA Exam IFM Introductory Derivatives Sample Question 39: How to create a ratio spread) Determine which of the following strategies creates a ratio spread, assuming all options are European.

(A) Buy a one-year call, and sell a three-year call with the same strike price.

(B) Buy a one-year call, and sell a three-year call with a different strike price.

(C) Buy a one-year call, and buy three one-year calls with a different strike price.

(D) Buy a one-year call, and sell three one-year puts with a different strike price.

(E) Buy a one-year call, and sell three one-year calls with a different strike price.

Solution. A ratio spread involves buying and selling options of the same type (call or put), same time to maturity, but different strike prices. Only Answer (E) fulfills these requirements. **(Answer: (E))** ☐

Example 3.3.8. (SOA Exam IFM Introductory Derivatives Sample Question 15: Maximum profit and loss of a ratio spread) The current price of a nondividend-paying stock is 40 and the continuously compounded risk-free interest rate is 8%. You enter into a short position on 3 call options, each with 3 months to maturity, a strike price of 35, and an option premium of 6.13. Simultaneously, you enter into a long position on 5 call options, each with 3 months to maturity, a strike price of 40, and an option premium of 2.78.

All 8 options are held until maturity.

Calculate the maximum possible profit and the maximum possible loss for the entire option portfolio.

	Maximum Profit	Maximum Loss
(A)	3.42	4.58
(B)	4.58	10.42
(C)	Unlimited	10.42
(D)	4.58	Unlimited
(E)	Unlimited	Unlimited

Solution. The initial cost to establish this position is $5 \times 2.78 - 3 \times 6.13 = -4.49$. Thus, you receive 4.49 upfront. This grows to $4.49e^{0.08(0.25)} = 4.58$ after 3 months. Working from left to right (because we are dealing with calls) and adjusting the slope of the payoff function as we pass through each strike price,, we can sketch the payoff diagram of the ratio spread as follows:

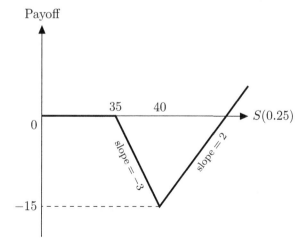

The maximum payoff and minimum payoff are unlimited and -15 (attained at $S(0.25) = 40$), respectively. It follows that the maximum profit and minimum profit are ⟨unlimited⟩ and $-15 + 4.58 = \boxed{-10.42}$, respectively. **(Answer: (C))** ☐

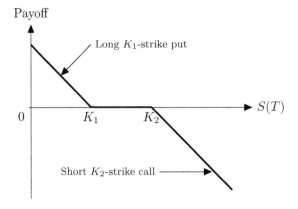

FIGURE 3.3.3
Payoff diagram of a long K_1-K_2 collar.

3.3.2 Collars

Composition.

A *collar* comprises a *long put* with strike price K_1 and a *short call* with a higher strike price K_2, where $K_1 \leq K_2$, and both options have the same underlying asset and maturity date. The difference between the strike prices of the call and put options, $K_2 - K_1$, is termed the *collar width*. If the positions in the put and call are reversed (i.e., a short K_1-strike put plus a long K_2-strike call), then a short collar is created.

Payoff.

The payoff of a long K_1-K_2 collar is given by

$$\underbrace{(K_1 - S(T))_+}_{\text{Long } K_1\text{-strike put}} \underbrace{-(S(T) - K_2)_+}_{\text{Short } K_2\text{-strike call}} = \begin{cases} K_1 - S(T), & \text{if } S(T) < K_1, \\ 0, & \text{if } K_1 \leq S(T) < K_2, \\ K_2 - S(T), & \text{if } K_2 \leq S(T). \end{cases}$$

Depicted in Figure 3.3.3, the payoff diagram of a collar shows that it is short with respect to the underlying asset. A short synthetic forward is a special case of a long collar with forward price $K_1 = K_2$.

To construct a collar, we need to pay $P(K_1) - C(K_2)$ at time 0. In general, the financing cost can be positive or negative, depending on the relative locations of K_1 and K_2, and the model that is used to price the call and put (see the "Zero-cost collars" paragraph on page 89).

Collared stock.

In practice, collars are often used to effect an insurance strategy. Suppose that we are holding a stock and would like to limit the range of possible payoffs to a specific range, say between K_1 and K_2, where $K_1 < K_2$. In this case, we can couple the long position in the underlying stock with a long collar. The resulting portfolio, called a *collared stock*, has a total payoff of

$$\text{Payoff} = S(T) + \underbrace{(K_1 - S(T))_+ - (S(T) - K_2)_+}_{\text{Long collar}} = \begin{cases} K_1, & \text{if } S(T) < K_1, \\ S(T), & \text{if } K_1 \leq S(T) < K_2, \quad (3.3.1) \\ K_2, & \text{if } K_2 \leq S(T), \end{cases}$$

which always lies between K_1 and K_2. Compared to a mere long stock position, a collared stock insures us against downside risk because of the long put position. Meanwhile, the sale of a call helps finance the purchase of the put, but undesirably rules out the possibility of a huge gain when $S(T)$ is large enough.

Example 3.3.9. (SOA Exam IFM Introductory Derivatives Sample Question 59: Payoff diagram of a long collared stock) An investor has a long position in a nondividend-paying stock, and, additionally, has a long collar on this stock consisting of a 40-strike put and 50-strike call.

Determine which of these graphs represents the payoff diagram for the overall position at the time of expiration of the options.

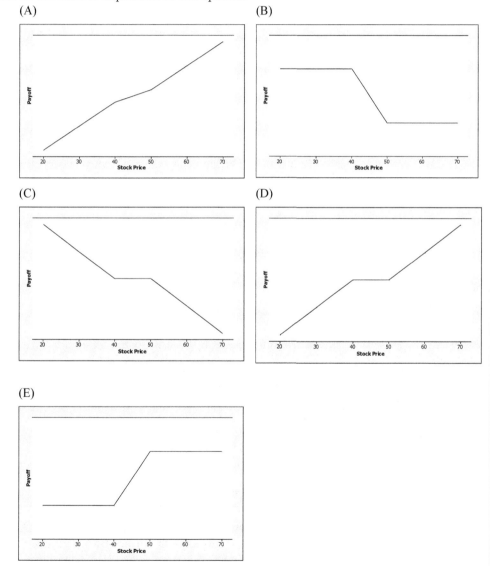

(A) (B) (C) (D) (E)

Solution. We determine the shape of the collared stock by looking at the slope of the payoff function over different stock price ranges:

Stock price range	Slope		
	$[0, 40)$	$[40, 50)$	$[50, \infty)$
Long stock	$+1$	$+1$	$+1$
Long 40-strike put	-1	0	0
Short 50-strike call	0	0	-1
Total	0	$+1$	0

Only (E) is consistent with the shape of the payoff function. (**Answer: (E)**) □

Remark. The payoff of a long collared stock has the same shape and therefore the same profit as that of a long bull spread.

Example 3.3.10. (Stock vs collared stock) Determine which of the following statements about a long stock and a long K_1-K_2 collared stock is/are always correct.

 I. Both are long with respect to the underlying stock.

 II. The long collared stock requires a lower amount of initial investment than the long stock.

 III. The long collared stock has a higher payoff than the long stock when the stock price at expiration is sufficiently high.

(A) I only

(B) II only

(C) III only

(D) I and III only

(E) The correct answer is not given by (A), (B), (C), or (D)

Solution.   I. True. Both benefit from increases in the stock price.

 II. False. The initial investment of a long stock is $S(0)$, while that of a collared stock is $S(0) + P(K_1) - C(K_2)$. In general, $P(K_1) - C(K_2)$ can be positive, zero, or negative, so we are not sure which costs more in terms of initial investment.

 III. False. When $S(T) \geq K_2$, the payoff of the collared stock is constant at K_2, which is less than the payoff of the long stock being $S(T)$.
   (**Answer: (A)**)   □

Example 3.3.11. (SOA Exam IFM Introductory Derivatives Sample Question 60: A collared stock in a business context − I) Farmer Brown grows wheat, and will be selling his crop in 6 months. The current price of wheat is 8.50 per bushel. To reduce the risk of fluctuation in price, Brown wants to use derivatives with a 6-month expiration date to sell wheat between 8.60 and 8.80 per bushel. Brown also wants to minimize the cost of using derivatives.

The continuously compounded risk-free interest rate is 2%.
Which of the following strategies fulfills Farmer Brown's objectives?

(A) Short a forward contract

(B) Long a call with strike 8.70 and short a put with strike 8.70

(C) Long a call with strike 8.80 and short a put with strike 8.60

(D) Long a put with strike 8.60

(E) Long a put with strike 8.60 and short a call with strike 8.80

Solution. In the absence of any hedging strategy, Brown's payoff per bushel of wheat is $S(0.5)$, the 6-month price of wheat. By (3.3.1), coupling his position with a long 8.6-8.8 collar (i.e., a long 8.6-strike put and a short 8.8-strike call) allows Brown to lock in a selling price between 8.6 and 8.8. (**Answer: (E)**)   □

Remark. The current price of wheat and the continuously compounded risk-free interest rate play no role in this question.

Example 3.3.12. (SOA Exam IFM Introductory Derivatives Sample Question 3: A collared stock in a business context – II) Happy Jalapenos, LLC has an exclusive contract to supply jalapeno peppers to the organizers of the annual jalapeno eating contest. The contract states that the contest organizers will take delivery of 10,000 jalapenos in one year at the market price. It will cost Happy Jalapenos 1,000 to provide 10,000 jalapenos and today's market price is 0.12 for one jalapeno. The continuously compounded risk-free interest rate is 6%.

Happy Jalapenos has decided to hedge as follows:

Buy 10,000 0.12-strike put options for 84.30 and sell 10,000 0.14-stike call options for 74.80. Both options are one-year European.

Happy Jalapenos believes the market price in one year will be somewhere between 0.10 and 0.15 per jalapeno.

Determine which of the following intervals represents the range of possible profit one year from now for Happy Jalapenos.

(A) −200 to 100

(B) −110 to 190

(C) −100 to 200

(D) 190 to 390

(E) 200 to 400

Solution. The future value of the long collar investment is $(84.3 - 74.8)e^{0.06} = 10.09$.

- The minimum profit is $\underbrace{0.12}_{\because \text{Floor}} \times 10,000 - \underbrace{1,000}_{\text{Cost}} - \underbrace{10.09}_{\text{FV of investment}} = \boxed{189.91}$

- The maximum profit is $\underbrace{0.14}_{\because \text{Covered call}} \times 10,000 - 1,000 - 10.09 = \boxed{389.91}$

(**Answer: (D)**)   □

Example 3.3.13. (SOA Exam IFM Introductory Derivatives Sample Question 43: Short collared stock) You are given:

(i) An investor short-sells a nondividend-paying stock that has a current price of 44 per share.

(ii) This investor also writes a collar on this stock consisting of a 40-strike European put option and a 50-strike European call option. Both options expire in one year.

(iii) The prices of the options on this stock are:

Strike Price	Call option	Put option
40	8.42	2.47
50	3.86	7.42

(iv) The continuously compounded risk-free interest rate is 5%.

(v) Assume there are no transaction costs.

Calculate the maximum profit for the overall position at expiration.

(A) 2.61

(B) 3.37

(C) 4.79

(D) 5.21

(E) 7.39

Solution 1 (Standard). The written collar consists of a short 40-strike put and a long 50-strike call. The initial cost of the short collared stock is $-S(0) - P(40) + C(50) = -44 - 2.47 + 3.86 = -42.61$, which grows to $-42.61e^{0.05} = -44.7947$ in one year.

Because the short collared stock is a short position, its maximum profit is attained at the smallest 1-year stock price which is $S(1) = 0$. With the maximum payoff being $-0 - (40 - 0)_+ + (0 - 50)_+ = -40$, the maximum profit is $-40 - (-44.7947) = \boxed{4.7947}$.
(**Answer: (C)**)   □

Solution 2. In terms of profit, the short collared stock is equivalent to a 40-50 put bear spread. Hence the maximum profit is achieved when $S(1) \leq 40$ and equals

$$10 - (\underbrace{7.42}_{\text{Buy 50-strike put}} \underbrace{-2.47}_{\text{Sell 40-strike put}}) \times e^{0.05} = \boxed{4.7962}. \quad (\textbf{Answer: (C)})$$

□

Remark. (i) Solution 2 is shorter and does not require the knowledge of the values of $S(0)$ and the call premiums.

(ii) The current stock price and the continuously compounded interest rate r can be deduced from the table in (iii) via two applications of put-call parity.

Zero-cost collars.

A *zero-cost collar* is a special collar for which the prices of the purchased put and the written higher-strike call exactly offset each other, resulting in a zero net investment. Two questions concerning a zero-cost collar arise naturally. First, does a zero-cost collar always exist? That is, for any option pricing model, can one always find some strike prices K_1 and K_2 with $K_1 \leq K_2$ such that the K_1-K_2 collar is costless? Second, is the design of a zero-cost collar unique, given that one exists? That is, can we find one or more pairs of strike prices (K_1, K_2) satisfying the zero-cost requirement? We answer these two questions in turn.

Existence. With respect to the existence of zero-cost collars, the answer is simple. A T-year collar with $K_1 = K_2 = F_{0,T}$ is trivially a zero-cost collar. This is a direct consequence of put-call parity:

$$C(F_{0,T}, T) - P(F_{0,T}, T) = \text{PV}_{0,T}(F_{0,T} - F_{0,T}) = 0,$$

or $C(F_{0,T}, T) = P(F_{0,T}, T)$. In fact, such a collar is identical to a short genuine forward contract, which entails zero investment by definition.

Uniqueness. The uniqueness issue is more complex. An important result is the following inequality relating the put strike and call strike of a zero-cost collar, and the forward price:

$$\boxed{K_1 \leq F_{0,T} \leq K_2.} \tag{3.3.2}$$

To prove this, we assume by way of contradiction that $K_1 > F_{0,T}$. Then put-call parity implies that

$$C(K_1) - P(K_1) = \text{PV}_{0,T}(F_{0,T} - K_1) < 0.$$

Assuming for the time being that call option prices are non-increasing in the strike price (see page 348 in Chapter 9), as intuition suggests, we have

$$P(K_1) > C(K_1) \geq C(K_2) \quad \text{for all } K_2 \geq K_1.$$

In other words, it is impossible to find $K_2 \geq K_1$ such that $P(K_1) = C(K_2)$. Similarly, if $K_2 < F_{0,T}$, then $P(K_1) = C(K_2)$ cannot be true for all $K_1 \leq K_2$. With (3.3.2), we can first fix the put strike $K_1(\leq F_{0,T})$, then find the call strike $K_2(\geq F_{0,T})$ such that $C(K_2) = P(K_1)$ (see Figure 3.3.4). The existence of K_2 is guaranteed by the continuity of the call price with respect to the strike price (see Property 2 in Subsection 9.3.1), the intermediate value theorem in calculus, as well as the fact that $C(F_{0,T}) = P(F_{0,T}) \geq 0$ and $\lim_{K \to \infty} C(K) = 0$. As K_1 can be any number less than $F_{0,T}$, there are infinitely many zero-cost collars that can be constructed using different pairs of strike prices (K_1, K_2).

Example 3.3.14. (SOA Exam IFM Introductory Derivatives Sample Question 1: Some facts about a zero-cost collar) Determine which statement about zero-cost purchased collars is FALSE.

(A) A zero-width, zero-cost collar can be created by setting both the put and call strike prices at the forward price.

(B) There are an infinite number of zero-cost collars.

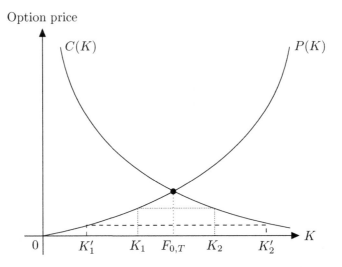

FIGURE 3.3.4
Illustration of the construction of two different zero-cost collars: a K_1-K_2 zero-cost collar and a K_1'-K_2' zero-cost collar.

(C) The put option can be at-the-money.

(D) The call option can be at-the-money.

(E) The strike price on the put option must be at or below the forward price.

Solution. Answers (A), (B), and (E) have been discussed above. If we assume that the underlying asset pays no dividends, then (D) must be false. This is because $K_2 \geq F_{0,T} = S(0)e^{rT} > S(0)$, i.e., $K_2 = S(0)$ is impossible.   □

Remark. If the underlying asset pays dividends, then (C) and (D) are both *possible* (not necessarily true). This is because there is no definite order between $F_{0,T}$ and $S(0)$—it is possible for $F_{0,T}$ to be less than $S(0)$ if the present value of the dividends is large enough.

3.4   Volatility Speculation

The option strategies in Section 3.3 (with the exception of box spread) are all directional in nature in the sense that they benefit from movements of the price of the underlying asset in a particular direction. In this section, we study how options can be fused so that the resulting portfolio is non-directional with respect to the underlying asset. These option strategies are valuable to investors who would be adversely affected or would like to speculate on vigorous movements in the price of the underlying asset in either direction in the future.

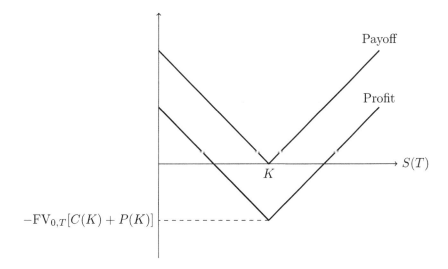

FIGURE 3.4.1
The payoff and profit diagrams of a long K-strike straddle.

3.4.1 Straddles

Composition.

A *straddle* is the combination of a long call and a long put with the same underlying asset, strike price K, and expiration time T. Often the call and put are at-the-money. This position is non-directional in nature, because if the asset price rises, the straddle benefits from the purchased call; if the asset price drops, the purchased put yields a positive payoff. This two-sided protection is achieved at the cost of paying both the call premium and put premium upfront.

Payoff and profit.

The payoff of a long K-strike straddle is

$$\underbrace{(S(T) - K)_+}_{\text{Long call}} + \underbrace{(K - S(T))_+}_{\text{Long put}} = \begin{cases} K - S(T), & \text{if } S(T) < K \\ S(T) - K, & \text{if } S(T) \geq K \end{cases}$$

$$= \boxed{|S(T) - K|}.$$

As shown in Figure 3.4.1, the payoff of a straddle is V-shaped, with the tip situated at the common strike price K. The two straight lines emanating from K can be thought of as the two legs of a man who "straddles" on the plane, hence the name "straddle" for this position. The more the terminal asset price $S(T)$ differs from the strike price K, the higher the payoff of the straddle will be. Holding a straddle can therefore be viewed as a bet on the volatility of the underlying asset being higher than that perceived by the market. The profit diagram is obtained by subtracting from the payoff diagram the future value of the total initial investment, which is the sum of the call price and the put price.

Example 3.4.1. (Break-even asset prices) In Figure 3.4.1, identify all possible asset price(s) at expiration such that the profit of a long K-strike straddle is zero.

Solution. Let $\Delta = \mathrm{FV}_{0,T}[C(K) + P(K)]$. Then the two values of $S(T)$ such that the profit of a long K-strike straddle is zero are $\boxed{K - \Delta}$ and $\boxed{K + \Delta}$. $\square$

Example 3.4.2. (Calculating the price of a straddle by put-call parity) You are given the following information:

(i) The current price of the stock is 45 per share.

(ii) The stock pays dividends continuously at a rate proportional to its price. The dividend yield is 2%.

(iii) The price of a 9-month at-the-money European call option on the stock is 3.

(iv) The continuously compounded risk-free interest rate is 4%.

Calculate the current price of a 9-month 45-strike European straddle.

Solution. The (long) 9-month 45-strike straddle comprises a long 9-month 45-strike call and a long 9-month 45-strike put. By put-call parity, the price of the put is

$$P(45) = C(45) - S(0)\mathrm{e}^{-\delta T} + K\mathrm{e}^{-rT} = 3 - 45\mathrm{e}^{-0.02(0.75)} + 45\mathrm{e}^{-0.04(0.75)} = 2.340012.$$

The price of the straddle is $C(45) + P(45) = \boxed{5.3400}$. $\square$

Written straddle.

A short straddle is the opposite of a long straddle. It involves simultaneously selling a call and a put with the same underlying asset, strike price, and expiration date. The payoff has an inverted V-shape. The profit is limited to the premiums of the call and put accumulated with interest, but the loss can be huge if the price of the underlying asset goes up or down substantially. Therefore, a short straddle position is highly risky due to unlimited potential loss. An investor may take a short straddle position if he/she strongly believes that the volatility of the underlying asset is smaller than the market's assessment.

3.4.2   Strangles

Motivation.

If one believes that the market is highly volatile so that $S(T)$ would be far away from K, a long straddle can be used. An obvious objection to the use of a straddle is the high investment required. A strangle is the more economical alternative to a straddle involving the purchase of out-of-the-money calls and puts in place of at-the-money ones as are typically used in a straddle. More specifically, a long K_1-K_2 *strangle* is established by a long put with the lower strike price K_1 and a long call with the higher strike price K_2, with $K_1 \leq S(0) \leq K_2$. The K_1-strike put is cheaper than the at-the-money put, and the K_2-strike call costs less than the at-the-money call, so a strangle requires a lower initial investment to set up than that of a straddle while allowing its holder to protect himself from or to speculate on the volatility of the underlying asset.

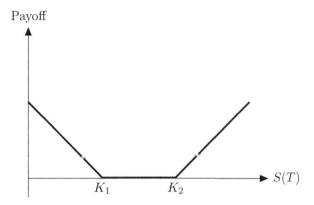

FIGURE 3.4.2
Payoff diagram of a long K_1-K_2 strangle.

Payoff.

The payoff of a strangle, sketched in Figure 3.4.2, is akin to a horrendous "strangle" (imagine that the base between K_1 and K_2 is a cord to be put around someone's throat!) shares characteristics similar to that of a straddle in the sense that both increase when $S(T)$ deviates significantly from K, but the payoff of a strangle has a flattened base on the $S(T)$-axis when $S(T)$ is between the two strike prices.

Example 3.4.3. (SOA Exam IFM Introductory Derivatives Sample Question 16: Straddle vs strangle) The current price of a nondividend-paying stock is 40 and the continuously compounded risk-free interest rate is 8%. The following table shows call and put option premiums for three-month European options of various exercise prices:

Exercise price	Call premium	Put premium
35	6.13	0.44
40	2.78	1.99
45	0.97	5.08

A trader interested in speculating on volatility in the stock price is considering two investment strategies. The first is a 40-strike straddle. The second is a strangle consisting of a 35-strike put and a 45-strike call.

Determine the range of stock prices in 3 months for which the strangle outperforms the straddle.

(A) The strangle never outperforms the straddle.

(B) $33.56 < S(T) < 46.44$

(C) $35.13 < S(T) < 44.87$

(D) $36.57 < S(T) < 43.43$

(E) The strangle always outperforms the straddle.

Solution. To begin with, the 40-strike straddle comprises a long 40-strike call and a long 40-strike put. It costs $2.78 + 1.99 = 4.77$, which grows to $4.77e^{0.08/4} = 4.8664$ in 3 months. The 35-45 strangle consists of a long 35-strike put and a long 45-strike call. It costs $0.97 + 0.44 = 1.41$, which grows to $1.41e^{0.08/4} = 1.4385$ in 3 months.

We present two solutions for solving this example.

- *Geometric solution:* The payoff diagrams of the straddle and strangle are sketched below (the diagrams are not drawn to scale):

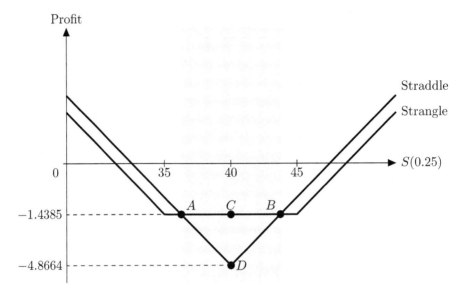

By inspection, the strangle outperforms the straddle when $S(0.25)$ lies in the interval AB. To find the horizontal coordinates of the two points, A and B, observe that the slope of line BD is 1, so that triangle BCD is an isosceles triangle with $BC = CD = -1.4385 - (-4.8664) = 3.4279$. Thus the horizontal coordinate of point B is $40 + 3.4279 = 43.4279$. Similarly, the horizontal coordinate of point A is $40 - 3.4279 = 36.5721$. The required stock price range is $S(0.25) \in \boxed{(36.5721, 43.4279)}$. (**Answer: (D)**)

- *Algebraic solution:* Comparing the profit of the strangle to that of the straddle, we solve the inequality

$$(S(0.25) - 45)_+ + (35 - S(0.25))_+ - 1.4385 > |S(0.25) - 40| - 4.8664. \quad (3.4.1)$$

and distinguish three cases:

Case 1. If $S(0.25) \geq 45$, then LHS $= S(0.25) - 45 - 1.4385 = S(0.25) - 46.4385$, while RHS $= S(0.25) - 44.8664 >$ LHS. Therefore, (3.4.1) has no solution.

Case 2. If $S(0.25) \leq 35$, then LHS $= 35 - S(0.25)$, while RHS $= (40 - S(0.25)) - 4.8664 = 35.1336 - S(0.25) >$ LHS. Again, (3.4.1) has no solution.

Case 3. It remains to investigate inequality (3.4.1) when $S(0.25) \in (35, 45)$, in which case it reduces to

$$-1.4385 > |S(0.25) - 40| - 4.8664,$$

or $S(0.25) \in \boxed{(36.5721, 43.4279)}$. (**Answer: (D)**)   □

Remark. (i) The current stock price is not used anywhere in the solution.

(ii) If I were the setter of this exam question, I might not be so kind to give you the value of r—You can deduce it from the given table of option prices!

(iii) This example reveals that there are relative merits and demerits of a strangle and straddle. While a strangle entails a lower amount of initial investment and carries a higher profit when the terminal stock price stays near $S(0)$, the terminal price at expiration has to move further away from K in order to result in a positive profit.

Example 3.4.4. (How serious is the student's mistake?) In the Midterm Exam of your favorite derivatives pricing course, a really "clever" student erroneously constructed a K_1-K_2 strangle by using *in-the-money* options. Specifically, he bought a call option with strike price K_1, and bought a put option with strike price K_2, where $K_1 < S(0) < K_2$.

Determine which of the following statements about the student's "strangle" and a correct K_1-K_2 strangle is/are correct.

I. Both positions are long with respect to the underlying asset.

II. Both positions have the same payoff at expiration.

III. Both positions have the same profit.

(A) I only

(B) II only

(C) III only

(D) II and III only

(E) The correct answer is not given by (A), (B), (C), or (D)

Solution. We first determine the payoff function of the student's "strangle":

$$\text{Payoff} = (K_2 - S(T))_+ + (S(T) - K_1)_+ = \begin{cases} K_2 - S(T), & \text{if } S(T) < K_1, \\ K_2 - K_1, & \text{if } K_1 \leq S(T) < K_2, \\ S(T) - K_1, & \text{if } K_2 \leq S(T). \end{cases}$$

The payoff diagram is sketched in Figure 3.4.3. It can be seen that the payoff of the student's "strangle" is always greater than that of a genuine K_1-K_2 strangle by $K_2 - K_1$. However, because the two payoff functions are parallel, the two positions actually share the same profit. In other words, if the student was asked to calculate the profit of the genuine strangle at a particular $S(T)$ in the Midterm Exam, the student would get the correct answer based on his own "strangle!" **(Answer: (C))** ☐

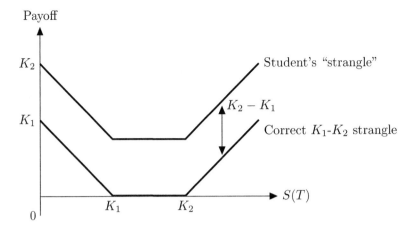

FIGURE 3.4.3
The payoff diagrams of the student's "strange" and a genuine strangle in Example 3.4.4.

3.4.3 Butterfly Spreads

Motivation.

If you believe that the market volatility is low so that $S(T)$ would be close to a certain level, say K_2, you may take a short position in a K_2-strike straddle. As we have seen in Subsection 3.4.1, the most obvious objection to a written straddle is that you will be exposed to an unlimited loss potential. To limit your losses in case you are wrong, you may insure your written straddle with:

- A long out-of-the-money, say K_3-strike call, which provides protection on the upside.

- A long out-of-the-money, say K_1-strike put, which offers protection on the downside.

The resulting portfolio consisting of a short K_2-strike straddle and a long K_1-K_3 strangle is called a K_1-K_2-K_3 *butterfly spread*, which derives its name from the fact that its payoff function resembles a butterfly (exercise some imagination, please!). Graphed in Figure 3.4.4, the payoff is the greatest when the terminal stock price is closest to K_2. Even if your belief about low volatility turns out to be incorrect and the stock price does change inexorably, your loss is still limited.

Example 3.4.5. (SOA Exam IFM Introductory Derivatives Sample Question 8) Joe believes that the volatility of a stock is higher than indicated by market prices for options on that stock. He wants to speculate on that belief by buying or selling at-the-money options.

Determine which of the following strategies would achieve Joe's goal.

(A) Buy a strangle

(B) Buy a straddle

(C) Sell a straddle

(D) Buy a butterfly spread

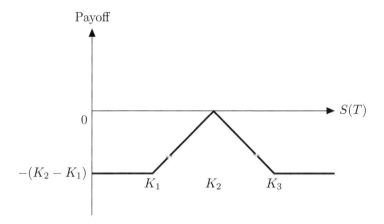

FIGURE 3.4.4
Payoff diagram of a long K_1-K_2-K_3 butterfly spread constructed by a short K_2-strike strad-dle coupled with a long K_1-K_3 strangle.

(E) Sell a butterfly spread

Solution. Only straddles use at-the-money options and buying is correct for this spec-ulation. **(Answer: (B))** □

Other methods of construction.

As discussed in the preceding paragraph, a long K_1-K_2-K_3 butterfly spread can be con-structed by a short K_2-strike straddle along with a long K_1-K_3 strangle. Alternative, indeed more popular, ways of construction using solely calls and puts include:

1. Long 1 K_1-strike call, 2 short K_2-strike calls and long 1 K_3-strike call

2. Long 1 K_1-strike put, 2 short K_2-strike puts and long 1 K_3-strike put

3. Long 1 K_1-K_2 (call/put) bull spread and long 1 K_2-K_3 (call/put) bear spread

The payoff functions of all of these constructions have the same general shape as Figure 3.4.4. As a consequence, all of them lead to the same profit.

As an illustration, let's examine the construction solely by calls and proceed from left to right (why not from right to left?).

Case 1. If $S(T) < K_1$, then none of the three call options pay off. The payoff of the entire portfolio is constant at zero.

Case 2. If $K_1 \leq S(T) < K_2$, then only the K_1-strike call pays off. As we are long 1 K_1-strike call, the portfolio payoff increases by 1 for every unit increase in $S(T)$. In other words, the payoff function after passing through K_1 is upward tilting with a slope of $+1$.

Case 3. If $K_2 \leq S(T) < K_3$, then the K_2-strike call, in addition to the K_1-strike call, also pays off. Due to the short 2 K_2-strike calls, the slope of the payoff function in this range of stock price is decreased by 2 to $-1 (= 1 - 2)$. The payoff starts to become downward sloping.

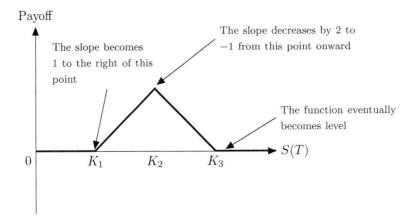

FIGURE 3.4.5
Payoff diagram of a long K_1-K_2-K_3 call (or put) butterfly spread.

Case 4.   If $K_3 \leq S(T)$, then all three kinds of call options pay off. In this range of stock
price, the slope of the payoff function is increased from -1 to $-1 + 1 = 0$. Thus
the payoff eventually levels off and becomes constant at zero again.

These considerations are visualized in Figure 3.4.5, which displays the payoff diagram of such
a long K_1-K_2-K_3 call butterfly spread. Observe that unlike Figure 3.4.4, the payoff function
in Figure 3.4.5 is always non-negative and is strictly positive when $S(T) \in (K_1, K_3)$. By
the no-arbitrage principle, the initial cost of setting up the call butterfly spread, which is
$C(K_1) + C(K_3) - 2C(K_2)$, must be non-negative. Because K_1 and K_3 are arbitrary numbers,
this non-negativity condition is a remarkably strong no-arbitrage restriction on call prices.
In Subsection 9.3.1, we will see that this condition is a special form of the so-called convexity
of option prices.

**Example 3.4.6. (SOA Exam IFM Introductory Derivatives Sample Question
9: Possible composition of a butterfly spread)** Stock ABC has the following
characteristics:

- The current price to buy one share is 100.

- The stock does not pay dividends.

- European options on one share expiring in one year have the following prices:

Strike Price	Call option price	Put option price
90	14.63	0.24
100	6.80	1.93
110	2.17	6.81

A butterfly spread on this stock has the following profit diagram.

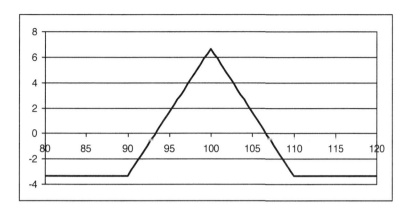

The continuously compounded risk-free interest rate is 5%.

Determine which of the following will NOT produce this profit diagram.

(A) Buy a 90 put, buy a 110 put, sell two 100 puts

(B) Buy a 90 call, buy a 110 call, sell two 100 calls

(C) Buy a 90 put, sell a 100 put, sell a 100 call, buy a 110 call

(D) Buy one share of the stock, buy a 90 call, buy a 110 put, sell two 100 puts

(E) Buy one share of the stock, buy a 90 put, buy a 110 call, sell two 100 calls

Ambrose's comments:

(A) and (B) refer to the standard ways to construct a butterfly spread by calls and puts. (C) corresponds to a 90-100 long put bull spread and a 100-110 long call bear spread. What about (D) and (E)?

Solution. Consider the shape of the payoff function of Position (D):

Stock price range	Slope			
	$[0, 90)$	$[90, 100)$	$[100, 110)$	$[110, \infty)$
Long stock	$+1$	$+1$	$+1$	$+1$
Long one 90-strike call	0	$+1$	$+1$	$+1$
Long one110-strike put	-1	-1	-1	0
Short two 100-strike puts	$+2$	$+2$	0	0
Total	$+2$	$+3$	$+1$	$+2$

This payoff function is always increasing, not in the shape of a butterfly spread. **(Answer: (D))** $\square$

Example 3.4.7. (SOA Exam IFM Introductory Derivatives Sample Question 67: Payoff calculations) Consider the following investment strategy involving put options on a stock with the same expiration date.

(i) Buy one 25-strike put

(ii) Sell two 30-strike puts

(iii) Buy one 35-strike put

Calculate the payoffs of this strategy assuming stock prices (i.e., at the time the put options expire) of 27 and 37, respectively.

(A) −2 and 2

(B) 0 and 0

(C) 2 and 0

(D) 2 and 2

(E) 14 and 0

Solution. When $S(T) = 27$, the payoff is $-2(30-27)+(35-27) = \boxed{2}$. When $S(T) = 37$, the payoff is $\boxed{0}$ because none of the puts will be exercised. (**Answer: (C)**)   □

Remark. You may also draw a payoff diagram to answer this question.

Asymmetric butterfly spreads.

It is not necessary for the peak of a butterfly spread to lie mid-way between the two extreme strike prices, K_1 and K_3. By varying the number of options you buy at K_1 and K_3, it is possible to make the peak tilted to the left or right and the butterfly spread *asymmetric*. The slope adjustment technique featured in this chapter continues to be useful for understanding the composition and payoff structure of such asymmetric butterfly spreads.

Example 3.4.8. (Given its payoff diagram, find the cost of an asymmetric butterfly spread) The payoff diagram of a certain investment strategy involving 2-year European put options on a stock is shown on the right.

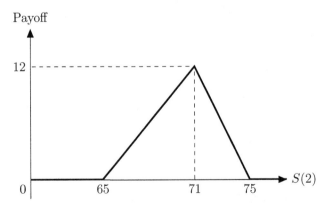

You are given:

(i)

Strike Price	Put Option Price
65	1.5
71	4.0
75	6.0

(ii) The continuously compounded risk-free interest rate is 1.5%.

Calculate the profit on the investment strategy assuming that the stock price at expiration is 70.

Solution. To determine how the investment strategy, which is just an asymmetric butterfly spread, can be constructed by the given puts, observe that the slope of the upward sloping straight line is $(12 - 0)/(71 - 65) = 2$ and the slope of the downward sloping straight line is $(0 - 12)/(75 - 71) = -3$. Going from right to left, we deduce that three 75-strike puts should be bought (so that the slope of the payoff function between 71 and 75 is -3), five 71-strike puts should be sold (so that the slope between 65 and 71 is $-3 + 5 = 2$), and two 65-strike puts should be bought (so that the slope to the left of 65 is $2 - 2 = 0$) to construct the given butterfly spread. The initial investment required is

$$3P(75) - 5P(71) + 2P(65) = 3(6.0) - 5(4.0) + 2(1.5) = 1.$$

When the 2-year stock price is 70, the payoff of the butterfly spread, by linear interpolation, is $12 \times 5/6 = 10$. Therefore, the profit on the butterfly spread is $10 - e^{0.015(2)} = \boxed{8.9695}$. □

Remark. For any butterfly spread, symmetric or asymmetric, the number of options to buy ($2 + 3 = 5$ in this example) must equal the number of options to sell (equal to 5 in this example). This ensures that the slope of the overall payoff function must be zero when the terminal asset price is sufficiently small (less than 65 in this example) or sufficiently large (greater than 75 in this example).

3.5   Problems

Problem 3.5.1. (Warm-up true-or-false items) Determine whether each of the following positions has an unlimited loss potential from adverse price movements in the underlying asset, regardless of the initial premium received.

(A) Long forward

(B) Short naked call

(C) Long collared stock

(D) Short straddle

(E) Long butterfly spread

Basic insurance strategies

Problem 3.5.2. (Simple profit calculation for a floor) You buy a stock at $300 and buy an at-the-money 9-month European put option on the stock at a price of $15.

The continuously compounded risk-free interest rate is 5%.

Calculate your 9-month profit if the 9-month stock price is $280.

Problem 3.5.3. (Considerations that go into the construction of a floor) You are selecting among various put options with different strike prices to hedge a long asset position.

Which of the following statements is true? Give your reasoning.

(A) Higher-strike puts cost more and provide higher floors.

(B) Higher-strike puts cost less and provide higher floors.

(C) Lower-strike puts cost more and provide higher floors.

(D) Lower-strike puts cost less and provide higher floors.

(E) The strike price does not matter at all.

Problem 3.5.4. (Hedging an implicit short position) Supway is a sandwich shop, one of its main production inputs being wheat.

Determine whether each of the following risk management techniques can hedge the financial risk faced by Supway arising from the price of wheat that it *buys*.

(A) Long forward on wheat

(B) Long call option on wheat

(C) Short put option on wheat

(D) Long bear spread on wheat

(E) Short collar on wheat

Problem 3.5.5. (Combining different insurance strategies – I) Assume the same underlying stock, same time to expiration, and same strike price for all derivatives in this problem.

Which of the following must have the same profit as a floor coupled with a cap? Give your reasoning.

(A) Long stock

(B) Short stock

(C) Long straddle

(D) Short straddle

(E) None of the above

Problem 3.5.6. (Combining different insurance strategies – II) Assume the same underlying stock, same time to expiration, and same strike price for all derivatives in this problem.

Which of the following must have the same profit as a floor coupled with a written covered call? Give your reasoning. There can be more than one answer.

(A) Long stock

(B) Short stock

(C) Long forward

(D) Short forward

(E) Long straddle

(F) Short straddle

Put-call parity

Problem 3.5.7. (Synthetic short forward) The current price of a nondividend-paying stock is 1,000 and the continuously compounded risk-free interest rate is 5%.

Richard wants to lock in the ability to *sell* a unit of this stock in six months for a price of 1,020. He can do this by buying or selling 6-month 1,020-strike European put and call options on the stock.

Determine the positions in the put and call options that Richard should take to achieve his objective and calculate the cost today of establishing this position.

Problem 3.5.8. (Long/short synthetic forward with discrete dividends) The current price of a stock is 100, and the continuously compounded risk-free interest rate is 10%. A $2.5 dividend will be paid every quarter, with the first dividend occurring 2 months from now.

Roger uses a K-strike European call option and a K-strike European put option on the same stock to create a synthetic six-month short forward. The initial investment is 5.3.

Calculate K.

Problem 3.5.9. (Comparing a short call with a long synthetic forward) The current price of a nondividend-paying stock is 60 and the continuously compounded risk-free interest rate is 6%.

Actuary A writes a 1-year 70-strike call option whose price is 1.50. Actuary B enters into a 1-year synthetic long forward which permits him to buy the stock for 65 in one year.

It is known that Actuary B's profit from his synthetic long forward is twice as large as Actuary A's profit from his written call option.

Determine all possible 1-year price(s) of the stock.

Problem 3.5.10. (The strike price such that the call and put prices coincide) You are given that one-year 15-strike European call and put premiums on a share of Iowa Inc. are 6.46 and 0.75, respectively.

The stock pays dividends continuously at a rate proportional to its price. The dividend yield is 2%. The effective annual interest rate is 5%.

Determine the strike price at which the call and put premiums on a share of Iowa Inc. will be equal.

Problem 3.5.11. (Given a table of option prices) Stock X pays dividends continuously at a rate proportional to its price. The dividend yield is 2%.

You are given the following option prices for European puts and calls, all written on stock X and with two years to expiration:

Strike Price	Put Price	Call Price
98	0.4394	14.3782
100	0.6975	12.7575

Calculate the current price of stock X.

Problem 3.5.12. (Profit comparison) The current price of stock ABC is 40. Stock ABC pays dividends continuously at a rate proportional to its price. The dividend yield is 2%.

You are given the following premiums of one-year European call options on stock ABC for various strike prices:

Strike	Call premium
35	7.24
40	4.16
45	2.62

The effective annual risk-free interest rate is 8%.

Let $S(1)$ be the price of the stock one year from now.

Determine the range for $S(1)$ such that a 35-strike short put produces a higher profit than a 45-strike short put, but a lower profit than a 40-strike short put.

(Note: All put positions being compared are short.)

Problem 3.5.13. (Absolute value of the difference between call/put premiums) Let $C(K)$ and $P(K)$ be the premiums of three-month K-strike European call and put options on the same stock, respectively. You are given that $|C(60) - C(65)| = 3$ and the continuously compounded risk-free interest rate is 5%.

Calculate $|P(60) - P(65)|$.

Problem 3.5.14. [HARDER!] (Parity arbitrage with discrete dividends) You are given the following information:

(i) The current price of stock Y is 30.

(ii) Dividends of 1 per unit of stock will be paid in two months and in eight months.

(iii) The continuously compounded risk-free interest rate is 6%.

(iv) The price of a 1-year 32-strike European call option on stock Y is 3.

(v) The price of a 1-year 32-strike European put option on stock Y is 4.

Describe how you could earn arbitrage profits with actions taken *at time 0 only*.

Spreads and collars

Problem 3.5.15. (How to create a bear spread by calls and puts?) Determine whether each of the following strategies creates a long bear spread, assuming that all options are European and on the same underlying asset.

(A) Buy a 45-strike call and sell a 50-strike call.

(B) Buy a 45-strike put and sell a 50-strike put.

(C) Buy a call with a price of 6 and sell a call with a price of 10.

(D) Buy a put with a price of 6 and sell a put with a price of 10.

Problem 3.5.16. (Comparing a bull spread with a bear spread) The current price of a nondividend-paying stock is 40 and the continuously compounded risk-free interest rate is 8%. The following table shows call option premiums for 3-month European options of various exercise prices:

Exercise Price	Call Premium
35	6.13
40	2.78
45	0.97

Student A constructs a 35-45 bull spread using call options.
Student B constructs a 40-45 bear spread using put options.
Determine the three-month stock price such that Student A and Student B have the same profit, and the value of the common profit.

Problem 3.5.17. (Call/put bear spread) You are given that the price of a 70-strike call option is 8.3 and the price of a 80-strike call option is 2.7, where both options expire in one year and have the same underlying asset.

The continuously compounded risk-free interest rate is 6%.

You create a one-year 70-80 long bear spread using put options. If the amount of profit from the bear spread is 4, calculate the 1-year stock price.

(Hint: Does it matter whether the bear spread is a call bear spread or a put bear spread?)

Problem 3.5.18. (Break-even price for a call bear spread) An investor wrote a 45-strike European call option on an index with three years to expiration. The premium for this option was 4.

The investor also bought a 55-strike European call option on the same index with three years to expiration. The premium for this option was 2.5.

The continuously compounded risk-free interest rate is 2%.

Calculate the index price at expiration that will allow the investor to break even.

Problem 3.5.19. (Break-even price for a bear spread) The following table shows the premiums of European call and put options having the same nondividend-paying stock, the same time to expiration but different strike prices:

Strike Price	Call Premium	Put Premium
50	8.4	0.8
60	2.6	4.7

An investor constructs a 50-60 long bear spread using the above options and breaks even at expiration.

Calculate the amount that the stock price at expiration should move from its current level.

Problem 3.5.20. (Different ways of constructing box spreads) Consider a box spread based on two distinct strike prices K and L with $K < L$ that is used to lend money, so that there is a positive cost to this transaction up front, but a guaranteed positive payoff at expiration.

Determine which of the following sets of transactions is equivalent to this type of box spread.

 I. Buy a K-strike put, buy a L-strike call, sell a K-strike call, and sell a L-strike put

 II. Buy a K-L call bull spread and sell a K-L put bull spread

 III. Buy a K-L strangle and sell a K-L strangle

(A) I only

(B) II only

(C) III only

(D) I, II, and III

(E) The correct answer is not given by (A), (B), (C), or (D)

Problem 3.5.21. (Does the option box permit arbitrage?) You are given the following information about four European options on the same underlying asset:

(i) The price of a 25-strike 1-year call option is 6.85.

(ii) The price of a 35-strike 1-year call option is 1.77.

(iii) The price of a 25-strike 1-year put option is 0.63.

(iv) The price of a 35-strike 1-year put option is 5.06.

The continuously compounded risk-free interest rate is 6%.

Describe actions you could take at time 0 using only appropriate bull/bear spread(s) and/or zero-coupon bond(s) to earn arbitrage profits at time 0. Specify the contractual details of the bull/bear spread(s) and zero-coupon bond(s) you use clearly.

Problem 3.5.22. (Profit of a collared stock) The following table shows the prices of European call and put options with the same underlying asset, time to expiration, but different strike prices:

Strike Price	Call Premium	Put Premium
17	5.16	1.35
23	2.45	4.36

Calculate the profit on a 17-23 long collared stock position if the ending stock price is 22.

Problem 3.5.23. [HARDER!] (Comparing a bear spread with a collar in terms of profit) The current price of a stock is 50. The stock will pay a single dividend of 0.75 in one month. The continuously compounded risk-free interest rate is 6%.

The following table shows the premiums of 6-month European call options on the stock:

Strike Price	Call Premium
50	5.19
60	1.96

Let S be the price of the stock six months from now.

Determine the range of values of S such that a long 50-60 6-month European bear spread outperforms (in terms of profit) a long 50-60 6-month European collar.

Problem 3.5.24. (Zero-cost collar: Forward price inequality) Apple expects to sell pork bellies 3 months from now. The current 3-month forward price for pork belly is 4 per ton. Apple has decided to buy a zero-cost collar to reduce his exposure to the price of pork belly.

Determine whether each of the following options could be a *possible* component of Apple's collar.

(A) A written call with a strike price of 4.2

(B) A written put with a strike price of 3.9

(C) A purchased call with a strike price of 4.1

(D) A purchased put with a strike price of 3.9

(E) A purchased put with a strike price of 4.0

Problem 3.5.25. (Some basic facts about zero-cost collars: Motivation, existence, uniqueness, and moneyness of call and put) Determine whether each of the following statements about zero-cost purchased collars on stocks is true or false. Assume that options at all positive strike prices are available for trading. Note that you are not given whether the underlying stock pays dividends or not.

(A) A long zero-cost collar is a bet on the price of the underlying stock going up in the future.

(B) A zero-cost collar can always be constructed by some pair of call and put options.

(C) A zero-cost collar, if it exists, can be constructed by one and only one pair of call and put options.

(D) The put option can be at-the-money.

(E) The call option must be out-of-the-money.

Problem 3.5.26. [HARDER!] (The fair price of a strange derivative – I) The payoff of a derivative contract maturing in one year is given by

$$\text{Payoff} = \begin{cases} 3S(1) - 40, & \text{if } 0 \leq S(1) < 10, \\ 4S(1) - 50, & \text{if } 10 \leq S(1) < 20, \\ S(1) + 10, & \text{if } S(1) \geq 20, \end{cases}$$

where $S(1)$ is the one-year price of the underlying nondividend-paying stock.

You are given:

(i) The continuously compounded risk-free interest rate is 5%.

(ii) The current stock price is 15.

(iii) The price of a 10-strike 1-year call option is 5.52.

(iv) The price of a 20-strike 1-year call option is 0.38.

Calculate the fair price of the above derivative.

(Hint: It will help if you first sketch the payoff diagram of the derivative. Then work from left to right and deduce how many call options should be bought/sold at the two strike prices.)

Problem 3.5.27. [HARDER!] (The fair price of a strange derivative – II) The payoff of a special 1-year derivative on a nondividend-paying stock is described by the following piecewise linear function:

$$\text{Payoff} = \begin{cases} -S(1) + 25, & \text{if } 0 \leq S(1) < 5, \\ 2S(1) + 10, & \text{if } 5 \leq S(1) < 10, \\ 4S(1) - 10, & \text{if } 10 \leq S(1) < 15, \\ -2S(1) + 80, & \text{if } 15 \leq S(1), \end{cases}$$

where $S(1)$ is the one-year price of the stock.

You are given:

(i) The continuously compounded risk-free interest rate is 4%.

(ii) The current stock price is 10.

(iii) The price of a 5-strike 1-year European put option is 0.11.

(iv) The price of a 10-strike 1-year European put option is 1.75.

(v) The price of a 15-strike 1-year European put option is 9.52.

(vi) The price of the above derivative is 3.60.

Describe actions you could take to exploit an arbitrage opportunity using only the above put options, stocks and zero-coupon bonds, and calculate the present value of the arbitrage profit (per unit of stock).

(Hint: Decide whether you should work from left to right or from right to left, and deduce how many put options should be bought/sold at the three strike prices.)

Volatility speculation

Problem 3.5.28. (Given the profit/loss of a straddle) The stock of Iowa Actuarial Corporation has been trading in a narrow range around its current price of 45 per share for months. Dividends of 2 are payable quarterly, with the first dividend payable one month from now. The continuously compounded risk-free rate of interest is 6%.

You are convinced that the stock price will remain in a narrow range around 45 over the next three months. To take advantage of your belief, you speculate on the volatility of the stock through an appropriate position in a 3-month 45-strike straddle.

It turns out that you realize a loss (i.e , your profit is negative) if the 3-month stock price moves by more than 8.52 in either direction from its current level of 45.

Calculate the price of a 3-month 45-strike put option.

Problem 3.5.29. (How volatile does the stock have to be?) The price of a stock has hovered around its current price of 70 for several months, but you believe that the stock price will break far out of that range over the next 4 months, without knowing whether it will go up or down. You have decided to take advantage of your conviction through an appropriate position in a 4-month 70-strike European straddle.

You are given:

(i) The stock pays dividends continuously at a rate proportional to its price. The dividend yield is 2%.

(ii) The price of a 4-month at-the-money European call option on the stock is 7.

(iii) The continuously compounded risk-free interest rate is 8%.

Determine how far the price of the stock has to move in either direction from its current level so that you make a profit on your straddle after 4 months.

Problem 3.5.30. (Bull spread vs straddle) The current price of stock ABC is 40. Stock ABC pays dividends continuously at a rate proportional to its price. The dividend yield is 3%. The continuously compounded risk-free interest rate is 6%.

The following table shows the premiums of two-year put options on stock ABC of various strike prices:

Strike Price	Put Premium
35	0.44
40	1.99
45	5.08

Let $S(2)$ be the price of stock ABC two years from now.

Determine the range of values of $S(2)$ for which a 35-45 long bull spread outperforms a 40-strike long straddle, both of which are on stock ABC and expire in two years.

(Suggestion: Sketch the payoff diagrams of the bull spread and the straddle.)

Problem 3.5.31. [HARDER!] (Butterfly spread, the bloodshed – I! Some theoretical questions) Determine whether each of the following statements about butterfly spread is true or false.

(A) A long butterfly spread is a bet on the volatility of the underlying asset being higher than that perceived by the market.

(B) Combining a long K_1-K_2 bear spread combined with a long K_2-K_3 bull spread produces a short K_1-K_2-K_3 butterfly spread.

(C) A long butterfly spread created using call options has the same cost as an otherwise identical long butterfly spread created using put options.

(D) The maximum profit of a butterfly spread can be made arbitrarily large by appropriately scaling up the number of options to buy and sell.

(Note: This problem, which is not as easy as it may seem, tests the theoretical understanding of butterfly spread, particularly its motive and construction. The result of part (c) will be useful in 9.3.1.)

Problem 3.5.32. [HARDER!] (Butterfly spread, the bloodshed – II!) The current price of stock XYZ is 1,000, and the continuously compounded risk-free interest rate is 8%. A dividend will be paid at the end of every 2 months over the next year, with the first dividend occurring 2 months from now. The amount of the first dividend is $1^2 = 1$, that of the second dividend is $2^2 = 4$. In general, the amount of the nth dividend is n^2, for $n = 1, 2, \ldots, 6$.

The following table shows the premiums of 1-year call options on stock XYZ of various exercise prices:

Exercise Price	Call Premium
950	120.41
1,020	80.28
1,050	71.80

You create a one-year 950-1,020-1,050 long asymmetric butterfly spread with the following characteristics:

• The maximum payoff of 210 is attained when the stock price at expiration is 1,020.

• The payoff is strictly positive only when the stock price at expiration is strictly between 950 and 1,050.

• Only put options are used in the construction of the butterfly spread.

Calculate all possible 1-year stock price(s) that result(s) in a profit of 15.

(Hint: The calculations can be tedious, but...they need not be, if you make good use of the results in the previous problem.)

Problem 3.5.33. (Butterfly spread: Given its desired characteristics) The following table shows the premiums of European call and put options having the same underlying stock, the same time to expiration but different strike prices:

Strike price	Call premium	Put premium
20	3.59	2.64
23	2.45	4.36
25	1.89	5.70

You use the above call and put options to construct an asymmetric butterfly spread with the following characteristics:

(i) The maximum payoff of 6 is attained when the stock price at expiration is 23.

(ii) The payoff is strictly positive as long as the stock price at expiration is strictly between 20 and 25.

Calculate your profit from the asymmetric butterfly spread if the stock price at expiration is 21.

(Hint: First sketch the payoff diagram of the butterfly spread from the descriptions in (i) and (ii). Then construct the butterfly spread using any method you please — why is this possible?)

Problem 3.5.34. (Given its payoff diagram, find the maximum profit of an unfamiliar derivative) The payoff diagram of an investment strategy involving 3-year European put options on a stock is shown below:

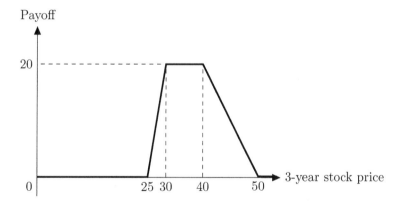

You are given:

(i)

Strike Price	Put Option Price
25	0.26
30	0.87
40	3.98
50	9.66

(ii) The continuously compounded risk-free interest rate is 2.5%.

Calculate the maximum possible profit of this investment strategy.

Part II

Pricing and Hedging of Derivatives

4

Binomial Option Pricing Models

Chapter overview: This chapter marks the beginning of *option pricing*, which is the central theme of Part II of this book. Unlike the pricing of prepaid forwards and forwards, option pricing is inherently model-dependent, i.e., different models of the stock price give rise to different option prices. Consider, for instance, the pricing of a K-strike call. Loosely speaking, the price of the call depends on the thickness of the right tail of the stock price distribution beyond the strike price K. The more likely the stock stays above K at expiration, the higher the call price. Similar considerations apply to a put, for which the left tail of the stock price distribution plays a pivotal role.

In this book, you will learn two broad classes of option pricing models, namely, the discrete-time binomial option pricing model (Chapter 4 and part of Chapter 8) and the continuous-time Black-Scholes model (Chapters 5 to 8). The discrete-time binomial model is intuitively easy to understand and, albeit simplistic to some extent, gives you the valuable intuition that carries over to the more complicated Black-Scholes model. More importantly, all derivatives can be priced and hedged in this discrete-time framework, at least in theory. The binomial model therefore forms the ideal starting point for understanding option pricing.

This chapter describes the mechanics of valuing options in the binomial option pricing model. Section 4.1 sets the stage, introducing the one-period binomial model and illustrating many of the essential ideas that underlie option pricing in general. In particular, the conceptually and practically important methods of pricing by replication and risk-neutral pricing are presented with brief discussions on their pros and cons. The valuation task is continued in Section 4.2, where we extend the simple one-period model to multiple periods, and the method of risk-neutral pricing remains to thrive. American options are treated in Section 4.3, in which we will see that pricing European options and pricing American options share essentially the same fundamental ideas. Options on other underlying assets such as exchange rates and futures are studied in Section 4.4. Finally, Section 4.5 reconciles the apparent inconsistency between risk-neutral valuation and traditional discounted cash flow valuation.

4.1 One-period Binomial Trees

4.1.1 Pricing by Replication

Model setting.

Suppose that the current time is 0. Consider a primitive market comprising a risky stock and a risk-free (zero-coupon) bond (also known as a bank account) earning a continuously compounded risk-free interest rate of r. The stock has a current price of S_0 and pays dividends continuously at a rate proportional to its price, with a dividend yield[i] of δ.

In a one-period binomial stock price model whose duration is h[ii] (in years), the stock price at the end of the period is assumed to take only two possible values (hence the term "binomial"):

$$\boxed{S_u := u \times S_0 \quad \text{or} \quad S_d := d \times S_0,}$$

where the subscripts "u" and "d" on "S" suggest "up" and "down,"[iii] respectively, and the proportional constants u and d are the multiplicative factors applied to the initial stock price to form the time-h stock price. Between time 0 and time h, the stock price stays put at the initial stock price. Because the stock price in h years is a two-point (hence binomial) random variable and the evolution of the stock price, depicted in Figure 4.1.1, resembles a tree[iv], we also refer to a binomial stock price model as a *binomial tree model*. To visualize the development of future stock prices, tree diagrams such as the one in Figure 4.1.1 will be intensively used in this chapter. For expository convenience, we say that we are in the u *node* (resp. d *node*) of the binomial tree when the time-h stock price is S_u (resp. S_d).[v] The initial node is nothing but a synonym for the initial time.

In the context of this one-period binomial model, we are interested in a generic European derivative (e.g., call, put, bull spread, straddle, etc.) maturing at time h written on the above stock with the following payoff structure:

- If the stock price at time h is S_u, then the payoff of the derivative is V_u.

- If the stock price at time h is S_d, then the payoff of the derivative is V_d.

Our principal objective is to determine the time-0 price (or current price) of this derivative, denoted by V_0.

Pricing by replication.

As with forwards, the fundamental idea behind pricing options is replication: to invest in assets available in the market to replicate the payoffs of the security of interest. In the current binomial tree setting, we have two assets at our disposal: (1) the risky stock; and (2) the risk-free bond. If we can invest in these two assets in such a way that the payoff of the resulting portfolio is always identical to that of the concerned derivative, then the no-arbitrage principle mandates that the fair price of the derivative coincides with the cost of setting up this portfolio.

Mathematically, suppose that our portfolio consists of Δ shares of the stock and $\$B$ in the risk-free bond (i.e., lending $\$B$; you may regard B as "bonds"). The cost of this

---

[i] In the binomial tree framework, we will only treat stocks paying continuous proportional dividends. Binomial tree models involving stocks that pay discrete dividends are studied, for example, in Section 11.4 of McDonald (2013).

[ii] In binomial tree models, the symbol h typically denotes the length of each period. In a one-period model, it also equals the entire duration of the tree T.

[iii] Associating "d" with "down" can occasionally be misleading. As we will see in (4.1.6), it is possible that $d \geq 1$ if r is substantially larger than δ.

[iv] The tree diagram, although aesthetically appealing, may encourage the misguided impression that the evolution of the stock price is uniform over time. In reality, the stock price stays put at S_0 and moves momentarily to either S_u or S_d at the end of the period.

[v] In a break with the notation in other parts of this book, where we write $S(t)$ as the time-t price of the underlying stock, in binomial tree models we distinguish stock prices by their locations in the binomial tree (e.g., S_0, S_u, S_d), with our attention paid to not only the time point of interest, but also the particular stock node in question.

Time 0   Time h

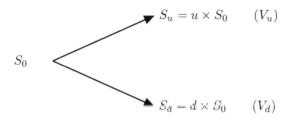

FIGURE 4.1.1
A generic one-period binomial stock price model. The derivative payoffs are shown in parentheses.

portfolio is $\Delta S_0 + B$, and its payoff at time h is

$$\begin{cases} (\Delta e^{\delta h})S_u + Be^{rh}, & \text{if the time-}h \text{ stock price is } S_u, \\ (\Delta e^{\delta h})S_d + Be^{rh}, & \text{if the time-}h \text{ stock price is } S_d. \end{cases}$$

Note that Δ shares of the stock at time 0 grow, because of the reinvestment of dividends, to $\Delta e^{\delta h}$ shares at time h. We try to choose Δ and B so that the payoff of the portfolio matches that of the derivative under all circumstances, i.e., Δ and B satisfy

$$\begin{cases} \Delta S_u e^{\delta h} + Be^{rh} = V_u \\ \Delta S_d e^{\delta h} + Be^{rh} = V_d \end{cases}.$$

This 2×2 linear system in the two variables Δ and B can be easily solved to yield[vi]

$$\boxed{\Delta = e^{-\delta h}\left(\frac{V_u - V_d}{S_u - S_d}\right) = e^{-\delta h}\left[\frac{V_u - V_d}{S_0(u - d)}\right], \quad B = e^{-rh}\left(\frac{uV_d - dV_u}{u - d}\right).}$$ (4.1.1)

With these choices of Δ and B, the portfolio successfully replicates the derivative in terms of payoff regardless of the level of the time-h stock price. For this reason, we call this portfolio the *replicating portfolio*, represented by (Δ, B). Consequently, the fair time-0 price of the derivative should equal the cost of the replicating portfolio, which in turn is

$$\boxed{V_0 = \Delta S_0 + B,}$$ (4.1.2)

with Δ and B given in (4.1.1). Our pricing task is thus completed!

Example 4.1.1. (CAS Exam 3 Spring 2007 Question 16: Calculating Δ) A nondividend-paying stock, S, is modeled by the binomial tree shown below.

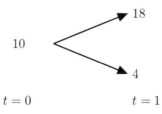

[vi]We assume that $S_u \neq S_d$, so that the stock is genuinely risky.

A European call option on S expires at $t = 1$ with strike price $K = 12$.
Calculate the number of shares of stock in the replicating portfolio for this option.

(A) Less than 0.3

(B) At least 0.3, but less than 0.4

(C) At least 0.4, but less than 0.5

(D) At least 0.5, but less than 0.6

(E) At least 0.6

Solution. Note that the payoff of the call is either $C_u = (18 - 12)_+ = 6$ or $C_d = (4 - 12)_+ = 0$. By (4.1.1), the number of shares of stock in the replicating portfolio for the call is

$$\Delta = e^{-\delta h}\left(\frac{C_u - C_d}{S_u - S_d}\right) = \frac{6 - 0}{18 - 4} = \boxed{0.4286}. \quad \textbf{(Answer: (C))}$$

$\square$

Example 4.1.2. (CAS Exam 8 Spring 2004 Question 27: Calculating call price) The price of a nondividend-paying stock is currently $50.00. It is known that at the end of two months, it will be either $54.00 or $46.00. The risk-free interest rate is 9.0% per annum with continuous compounding.

Calculate the value of a two-month European call option with a strike price of $48 on this stock. Show all work.

Solution. Note that the tree parameters are $u = 54/50 = 1.08$ and $d = 46/50 = 0.92$, and the payoff of the call is either $C_u = (54 - 48)_+ = 6$ or $C_d = (46 - 48)_+ = 0$. By (4.1.1),
$$\Delta = e^{-\delta h}\left(\frac{C_u - C_d}{S_u - S_d}\right) = \frac{6 - 0}{54 - 46} = 0.75$$
and
$$B = e^{-rh}\left(\frac{uC_d - dC_u}{u - d}\right) = e^{-0.09(2/12)}\left[\frac{-0.92(6)}{1.08 - 0.92}\right] = -33.9864.$$
The call price is
$$C_0 = \Delta S_0 + B = 0.75(50) - 33.9864 = \boxed{3.5136}.$$

$\square$

Example 4.1.3. (SOA Exam MFE Spring 2007 Question 14: Our old friend from Chapter 3 – straddle!) For a one-year straddle on a nondividend-paying stock, you are given:

(i) The straddle can only be exercised at the end of one year.

(ii) The payoff of the straddle is the absolute value of the difference between the strike price and the stock price at expiration date.

(iii) The stock currently sells for $60.

(iv) The continuously compounded risk-free interest rate is 8%.

(v) In one year, the stock will either sell for $70 or $45.

(vi) The option has a strike price of $50.00.

Calculate the current price of the straddle.

(A) $0.90

(B) $4.80

(C) $9.30

(D) $14.80

(E) $15.70

Ambrose's comments:

Using a binomial tree, it is easy to value any derivatives, not necessarily calls or puts, however complicated their payoffs are. Even a straddle can be easily handled!

Solution. Consider a replicating portfolio consisting of Δ shares of stock and B amount of cash. Matching the payoffs in the up and down scenarios, we solve

$$\begin{cases} 70\Delta + e^{0.08}B = |50 - 70| = 20, \\ 45\Delta + e^{0.08}B = |50 - 45| = 5, \end{cases}$$

which gives $\Delta = 0.6$ and $B = -20.3086$. The current price of the derivative is the same as the portfolio value at time 0, which is

$$60(0.6) + (-20.3086) = \boxed{15.6914}. \quad \textbf{(Answer: (E))}$$

$\square$

Remark. We will revisit this example in the next subsection using an alternative method.

4.1.2 Risk-neutral Pricing

Pricing formula viewed as a discounted expectation.

Pricing via replication is the fundamental, economically correct way of determining the fair price of a derivative. The actual computations, as you experienced in Subsection 4.1.1, can be cumbersome and lack insights. It turns out that the pricing formula given in equation (4.1.2) can be rearranged into a form that admits a very useful probabilistic interpretation.

To see this, we plug in the expressions for Δ and B into the pricing formula, yielding

$$
\begin{aligned}
V_0 &= \Delta S_0 + B \\
&= \mathrm{e}^{-\delta h}\left[\frac{V_u - V_d}{S_0(u-d)}\right]S_0 + \mathrm{e}^{-rh}\left(\frac{uV_d - dV_u}{u-d}\right) \\
&= \mathrm{e}^{-rh}\left[\left(\frac{\mathrm{e}^{(r-\delta)h} - d}{u-d}\right)V_u + \left(\frac{u - \mathrm{e}^{(r-\delta)h}}{u-d}\right)V_d\right].
\end{aligned}
$$

If we define

$$
p^* = \frac{\mathrm{e}^{(r-\delta)h} - d}{u-d}, \tag{4.1.3}
$$

then the time-0 price of the derivative can be written more compactly as

$$
\boxed{V_0 = \mathrm{e}^{-rh}[p^* V_u + (1-p^*)V_d].} \tag{4.1.4}
$$

We will show shortly that p^* lies between 0 and 1 for the binomial model to make sense (see Subsection 4.1.3). For now, it is enough to notice that: If we "regard" p^* as the "probability" that the stock price will go up in h years, then $p^* V_u + (1-p^*)V_d$ resembles the "expected value" of the derivative payoff at expiration. (4.1.4) then says that the price of the derivative can be interpreted as the *expected derivative payoff, discounted at the risk-free interest rate*:

$$
V_0 = \mathrm{e}^{-rh}\mathbb{E}^*[\text{Payoff at expiration}]. \tag{4.1.5}
$$

Here we use the asterisk ($*$) to emphasize that the expectation $\mathbb{E}^*[\cdot]$ is computed under the assumption that p^* is the probability that the stock price will increase at the end of the period. It should be firmly kept in mind that p^* is generally not the same as the probability that the stock will go up in price *in the real world*. It is an artificial, mathematical quantity which, when thought of as a probability, allows one to associate equation (4.1.4) with a discounted expectation.

What is so special about p^*?

The probability-looking quantity p^* is commonly designated as the *risk-neutral probability* (also known as pseudo-probabilities) of an increase in the stock price (or simply an up move), and equations (4.1.4) or (4.1.5) as the *risk-neutral pricing formula*. To explain the terminology, it will be useful to look at how a *risk-neutral* investor behaves. Such an investor cares only about the expected return on an asset, not on its other characteristics such as riskiness. He/she will be indifferent between the risky stock and the risk-free bond if both of them earn the same rate of r. In fact, p^* and $1 - p^*$ are the unique "probabilities" such that this indifference prevails, with the initial price of the risky stock equal to the discounted expected stock price *with dividends*:

$$
S_0 = \mathrm{e}^{-rh} \times \mathrm{e}^{\delta h}[p^* S_u + (1-p^*)S_d] = \mathrm{e}^{-rh} \times \mathrm{e}^{\delta h}\mathbb{E}^*[\text{time-}h\text{ stock price}].
$$

This can be readily checked by

$$
\begin{aligned}
\mathrm{e}^{-rh} \times \mathrm{e}^{\delta h}[p^* S_u + (1-p^*)S_d] &= S_0\mathrm{e}^{-(r-\delta)h}[p^* u + (1-p^*)d] \\
&= S_0\mathrm{e}^{-(r-\delta)h}\left[u\left(\frac{\mathrm{e}^{(r-\delta)h} - d}{u-d}\right) + d\left(\frac{u - \mathrm{e}^{(r-\delta)h}}{u-d}\right)\right] \\
&= S_0\mathrm{e}^{-(r-\delta)h} \times \frac{(u-d)\mathrm{e}^{(r-\delta)h}}{u-d} \\
&= S_0.
\end{aligned}
$$

At all other probabilities, the investor would strictly prefer the risky stock or the risk-free bond. This unique association with risk-neutrality makes p^* be termed the risk-neutral probability.

Again, it is imperative to emphasize that we are not assuming that investors in practice are risk-neutral. Rather, risk-neutral pricing provides an interpretation of the pricing formula presented in (4.1.4), which in turn arises from the no-arbitrage hedging argument.

To recap, the use of the risk-neutral pricing formula involves a simple three-step procedure:

Step 1. Identify the risk-neutral probability p^* that the stock price will go up.

Step 2. Calculate the expected payoff of the derivative under the probabilities p^* and $1 - p^*$.

Step 3. Discount the expected payoff in Step 2 at the risk-free interest rate.

Example 4.1.4. (Example 4.1.3 revisited) Rework Example 4.1.3 using risk-neutral valuation.

Solution. The risk-neutral probability of an "up" movement is

$$p^* = \frac{60e^{0.08} - 45}{70 - 45} = 0.7999.$$

Using risk-neutral valuation, the current price of the straddle is

$$e^{-0.08}\left[20p^* + 5(1 - p^*)\right] = \boxed{15.6914}. \quad \textbf{(Answer: (E))}$$

□

An intuitive explanation of why risk-neutral pricing works.

Risk-neutral pricing is one of the deepest results in modern finance. Not only does it have no obvious connections to replication (which is the origin of risk-neutral pricing), it also transforms the pricing problem to the computation of the *expected present value* of the derivative payoff, a concept that is familiar to actuaries (perhaps you have computed expected present values in a life contingencies course!). However, this expected present value is puzzling in two respects:

(i) The expectation is taken with respect to the so-called risk-neutral probability measure, which is generally different from the real probability measure.

(ii) The discounting is performed using the risk-free interest rate, not a risk-adjusted rate that reflects the riskiness of the derivative.

Useful and elegant as it is, risk-neutral pricing is at the same time very perplexing and counterintuitive. The inevitable question is:

Why can we simply "pretend" that investors are risk-neutral and that the derivative earns the risk-free rate r? Why?

An intuitive explanation of the "validity" of risk-neutral pricing, though not possibly tested in a multiple-choice exam environment, is clearly in order from an educational point of view.

To better understand why risk-neutral pricing works, consider the following two worlds, both of which consist of the same set of assets having the same current prices and realizable future prices, but with different probabilities of attaining these future prices:

World 1. This is the real world in which we live. The real probability that the stock price will go up is p (see Section 4.5 for more discussions on how real probabilities can be used in the valuation procedure).

World 2. This is the risk-neutral world in which investors care only about expected returns, the probability that the stock price will go up is the risk-neutral probability p^*, and the stock earns the risk-free rate.

Now we perform the replication procedure described in Subsection 4.1.1 and obtain the two replicating portfolios in the two worlds. Because the composition of the replicating portfolios does not depend on the probabilities of attaining different future stock prices, the two worlds actually share the same replicating portfolio. The price of the derivative, which is equal to the cost of setting up the common replicating portfolio, turns out to be the same in the two worlds! The important implication is that to price the derivative in the real world, we may instead look at how much the derivative would cost in the risk-neutral world. This risk-neutral price takes the form of the discounted risk-neutral expectation

$$(V_0^{\text{real}} =)V_0^{\text{risk-neutral}} = \mathrm{e}^{-rT}\mathbb{E}^*[\text{Derivative terminal payoff}]$$

by the very definition of the risk-neutral world.

Can we forget about the method of replication?

Armed with the risk-neutral interpretation of the pricing formula, it becomes unnecessary to determine the replicating portfolio (Δ, B) using the non-intuitive formulas given in (4.1.1). All we have to do is to calculate the risk-neutral probability p^* using equation (4.1.3), the risk-neutral expected derivative payoff, and discount it at the risk-free rate. However, it does not mean the method of replicating portfolio has no value. In fact, the uses of replicating portfolios are (at least) three-fold:

(1) From a theoretical point of view, replicating portfolios form the basis for risk-neutral valuation. Without the method of replicating portfolios, the validity of risk-neutral pricing may become ill-founded.

(2) Practically, the method of replicating portfolios provides a *synthetic construction* of any derivative of interest. To be precise, any derivative is financially equivalent to buying Δ shares of the stock and investing $\$B$ in risk-free bonds, where Δ and B are given in (4.1.1). This synthetic construction provides the recipe for an arbitrage strategy when the observed market price deviates from the fair derivative price.

(3) The method of replicating portfolios works even for multinomial trees, in which the stock price can take multiple values at the end of the model period, whereas it is not immediately clear what the risk-neutral probabilities should be.

Example 4.1.5 below illustrates the second use and Example 4.1.6 the third use.

Example 4.1.5. (SOA Exam MFE Spring 2009 Question 3: Arbitraging a mispriced call) You are given the following regarding stock of Widget World Wide (WWW):

(i) The stock is currently selling for $50.

(ii) One year from now the stock will sell for either $40 or $55.

(iii) The stock pays dividends continuously at a rate proportional to its price. The dividend yield is 10%.

The continuously compounded risk-free interest rate is 5%.

   While reading the *Financial Post*, Michael notices that a one-year at-the-money European call written on stock WWW is selling for $1.90. Michael wonders whether this call is fairly priced. He uses the binomial option pricing model to determine if an arbitrage opportunity exists.

   What transactions should Michael enter into to exploit the arbitrage opportunity (if one exists)?

(A) No arbitrage opportunity exists.

(B) Short shares of WWW, lend at the risk-free rate, and buy the call priced at $1.90.

(C) Buy shares of WWW, borrow at the risk-free rate, and buy the call priced at $1.90.

(D) Buy shares of WWW, borrow at the risk-free rate, and short the call priced at $1.90.

(E) Short shares of WWW, borrow at the risk-free rate, and short the call priced at $1.90.

Solution. The replicating portfolio of the given call is defined by

$$\Delta = e^{-0.1}\left(\frac{5-0}{55-40}\right) = 0.3016$$

and

$$B = e^{-0.05}\left(\frac{1.1 \times 0 - 0.8 \times 5}{1.1 - 0.8}\right) = -12.6831.$$

It follows that the fair price of the call is $C_0^{\text{fair}} = \Delta S_0 + B = 2.3969$, which is higher than the observed price of 1.9. This means that the observed call is underpriced.

   To exploit the arbitrage opportunity, Michael should "buy low and sell high," i.e., purchase the call option at the observed price of $1.9 and short sell the replicating portfolio, which means shorting 0.3016 shares of the stock and lending $12.6831 at the risk-free rate, for $2.3969 **(Answer: (B))**. Currently, he receives $0.4969. One year from now, his overall payoff will be constant at

$$\underbrace{(S(1)-50)_+}_{\text{long observed call}} - \underbrace{(S(1)-50)_+}_{\text{short replicating portfolio}} = 0.$$

☐

Remark. A two-period version of this example is Problem 4.6.15.

   We end this subsection with an example involving a multinomial tree model, in which the asset price at expiration can take multiple values.

Example 4.1.6. (SOA Exam MFE Advanced Derivatives Sample Question 27: A three-state world) You are given the following information about a securities market:

- There are two nondividend-paying stocks, X and Y.

- The current prices for X and Y are both $100.

- The continuously compounded risk-free interest rate is 10%.

- There are three possible outcomes for the prices of X and Y one year from now:

Outcome	X	Y
1	$200	$0
2	$50	$0
3	$0	$300

Let C_X be the price of a European call option on X, and P_Y be the price of a European put option on Y. Both options expire in one year and have a strike price of $95.
   Calculate $P_Y - C_X$.

(A) $4.30

(B) $4.45

(C) $4.59

(D) $4.75

(E) $4.94

Ambrose's comments:

Although there are three possible outcomes, the philosophy of the method of replicating portfolio is exactly the same as that in the binomial model.

Solution. Instead of calculating P_Y and C_X separately, we view regard $P_Y - C_X$ as the time-0 price of the portfolio consisting of a long put on Y and a short call on X. The payoff of this "put-less-call" portfolio in one year is simply the payoff of the put on Y less the payoff of the call on X:

Outcome	Payoff of Put on Y	Payoff of Call on X	Total Payoff
1	$(95-0)_+ = 95$	$(200-95)_+ = 105$	-10
2	$(95-0)_+ = 95$	$(50-95)_+ = 0$	95
3	$(95-300)_+ = 0$	$(0-95)_+ = 0$	0

We now replicate the payoff of the portfolio using Δ_X shares of X, Δ_Y shares of Y, and B in the risk-free bond. Solving the 3×3 linear system

$$\begin{cases} e^{0.1}B + 200\Delta_X = -10, \\ e^{0.1}B + 50\Delta_X = 95, \\ e^{0.1}B + 300\Delta_Y = 0, \end{cases}$$

gives $\Delta_X = -0.7$, $B = 117.6289$ and $\Delta_Y = -13/30$. The cost of the portfolio then equals the cost of the replicating portfolio, which in turn is

$$P_Y - C_X = B + 100\Delta_X + 100\Delta_Y = \boxed{4.2955}. \quad (\textbf{Answer: (A)})$$

$\square$

Remark. Stocks X and Y can be said to be *mutually exclusive*, in that whenever one of them has a positive payoff, then the payoff of the other must be zero.

4.1.3   Constructing a Binomial Tree

In the previous subsections, we take the binomial tree as given and perform the valuation task. We now discuss how a binomial tree should be constructed (i.e., how the growth factors u and d should be appropriately chosen) taking stock market data into account.

No-arbitrage restrictions on u and d.

To preclude arbitrage opportunities, any reasonable binomial tree model should satisfy

$$d < e^{(r-\delta)h} < u. \tag{4.1.6}$$

In other words, u cannot be too small whereas d cannot be too large. To make sense of these restrictions, we first multiply every quantity by the initial stock price, giving

$$S_d = S_0 \times d < S_0 e^{(r-\delta)h} < S_0 \times u = S_u.$$

Putting the factor $e^{-\delta h}$ in the middle to both sides, we further rearrange (4.1.6) equivalently as

$$\underbrace{S_d e^{\delta h}}_{\text{② in } d \text{ node}} < \underbrace{S_0 e^{rh}}_{①} < \underbrace{S_u e^{\delta h}}_{\text{② in } u \text{ node}}. \tag{4.1.7}$$

This chain of inequalities then imply that neither of the following two strategies with the same initial investment of S_0 in the binomial stock market is always dominant over another:

① Using $\$S_0$ to buy a risk-free bond that earns the risk-free interest rate r.

② Using $\$S_0$ to buy the risky stock that pays dividends continuously at the rate of δ.

The time-h payoff of Strategy ① is always $S_0 e^{rh}$, while that of Strategy ② is $S_u e^{\delta h}$ in the u node or $S_d e^{\delta h}$ in the d node. Then (4.1.6), or equivalently, (4.1.7), says that Strategy ① is inferior to Strategy ② in the u node but superior to Strategy ② in the d node. In a fair market, no strategy is universally the winner!

The no-arbitrage conditions (4.1.6) effectively mean that the risk-neutral probability p^* is valued between 0 and 1:

$$0 = \frac{d-d}{u-d} < p^* = \frac{e^{(r-\delta)h} - d}{u-d} < \frac{u-d}{u-d} = 1.$$

This justifies viewing p^* as a probability.

Example 4.1.7. (SOA Course 6 Spring 2001 Multiple-Choice Question 3: What can r possibly be?) An arbitrage-free securities market model consists of a bank account and one security. The security price today is 100. The security price one year from now will be either 104 or 107.

    Determine which of the following can be the bank account interest rate.

(A) 0%

(B) 3%

(C) 5%

(D) 8%

(E) 10%

Solution. Since the securities market model is arbitrage-free, we can use (4.1.6), which says that (assume no dividends)

$$1.04 < e^r < 1.07,$$

or $\boxed{3.92\% < r < 6.77\%}$. **(Answer: (C))**   □

Volatility.

The tree parameters u and d control the size of the "jaw" of the binomial tree and their selection naturally depends on the riskiness of the stock price. A formal measure of the variability of the stock price is given by its *volatility*, which is defined as the *annualized* standard deviation of its continuously compounded returns. It is given mathematically by

$$\boxed{\sigma = \sqrt{\frac{1}{h}\mathrm{Var}\left[\ln\left(\frac{S(h)}{S(0)}\right)\right]} \overset{\text{(why?)}}{=} \sqrt{\frac{\mathrm{Var}[\ln S(h)]}{h}},}$$

where $S(h)$ is the time-h (random) stock price (equal to either S_u or S_d). Note that the division by $\sqrt{h}$ serves to annualize the volatility. In general, under the assumption of independent and identically distributed continuously compounded returns over disjoint time periods, an h-year volatility σ_h and the annual volatility σ are related via

$$\sigma = \frac{\sigma_h}{\sqrt{h}}.$$

A common method of construction: Forward tree.

One popular way of constructing a binomial tree is that of the *forward tree*,[vii] which requires setting u and d to be

$$\boxed{u = e^{(r-\delta)h+\sigma\sqrt{h}} \quad \text{and} \quad d = e^{(r-\delta)h-\sigma\sqrt{h}}.} \tag{4.1.8}$$

These choices of u and d naturally satisfy the no-arbitrage conditions (4.1.6).

[vii]The term "forward tree" comes from page 303 of McDonald (2013).

To understand why a forward tree is so called, recall from Section 2.3 that the h-year forward price of the stock is $F_{0,h} = S(0)e^{(r-\delta)h}$. Since

$$
\begin{aligned}
S_u &= S(0)e^{(r-\delta)h+\sigma\sqrt{h}} = F_{0,h}e^{\sigma\sqrt{h}}, \\
S_d &= S(0)e^{(r-\delta)h-\sigma\sqrt{h}} = F_{0,h}e^{-\sigma\sqrt{h}},
\end{aligned}
$$

uncertainty about the future stock price is introduced using the forward price $F_{0,h}$ as the benchmark, with $e^{\sigma\sqrt{h}}$ and $e^{-\sigma\sqrt{h}}$ serving as the multiplicative increment and decrement factors.

The forward tree has an inadvertent implication for the risk-neutral probability. We plug (4.1.8) into (4.1.3), yielding

$$
p^* = \frac{e^{(r-\delta)h} - d}{u - d} = \frac{e^{(r-\delta)h} - e^{(r-\delta)h-\sigma\sqrt{h}}}{e^{(r-\delta)h+\sigma\sqrt{h}} - e^{(r-\delta)h-\sigma\sqrt{h}}} = \boxed{\frac{1}{1 + e^{\sigma\sqrt{h}}}}. \tag{4.1.9}
$$

(Warning: $p^* \neq (1 + e^{-\sigma\sqrt{h}})^{-1}$!) This simplified form of the risk-neutral probability means that it is possible to find p^* without knowing u, d, r and δ. This can bring huge computational convenience in an exam setting. However, the fact that $\sigma > 0$ forces $p^* < 1/2 < 1 - p^*$, which may be viewed as a built-in bias in the forward tree model.

Example 4.1.8. (CAS Exam 3 Spring 2007 Question 15: Calculation of p^* in a forward tree) For a binomial option pricing model, you are given the following information:

- The current stock price is $110.

- The strike price is $100.

- The interest rate is 5% (continuously compounded)

- The continuous dividend yield is 3.5%.

- The volatility is 0.30.

- The time to expiration is 1 year.

- The length of period is 4 months.

Compute the risk-neutral probability of an increase in the stock price over one period.

(A) Less than 0.425

(B) At least 0.425, but less than 0.445

(C) At least 0.445, but less than 0.465

(D) At least 0.465, but less than 0.485

(E) At least 0.485

Solution. Although the question doesn't state it, a tree based on forward prices is assumed. By (4.1.9), the risk-neutral probability of an up move is

$$
p^* = \frac{1}{1 + e^{\sigma\sqrt{h}}} = \frac{1}{1 + e^{0.3\sqrt{4/12}}} = \boxed{0.4568}. \quad \textbf{(Answer: (C))}
$$

□

Remark. There is no need to compute u and d.

Example 4.1.9. (CAS Exam 3 Fall 2007 Question 19: Calculation of call price in a forward tree) A three-month European call is modeled by a single period binomial tree using the following parameters:

- Continuously compounded risk-free rate = 4%

- Dividend = 0

- Annual volatility = 15%

- Current stock price = 10

- Strike price = 10.5

Calculate the value of the call option.

(A) Less than 0.15

(B) At least 0.15, but less than 0.30

(C) At least 0.30, but less than 0.45

(D) At least 0.45, but less than 0.60

(E) At least 0.60

Solution. The tree parameters are

$$
\begin{aligned}
u &= e^{0.04(0.25)+0.15\sqrt{0.25}} = 1.088717, \\
d &= e^{0.04(0.25)-0.15\sqrt{0.25}} = 0.937067.
\end{aligned}
$$

The risk-neutral probability of an up move is

$$
p^* = \frac{1}{1+e^{\sigma\sqrt{h}}} = \frac{1}{1+e^{0.15\sqrt{0.25}}} = 0.481259.
$$

As $C_u = (10 \times 1.088717 - 10.5)_+ = 0.387170$ and $C_d = (10 \times 0.937067 - 10.5)_+ = 0$, the risk-neutral pricing formula says that

$$
C_0 = e^{-0.04(0.25)}(0.481259)(0.387170) = \boxed{0.1845}. \quad \textbf{(Answer: (B))}
$$

$\square$

Example 4.1.10. (CAS Exam 3 Fall 2007 Question 23: Given the call price, what is σ?) A one-year European call option is currently valued at 0.9645. The following parameters are given.

- Current stock price = 10

- Continuously compounded risk-free rate = 6%

- Continuously compounded dividend rate = 1%

- Strike price = 10

Using a single period binomial tree, calculate the implied volatility of the stock, assuming that it is greater than 5%.

(A) Less than 0.10

(B) At least 0.10, but less than 0.20

(C) At least 0.20, but less than 0.30

(D) At least 0.30, but less than 0.40

(E) At least 0.40

Solution. Because $\sigma > 0.05$, we have $d = e^{0.05-\sigma} < 1$, so the call does not pay off at the d node. Using the risk-neutral pricing formula,

$$0.9645 = e^{-rh}p^*(C_u - K)_+ = \frac{e^{-0.06}}{1+e^{\sigma}}(10e^{0.05+\sigma} - 10),$$

which results in $e^{\sigma} = 1.161834$ and $\sigma = \boxed{0.15}$. (**Answer: (B)**) □

Sidebar: How to estimate volatility given past stock price data?[viii]

From the above discussions, it goes without saying that σ is a very important input for constructing a binomial tree. As σ is not a directly observable quantity, how can the value of σ be selected?

Suppose that $n+1$ stock prices, $S(ih)$ for $i = 0, 1, 2, \ldots, n$ are observed, where h is length of time (in years) between adjacent observations. For example, if $h = 1/12$, then the stock prices are observed at consecutive monthly intervals. These $n+1$ successive stock prices are neither independent nor identically distributed—knowledge of the preceding stock price dictates the possible values that can be assumed by the succeeding stock price. However, the *observed* (non-annualized) continuously compounded returns, denoted by $r_1, r_2, \ldots, r_n$ and defined as

$$r_i = \ln \frac{S(ih)}{S[(i-1)h]}, \quad \text{for } i = 1, 2, \ldots, n,$$

are independent and identically distributed (i.i.d.)—they are either $\ln u$ or $\ln d$, independently of each other—with common variance

$$\text{Var}(r_1) = \text{Var}\left[\ln\left(\frac{S(h)}{S(0)}\right)\right] = \sigma^2 h. \tag{4.1.10}$$

Using these n i.i.d. observed returns, $r_1, r_2, \ldots, r_n$, we are back to the standard i.i.d. setting in statistics for estimating the unknown variance of a population.

Method-of-moments estimate of σ.

In statistics, there are a multitude of methods for estimating the value of an unknown parameter from observed data. The current setting is no exception. One conceptually very simple estimation method, based on Equation (4.1.10), is the *method of moments* (or MOM in short). The idea of MOM is to consider the sample version of (4.1.10), with the population

---

[viii]The estimation procedure here also applies to the Black-Scholes framework.

variance $\mathrm{Var}(r_1)$ replaced by its sample counterpart computable from the returns data, and solve for the resulting value of σ. In essence, we pick the value of σ in order to match the sample variance (this is why MOM is also known as moment matching) and designate such a value the estimate of σ.

To determine the MOM estimate of σ, we equate, by virtue of Equation (4.1.10), the sample variance of $r_1, r_2, \ldots, r_n$ and $\sigma^2 h$. The sample variance of the r_i's is

$$S_r^2 = \frac{1}{n-1} \sum_{i=1}^{n} (r_i - \bar{r})^2 = \frac{n}{n-1} \left(\overline{r^2} - \bar{r}^2 \right),$$

where $\bar{r}$ and $\overline{r^2}$ are the sample mean and sample second moment given by

$$\bar{r} = \frac{\sum_{i=1}^{n} r_i}{n} = \frac{1}{n} \ln \frac{S(nh)}{S(0)} \quad \text{and} \quad \underbrace{\overline{r^2} = \frac{\sum_{i=1}^{n} r_i^2}{n}}_{\text{not to be confused with } (\bar{r})^2 !}.$$

Hence the MOM estimate of σ is

$$\boxed{\hat{\sigma} = \sqrt{\frac{S_r^2}{h}}.}$$

Note that:

- The division by $n-1$ makes the sample variance of the r_i's unbiased for their true variance.

- The division by h serves to annualize the estimate of σ.

The estimate of σ obtained this way is also known as *historical volatility*, because it is based on historical stock prices and stock returns.

Example 4.1.11. (SOA Exam IFM Advanced Derivatives Sample Question 17: Estimating σ) You are to estimate a nondividend-paying stock's annualized volatility using its prices in the past nine months.

Month	Stock Price ($/share)
1	80
2	64
3	80
4	64
5	80
6	100
7	80
8	64
9	80

Calculate the historical volatility for this stock over the period.

(A) 83%

(B) 77%

(C) 24%

(D) 22%

(E) 20%

Solution. Let r_i be the continuously compounded monthly returns for the ith month. Then:

Month i	$r_i = \ln\{S(ih)/S[(i-1)h]\}$		
1	—		
2	$\ln(64/80)$	$=$	$\ln 0.8$
3	$\ln(80/64)$	$=$	$\ln 1.25$
4	$\ln(64/80)$	$=$	$\ln 0.8$
5	$\ln(80/64)$	$=$	$\ln 1.25$
6	$\ln(100/80)$	$=$	$\ln 1.25$
7	$\ln(80/100)$	$=$	$\ln 0.8$
8	$\ln(64/80)$	$=$	$\ln 0.8$
9	$\ln(80/64)$	$=$	$\ln 1.25$

Note that four of the r_i's are $\ln 1.25$ and the other four are $\ln 0.8 = -\ln 1.25$. In particular, their mean $\bar{r}$ is zero.

The (unbiased) sample variance of the non-annualized monthly returns is

$$\hat{\sigma}^2_{1/12} = \frac{1}{n-1}\sum_{i=1}^{n}(r_i - \bar{r})^2 = \frac{1}{7}\sum_{i=1}^{8}r_i^2 = \frac{8}{7}(\ln 1.25)^2,$$

and the estimated annualized volatility is

$$\hat{\sigma} = \frac{\hat{\sigma}_{1/12}}{\sqrt{1/12}} = \sqrt{12} \times \sqrt{\frac{8}{7}(\ln 1.25)^2} = \boxed{82.64\%}. \quad (\textbf{Answer: (A)})$$

$\square$

4.2 Multi-period Binomial Trees

An obvious objection to the single-period binomial tree model presented in the preceding section is that the stock price at expiration can only take two possible values. This restriction can be easily overcome by dividing the time to expiration into several periods, in each of which the stock price evolves according to a one-period binomial model. This process generates a more realistic multi-period binomial tree where the asset price at expiration can take any finite number of values, at the expense of substantially more intensive computations. In this section, we build upon the foundation laid in Section 4.1 and discuss the valuation of options on assets whose price movements are driven by multi-period binomial trees.

How to represent a multi-period binomial tree?

Figure 4.2.1 shows a two-period binomial stock price tree, in which the length of each period is h and the duration of the whole tree is $T = 2h$. At the end of the first period, there are two possible stock prices, namely S_u and S_d. At the end of the second period, each of these

Time 0   Time h Time $T = 2h$

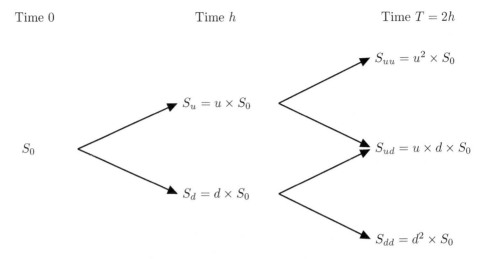

FIGURE 4.2.1
A generic two-period binomial stock price tree.

two prices can itself go up by u or down by d. In this two-period setting, the generic stock price symbol $S_{(\cdot)(\cdot)}$ is decorated by a subscript that represents chronologically the precise path through which the stock price passes over the two periods. For example, S_{uu} is the two-period stock price corresponding to the uu path (i.e., two consecutive up moves), and S_{ud} is the two-period stock price corresponding to an up move followed by a down move. Such notation will carry over to a general multi-period binomial tree.

One conspicuous feature of the stock price tree in Figure 4.2.1 is that the stock prices move up or down by the same factors u and d at the end of each period. The effect is $S_{ud} = S_{du}$, so there are only three distinct two-period stock prices. In general, a binomial tree in which an up move followed by a down move leads to the same price as a down move followed by an up move is said to be *recombining*. The use of a recombining binomial tree provides huge computational ease—an n-period binomial model only has $n + 1$ distinct terminal prices, in contrast to 2^n for a non-recombining one, and is an attempt to balance the flexibility and tractability of the stock price model. Unless otherwise stated, in this book we will always use a recombining binomial tree with the same growth factors u and d in each period.

The method of backward induction.

The key to valuing options using a multi-period binomial tree is to *work backward through the binomial tree*. We start from the option payoffs at expiration and determine the possible values of the option one period before expiration corresponding to different stock price paths. This valuation task involves only one-period binomial trees and therefore can be accomplished by what we learnt in Section 4.1. Then we use these option values one period before expiration to compute the option values two periods before expiration. This process is repeated recursively until we reach the initial node and obtain the price of the option at time 0. The essence of this backward induction procedure lies in reducing the multi-period valuation problem to a series of one-period valuation problems.

Let's use the following CAS exam question to illustrate the intricate ideas.

Example 4.2.1. (CAS Exam 3 Spring 2007 Question 14: First encounter with a two-period binomial tree) Consider the following information about a European call option on stock ABC:

- The strike price is $95.

- The current stock price is $100.

- The time to expiration is 2 years.

- The continuously compounded risk-free rate is 5% annually.

- The stock pays no dividends.

- The price is calculated using a 2-step binomial model where each step is one year in length.

The stock price tree is shown below:

		121
	110	
100		99
	90	
		81

Calculate the price of the call on stock ABC.

(A) Less than 13.50

(B) At least 13.50, but less than 14.00

(C) At least 14.00, but less than 14.50

(D) At least 14.50, but less than 15.00

(E) At least 15.00

Ambrose's comments:

This is the very first time we handle a multi-period binomial tree, so we will do this example from basics. Later we will introduce shortcuts that will save much of the work.

Solution. We break down our solution into three steps.

- *Step 1:* The first step of the solution is to identify the possible payoffs of the call option at expiration. They are

$$C_{uu} = (121 - 95)_+ = 26, \quad C_{ud} = (99 - 95)_+ = 4, \quad C_{dd} = (81 - 95)_+ = 0.$$

- *Step 2:* The second step, also the key step of the solution, lies in transforming the given two-period call into an equivalent one-period derivative (not a call). To this end, consider what happens in one year:

Case 1.   Suppose that the stock price goes up in one year to 110. From the perspective of the 110 node, the stock price can be either \$121 or \$99 one year thereafter. This constitutes a single-period binomial tree model with $u = 1.1$ and $d = 0.9$:

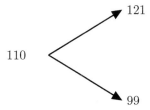

Since the risk-neutral probability of an up move is

$$p^* = \frac{e^{0.05(1)} - 0.9}{1.1 - 0.9} = 0.756355481,$$

by risk-neutral pricing the value of the call option at the 110 node is

$$C_u = e^{-0.05}[p^* C_{uu} + (1 - p^*)C_{ud}] = 19.6332.$$

Case 2.   Similarly, at the 90 node, we again have a single-period binomial tree model with the same parameters $u = 1.1$, $d = 0.9$ and $p^* = 0.756355$, but with different beginning and ending stock prices:

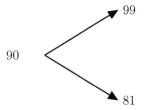

By risk-neutral pricing again, the value of the call option at the 90 node is
$$C_d = e^{-0.05}[p^*(99 - 95)_+ + (1 - p^*)(81 - 95)_+] = 2.8779.$$

At this point, we have reduced the two-year \$95-strike call option to a *one-year derivative* which pays $C_u = 19.6332$ if the stock price goes up in one year, and pays $C_d = 2.8779$ if the stock price goes down in one year.

• *Step 3:* In the third and final step of the solution, we price the equivalent one-year derivative using the single-period binomial model emanating from time 0:

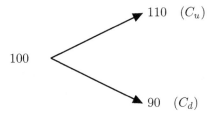

The required call price is

$$C_0 = e^{-0.05}[p^* C_u + (1 - p^*)C_d] = \boxed{14.7924}. \quad (\textbf{Answer: (D)})$$

□

Remark. If you love calculations, you might choose to determine the replicating portfolio at each node of the binomial tree:

Node	(Δ, B)
u	$(1, -90.3668)$
d	$(2/9, -17.1221)$
0	$(0.8378, -68.9841)$

You can check that the value of the replicating portfolio at time 0 is also equal to the call price you have just computed. Moreover, the composition of the replicating portfolio varies as the stock price tree unfolds.

An important shortcut for European options: From maturity date directly to time 0.

As Example 4.2.1 demonstrates, pricing an option in a multi-period binomial tree involves computing the value of the option at all intermediate nodes. These calculations can be computationally burdensome (although easily implemented by an Excel spreadsheet), especially when the binomial tree has three or more periods. Such pain, however, can be completely avoided in the case of *European* options, for which early exercise is not an "option." Instead of working backward through the entire binomial tree and computing all the intermediate option values, we simply calculate the expected payoff at expiration using *binomial risk-neutral probabilities* and discount this expected payoff directly back to time 0 at the risk-free interest rate without bothering with the interim stock prices.

To see this more concretely, consider a two-period binomial model in which the length of each period is h. The possible values of the option at the end of the first period are

$$V_u = \mathrm{e}^{-rh}[p^*V_{uu} + (1-p^*)V_{ud}] \quad \text{and} \quad V_d = \mathrm{e}^{-rh}[p^*V_{ud} + (1-p^*)V_{dd}].$$

Discounting the "risk-neutral" expectations of these two values back to time 0 yields

$$
\begin{aligned}
V_0 &= \mathrm{e}^{-rh}[p^*V_u + (1-p^*)V_d] \\
&= \mathrm{e}^{-rh}\{p^*\mathrm{e}^{-rh}[p^*V_{uu} + (1-p^*)V_{ud}] + (1-p^*)\mathrm{e}^{-rh}[p^*V_{ud} + (1-p^*)V_{dd}]\} \\
&= \mathrm{e}^{-2rh}[(p^*)^2 V_{uu} + 2p^*(1-p^*)V_{ud} + (1-p^*)^2 V_{dd}] \\
&\overset{(T=2h)}{=} \boxed{\mathrm{e}^{-rT}[(p^*)^2 V_{uu} + 2p^*(1-p^*)V_{ud} + (1-p^*)^2 V_{dd}]}.
\end{aligned}
$$
(4.2.1)

Since the risk-neutral probabilities of entering the uu node, ud node (Note: There are two paths to reach the ud node, namely up-down and down-up) and dd node are

$$(p^*)^2, \quad 2p^*(1-p^*), \quad \text{and} \quad (1-p^*)^2,$$

respectively, the preceding expression can be conveniently cast as

$$\boxed{V_0 = \mathrm{e}^{-rT}\mathbb{E}^*[\text{Payoff at expiration}],}$$
(4.2.2)

where the asterisk again signifies a risk-neutral expectation. By induction, it can be shown that (4.2.2) is true even for a general n-period binomial tree.

PRACTICAL NOTE

Given that (4.2.2) allows you to calculate the current price of the option directly from the terminal payoffs along with the binomial probabilities, this is the usual way you value a European option by hand calculation.

Example 4.2.2. (CAS Exam 3 Spring 2007 Question 17: A two-period binomial tree for a European put) For a two-year European put option, you are given the following information:

- The stock price is $35.

- The strike price is $32.

- The continuously compounded risk-free rate is 5%.

- The stock price volatility is 35%.

Using a binomial tree with annual valuations, calculate the price of this option.

(A) Less than $3.00

(B) At least $3.00, but less than $3.40

(C) At least $3.40, but less than $3.80

(D) At least $3.80, but less than $4.20

(E) At least $4.20

Solution. Prelude: The fact that the length of each period of the binomial tree is one year ($h = 1$) means that the binomial model is two-period ($n = 2$). Since there is no information about the dividends of the stock and how the binomial tree is constructed, we simply assume no dividends and a forward tree.

For a forward tree, we have

$$u = e^{0.05(1)+0.35\sqrt{1}} = 1.491825 \quad \text{and} \quad d = e^{0.05(1)-0.35\sqrt{1}} = 0.740818,$$

and the risk-neutral probability of an up move is

$$p^* = \frac{1}{1 + e^{0.35\sqrt{1}}} = 0.413382.$$

The possible payoffs at expiration of the put option are:

Terminal Stock Price	Terminal Put Payoff
$S_{uu} = 35u^2 = 77.8940$	$P_{uu} = (32 - 77.8940)_+ = 0$
$S_{ud} = 35ud = 38.6810$	$P_{ud} = (32 - 38.6810)_+ = 0$
$S_{dd} = 35d^2 = 19.2084$	$P_{dd} = (32 - 19.2084)_+ = 12.7916$

Using (4.2.1), we have

$$
\begin{aligned}
P_0 &= e^{-rT}[(p^*)^2 P_{uu} + 2p^*(1 - p^*)P_{ud} + (1 - p^*)^2 P_{dd}] \\
&= e^{-0.05(2)}[0 + 0 + (1 - 0.413382)^2(12.7916)] \\
&= \boxed{3.9830}. \quad \textbf{(Answer: (D))}
\end{aligned}
$$

□

Example 4.2.3. (A three-period binomial tree) You use the following information to construct a binomial forward tree for modeling the price movements of a stock:

(i) The length of each period is 4 months.

(ii) The current stock price is 41.

(iii) The stock's volatility is 30%.

(iv) The stock pays no dividends.

(v) The continuously compounded risk-free interest rate is 8%.

Calculate the price of a one-year at-the-money European call option on the stock.

Ambrose's comments:

The three-period risk-neutral pricing formula takes the form

$$e^{-rT}[(p^*)^3 V_{uuu} + 3(p^*)^2(1-p^*)V_{uud} + 3p^*(1-p^*)^2 V_{udd} + (1-p^*)^3 V_{ddd}].$$

Don't forget the factor of 3, which comes from $\binom{3}{1} = \binom{3}{2}$, in the second and third terms!

Solution. The tree parameters are

$$u = e^{0.08(1/3)+0.3\sqrt{1/3}} = 1.221246 \quad \text{and} \quad d = e^{0.08(1/3)-0.3\sqrt{1/3}} = 0.863693.$$

Then the terminal stock prices are

$$\begin{aligned}
S_{uuu} &= 41u^3 = 74.678110, \\
S_{uud} &= 41u^2 d = 52.814061, \\
S_{udd} &= 41ud^2 = 37.351308,
\end{aligned}$$

and there is no need to compute S_{ddd} (why?). With the risk-neutral probability of an up move being

$$p^* = \frac{1}{1 + e^{0.3\sqrt{1/3}}} = 0.456807,$$

the time-0 price of the required European call is

$$\begin{aligned}
e^{-0.08}[&0.456807^3(74.678110 - 41) \\
&+ 3(0.456807)^2(1 - 0.456807)(52.814061 - 41)] \\
= \;&\boxed{6.6720}.
\end{aligned}$$

$\square$

Example 4.2.4. (Given the call price tree, not the stock price tree!) For a two-period binomial stock price model, you are given:

(i) The length of each period is 1 year.

(ii) The (incomplete) price evolution of a 2-year European call option on the stock:

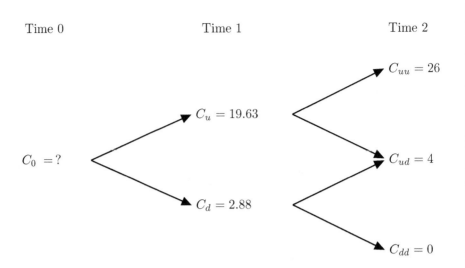

Calculate the current price of the call option.

Ambrose's comments:

We are not told anything about the stock (does it pay dividends?) or the call option (what is its strike price?). However, the given call price tree contains enough information for us to solve this problem.

Solution. Let's deduce the values of the risk-neutral probability p^* and the discount factor e^{-r} per period. To this end, we apply the risk-neutral pricing formula to C_d and C_u, yielding

$$\begin{cases} e^{-r}[p^*(4) + (1 - p^*)(0)] = 2.88 \\ e^{-r}[p^*(26) + (1 - p^*)(4)] = 19.63 \end{cases}.$$

The first equation implies that $e^{-r}p^* = 0.72$, which, when plugged into the second equation, leads to $e^{-r}(1 - p^*) = 0.2275$. Then it follows from a further application of the risk-neutral pricing formula that

$$C_0 = e^{-r}[p^*(19.63) + (1 - p^*)(2.88)] = 0.72(19.63) + 0.2275(2.88) = \boxed{14.79}.$$

$\square$

Remark. Although we do not need the individual values of p^* and e^{-r}, they can be found via $e^{-r} = e^{-r}p^* + e^{-r}(1 - p^*) = 0.9475$ and $p^* = 0.72/0.9475 = 0.7599$.

It is hard to conceive a pen-and-paper problem involving an n-period binomial tree with $n \geq 4$. If such catastrophic cases happen, chances are that the option will pay off only at a few nodes.

Example 4.2.5. (A 10-period tree!!) For a 10-period binomial stock price model, you are given:

(i) The length of each period is one year.

(ii) The current stock price is 1,000.

(iii) At the end of every year, the stock price will either increase by 5% or decrease by 5% in proportion.

(iv) The stock pays dividends continuously at a rate proportional to its price. The dividend yield is 1%.

(v) The continuously compounded risk-free interest rate is 2%.

Calculate the price of a 10-year 1,400-strike European call option on the stock.

Ambrose's comments:

A 10-period binomial model!? WTF!? ($\underline{W}$OW, $\underline{T}$HAT'S $\underline{F}$ANTASTIC!)

Solution. Although there are a total of 11 possible stock prices after 10 years, observe carefully that the call will be in-the-money only at two nodes. With $u = 1.05$, $d = 0.95$, and $K = 1,400$, note that the call will pay off only at

the u^{10} node, where $S_{u^{10}} = 1,000(1.05)^{10} = 1,628.8946 > 1,400$, and

the $u^9 d$ node, where $S_{u^9 d} = 1,000(1.05)^9(0.95) = 1,473.7618 > 1,400$,

but not at the $u^8 d^2$ node, where $S_{u^8 d^2} = 1,000(1.05)^8(0.95)^2 = 1,333.4035 < 1,400$, and other lower nodes.

The risk-neutral probability of an up move is

$$p^* = \frac{e^{(0.02-0.01)(1)} - 0.95}{1.05 - 0.95} = 0.600502.$$

By the risk-neutral pricing formula,

$$
\begin{aligned}
C &= e^{-rT}[(p^*)^{10} C_{u^{10}} + 10(p^*)^9(1-p^*)C_{u^9 d}] \\
&= e^{-0.02(10)}[0.600502^{10}(1,628.8946 - 1,400) \\
&\qquad\qquad + 10(0.600502)^9(1 - 0.600502)(1,473.7618 - 1,400)] \\
&= \boxed{3.592396}.
\end{aligned}
$$

$\square$

Remark. What if you are asked to price the otherwise identical put? It pays off at $11 - 2 = 9$ nodes, which means that the risk-neutral pricing formula will consists of 9 terms! See Problem 4.6.10.

Sidebar: A general representation of European option prices.

It is possible to give a closed-form expression for a European derivative in a general n-period recombining binomial tree model with constant u and d. For any fixed i between 0 and n, the probability of i up moves (and necessarily $n - i$ down-moves) over the n periods is

$$P_i := \binom{n}{i}(p^*)^i(1 - p^*)^{n-i}.$$

Then by risk-neutral pricing, the time-0 price of a K-strike European call option is

$$
\begin{aligned}
C &= \mathrm{e}^{-rT} \sum_{i=0}^{n} P_i(i)(S(0)u^i d^{n-i} - K)_+ \\
&= \mathrm{e}^{-rT} \sum_{i=0}^{n} \binom{n}{i} (p^*)^i (1-p^*)^{n-i} (S(0)u^i d^{n-i} - K)_+.
\end{aligned}
$$

This expression can be represented in a form that mimics the Black-Scholes pricing formula we shall learn in Chapter 6. We denote by n^* the smallest number of up moves that makes the call option in-the-money at expiration. Then the call price can be rewritten as

$$
\begin{aligned}
C &= \mathrm{e}^{-rT} \sum_{i=n^*}^{n} P_i(i)(S(0)u^i d^{n-i} - K) \\
&= S(0)\mathrm{e}^{-rT} \sum_{i=n^*}^{n} P_i(i)u^i d^{n-i} - K\mathrm{e}^{-rT} \sum_{i=n^*}^{n} P_i(i).
\end{aligned}
$$

The second term is the anticipated time-0 *cost* of exercising the call: the present value of the strike price K multiplied by the probability that the call will finish in-the-money. The first term measures the anticipated *benefit* from exercising the option and owning the stock. As you will see later, the Black-Scholes pricing formula takes a similar cost-benefit form:

$$
C = S(0) \times \text{term } 1 - K\mathrm{e}^{-rT} \times \text{term } 2,
$$

where "term 2" is again the risk-neutral probability that the call expires in-the-money.

4.3   American Options

Arguably the most powerful feature of the binomial tree models among all option pricing models is that it readily accommodates American options. Because of the possibility of early exercise, (4.2.2) based on discounting the payoffs at expiration is no longer applicable. As such, the options need not be held until expiration, making the term "terminal payoff" not well defined. In this section, we will illustrate how the method of backward induction can be employed to price American options.

Backward induction is inevitable...

To value American options using a binomial model, the end-of-tree shortcut as discussed on page 135 does not work and we have no choice but to work backward through the whole binomial tree and check whether early exercise is optimal at any intermediate node. The reason why we travel backward but not forward over time is that an early exercise decision made at a later node will increase the value of the option at an earlier node and may create a ripple effect along the tree, inducing early exercise at an earlier node.

In the presence of early exercise, the value of the option at a particular node is given by the *maximum* of:

(1) The *holding value* (i.e., the value of holding the American option until the end of the current period), which can be determined by risk-neutral valuation (or the method of replication, if you insist).

(2) The immediate exercise value, which is given by $(S - K)_+$ in the case of a call, and $(K - S)_+$ in the case of a put, where S is the price of the stock corresponding to that node of the binomial tree.

If $(1) < (2)$, then early exercise at that node is optimal, and the value of the option therein will become (2).

Example 4.3.1. (SOA Exam IFM Sample Question 4: American call I) For a two-period binomial model, you are given:

(i) Each period is one year.

(ii) The current price for a nondividend-paying stock is 20.

(iii) $u = 1.2840$, where u is one plus the rate of capital gain on the stock per period if the stock price goes up.

(iv) $d = 0.8607$, where d is one plus the rate of capital loss on the stock per period if the stock price goes down.

(v) The continuously compounded risk-free interest rate is 5%.

Calculate the price of an American call option on the stock with a strike price of 22.

(A) 0

(B) 1

(C) 2

(D) 3

(E) 4

Solution. The two-period binomial tree is constructed in Figure 4.3.1. The risk-neutral probability for the stock price to go up each year is

$$p^* = \frac{e^{0.05} - 0.8607}{1.2840 - 0.8607} = 0.450203.$$

Then we start from the end of the tree and travel backward to see if it pays to exercise the American call early.

- *At the u node:* The value of the call option is

$$\max \left\{ \underbrace{e^{-0.05}[p^*(10.9731) + (1 - p^*)(0.1028)]}_{=4.752922}, \underbrace{(25.680 - 22)_+}_{=3.680} \right\} = 4.752922.$$

Because the holding value of 4.752922 is greater than the exercise value of 3.680, it is more worthwhile to hold than to exercise the American call.

- *At the d node:* The value of the call option becomes

$$\max \left\{ \underbrace{e^{-0.05}[p^*(0.1028) + (1 - p^*)(0)]}_{=0.044023}, \underbrace{(17.214 - 22)_+}_{=0} \right\} = 0.044023.$$

Early exercise is again not optimal.

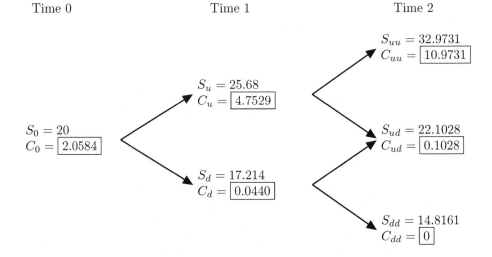

Time 0   Time 1   Time 2

FIGURE 4.3.1
The two-period binomial tree for Example 4.3.1.

- *At the initial node:* Finally, the call price is

$$\max\left\{\underbrace{e^{-0.05}[p^*(4.752922)+(1-p^*)(0.044023)]}_{=2.0584},\ \underbrace{(20-22)_+}_{=0}\right\} = \boxed{2.0584}.$$

(**Answer: (C)**)

$\square$

Remark. (i) Strictly speaking, point (iv) of the question should be written as "$d = 0.8607$, where d is one plus the *percentage change* in the stock price per period if the stock price goes down." When $d < 1$, the stock price drops and the rate of capital loss on the stock is a positive quantity.

(ii) It turns out that in the absence of dividends, early exercise of an American call is never optimal (see Section 9.4). Accordingly, the American call can be priced in the same way as its European counterpart.

Example 4.3.2. (SOA Exam MFE Spring 2009 Question 1: American call II)
You use the following information to construct a binomial forward tree for modeling the price movements of a stock. (This tree is sometimes called a forward tree.)

(i) The length of each period is one year.

(ii) The current stock price is 100.

(iii) The stock's volatility is 30%.

(iv) The stock pays dividends continuously at a rate proportional to its price. The dividend yield is 5%.

(v) The continuously compounded risk-free interest rate is 5%.

Calculate the price of a two-year 100-strike American call option on the stock.

(A) 11.40

(B) 12.09

(C) 12.78

(D) 13.47

(E) 14.16

Ambrose's comments:

This is a more involved version of the preceding example. You are required to construct the binomial tree, which is a forward tree, yourself. Also, early exercise is shown to be optimal at some point.

Solution. • *Step 1 (Constructing the forward tree):* In a forward tree, we have

$$u = \exp[(r - \delta)h + \sigma\sqrt{h}] = \exp[(0.05 - 0.05) \times 1 + 0.3\sqrt{1}] = e^{0.3},$$
$$d = \exp[(r - \delta)h - \sigma\sqrt{h}] = \exp[(0.05 - 0.05) \times 1 - 0.3\sqrt{1}] = e^{-0.3}.$$

The risk-neutral probability of an up move is

$$p^* = \frac{e^{(r-\delta)h} - d}{u - d} = \frac{e^{(0.05-0.05)\times 1} - e^{-0.3}}{e^{0.3} - e^{-0.3}} = 0.425557$$

or

$$p^* = \frac{1}{1 + e^{\sigma\sqrt{h}}} = \frac{1}{1 + e^{0.3}} = 0.425557.$$

• *Step 2 (Computing the time-1 option values):* The resulting stock prices are depicted in Figure 4.3.2. Now we calculate the option values at time 1, paying special attention to whether early exercise is optimal:

Node	Holding Value (Do the calculations!)	Exercise Value	Early Exercise?
u	$e^{-0.05}[p^* C_{uu} + (1 - p^*)C_{ud}] = 33.2796$	34.9859	Yes
d	$e^{-0.05}[p^* C_{ud} + (1 - p^*)C_{dd}] = 0$	0	No

Therefore, the values of the call options at time 1 are

$$C_u = 34.9859 \quad \text{and} \quad C_d = 0.$$

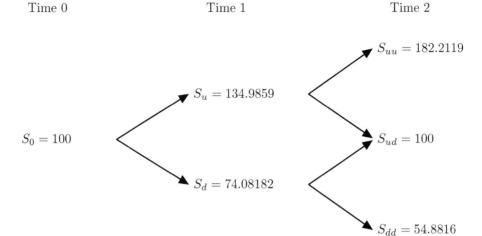

FIGURE 4.3.2
The two-period binomial forward tree for Example 4.3.2.

- *Step 3 (Time-0 price):* Now that the values of the call option at time 1 are available, the time-0 price of the call can be easily calculated as

$$
\begin{aligned}
C_0 &= \max\{e^{-0.05}[p^*C_u + (1 - p^*)C_d], 100 - 100\} \\
&= \max\{e^{-0.05}[0.425557(34.9859) + (1 - 0.425557)(0)], 0\} \\
&= \boxed{14.1624}. \quad \textbf{(Answer: (E))}
\end{aligned}
$$

$\square$

Remark. Don't forget to check whether early exercise is optimal at time 0.

Example 4.3.3. (SOA Exam MFE Spring 2007 Question 11: American put I) For a two-period binomial model for stock prices, you are given:

(i) Each period is 6 months.

(ii) The current price for a nondividend-paying stock is $70.00.

(iii) $u = 1.181$, where u is one plus the rate of capital gain on the stock per period if the price goes up.

(iv) $d = 0.890$, where d is one plus the rate of capital loss on the stock per period if the price goes down.

(v) The continuously compounded risk-free interest rate is 5%.

Calculate the current price of a one-year American put option on the stock with a strike price of $80.00.

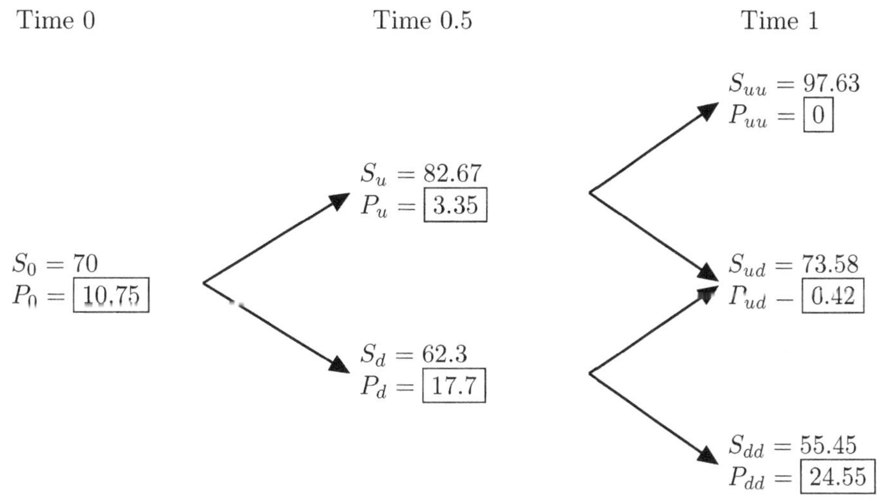

Time 0   Time 0.5   Time 1

$S_{uu} = 97.63$
$P_{uu} = \boxed{0}$

$S_u = 82.67$
$P_u = \boxed{3.35}$

$S_0 = 70$
$P_0 = \boxed{10.75}$

$S_{ud} = 73.58$
$P_{ud} = \boxed{0.42}$

$S_d = 62.3$
$P_d = \boxed{17.7}$

$S_{dd} = 55.45$
$P_{dd} = \boxed{24.55}$

FIGURE 4.3.3
The two-period binomial tree for Example 4.3.3.

(A) $9.75

(B) $10.15

(C) $10.35

(D) $10.75

(E) $11.05

Solution. With $u = 1.181$, $d = 0.89$, $h = 0.5$, $n = 2$ (i.e., two-period model) and $\delta = 0$, the risk-neutral probability of an increase in the stock price at the end of a period is

$$p^* = \frac{e^{0.05(0.5)} - 0.89}{1.181 - 0.89} = 0.4650.$$

The stock prices and put values are depicted in Figure 4.3.3. Note that early exercise is optimal at the d node, because the exercise value of $80 - 62.3 = 17.7$ exceeds the holding value of $e^{-0.05(0.5)}[p^*(6.42) + (1 - p^*)(24.55)] = 15.72$. The time-0 price of the American option is

$$P_0 = \max\{e^{-0.05(0.5)}[p^*(3.35) + (1 - p^*)(17.7)], 80 - 70\} = \boxed{10.75}. \quad (\textbf{Answer: (D)})$$

□

Example 4.3.4. (CAS Exam 3 Fall 2007 Question 18: American put II) You are given the following information about American options on a stock:

- The current stock price is 72.

- The strike price of the options is 80.

- The continuously compounded risk-free rate is 5%.

- Time to expiration is 1 year.

- Every six months, the stock price either increases by 25% or decreases by 15%.

Using a two-period binomial tree, calculate the price of an American put option.

(A) Less than 8

(B) At least 8, but less than 9

(C) At least 9, but less than 10

(D) At least 10, but less than 11

(E) At least 11

Solution. With $u = 1.25$, $d = 0.85$, $h = 0.5$, $n = 2$ (i.e., two-period model) and $\delta = 0$, the risk-neutral probability of an increase in the stock price at the end of one year is

$$p^* = \frac{e^{0.05(0.5)} - 0.85}{1.25 - 0.85} = 0.438288.$$

The stock prices and put values are depicted in Figure 4.3.4. Note that early exercise is optimal at the d node, because the exercise value of $80 - 61.2 = 18.8$ exceeds the holding value of $e^{-0.05(0.5)}[p^*(3.5) + (1 - p^*)(27.98)] = 16.8248$. The time-0 price of the American option is

$$P_0 = \max\{e^{-0.05(0.5)}[p^*(1.9175) + (1 - p^*)(18.8)], 80 - 72\} = \boxed{11.12}. \quad \textbf{(Answer: (E))}$$

$\square$

Example 4.3.5. (SOA Exam IFM Advanced Derivatives Sample Question 49: How high should the put strike be to induce immediate exercise?) You use the following information to construct a one-period binomial forward tree for modeling the price movements of a nondividend-paying stock. (The tree is sometimes called a forward tree.)

(i) The period is 3 months.

(ii) The initial stock price is $100.

(iii) The stock's volatility is 30%.

(iv) The continuously compounded risk-free interest rate is 4%.

At the beginning of the period, an investor owns an American put option on the stock. The option expires at the end of the period.

Determine the smallest integer-valued strike price for which an investor will exercise the put option at the beginning of the period.

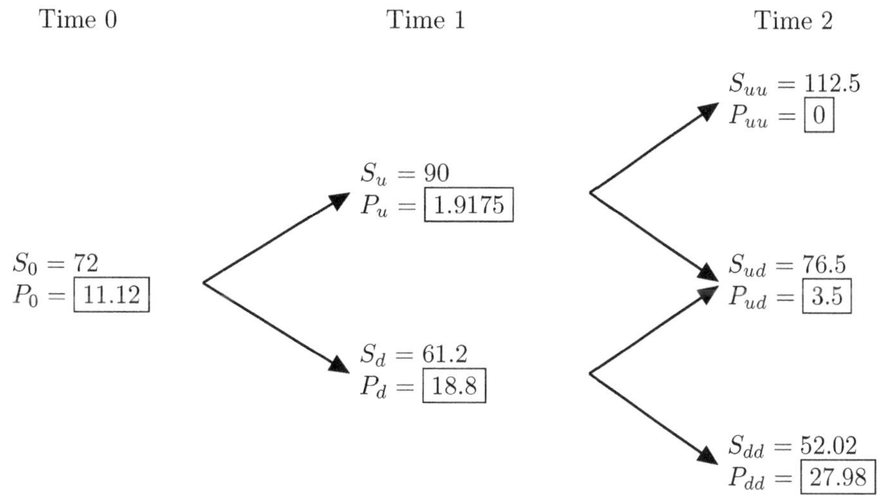

Time 0   Time 1   Time 2

$S_{uu} = 112.5$
$P_{uu} = \boxed{0}$

$S_u = 90$
$P_u = \boxed{1.9175}$

$S_0 = 72$
$P_0 = \boxed{11.12}$

$S_{ud} = 76.5$
$P_{ud} = \boxed{3.5}$

$S_d = 61.2$
$P_d = \boxed{18.8}$

$S_{dd} = 52.02$
$P_{dd} = \boxed{27.98}$

FIGURE 4.3.4
The two-period binomial tree for Example 4.3.4.

(A) 114

(B) 115

(C) 116

(D) 117

(E) 118

Solution. For a forward tree, we have

$$u = \exp[(r - \delta)h + \sigma\sqrt{h}] = \exp[(0.04)(0.25) + 0.3\sqrt{0.25}] = 1.173511,$$
$$d = \exp[(r - \delta)h - \sigma\sqrt{h}] = \exp[(0.04)(0.25) - 0.3\sqrt{0.25}] = 0.869358,$$
$$p^* = 1/(1 + e^{\sigma\sqrt{h}}) = 1/(1 + e^{0.3\sqrt{0.25}}) = 0.462570,$$

so that $S_u = S_0 \times u = 117.3511$ and $S_d = S_0 \times d = 86.9358$. In terms of the strike price K, the possible payoffs of the American put option are

$$P_u = (K - 117.3511)_+ \quad \text{and} \quad P_d = (K - 86.9358)_+.$$

In order that the investor exercises the put option at the beginning of the period, the strike price K should be such that

$$\underbrace{(K - S_0)_+}_{\text{Exercise value}} > \underbrace{e^{-rh}[p^* P_u + (1 - p^*)P_d]}_{\text{Holding value}},$$

or

$$(K - 100)_+ > 0.457968(K - 117.3511)_+ + 0.532082(K - 86.9358)_+. \qquad (4.3.1)$$

There are different good ways to solve this inequality:

- Try the five answer choices one by one and find the smallest K such that Inequality (4.3.1) is true.

- Distinguish various ranges of values of K so that Inequality (4.3.1) holds. Because the right-hand side of Inequality (4.3.1) is non-negative, we can restrict our attention to $K > 100$.

   Case 1. If $K > 117.3511$, then the left-hand side of Inequality (4.3.1) is $K - 100$, whereas its right-hand side is $e^{-0.01}K - 110 = 0.99005K - 110$, so Inequality (4.3.1) is true.

   Case 2. If $100 < K \leq 117.3511$, then we solve

$$K - 100 > 0.532082(K - 86.9325),$$

   resulting in $K > 114.8594$. The smallest integer-valued K is therefore $\boxed{115}$. (**Answer: (B)**)   □

Remark. It can be shown that if an American put on a nondividend-paying stock in a one-period binomial tree (not necessarily a forward tree) pays off at both the u and d nodes, then it must be optimally exercised at the beginning of the period. To see this, we determine the holding value of the put option, which is

$$e^{-rT}[p^*(K - S_u) + (1 - p^*)(K - S_d)] = e^{-rT}\{K - [p^*u + (1 - p^*)d]S_0\}.$$

The weighted average $p^*u + (1 - p^*)d$ can be evaluated as

$$\frac{e^{(r-\delta)h} - d}{u - d} \times u + \frac{u - e^{(r-\delta)h}}{u - d} \times d = e^{(r-\delta)h}.$$

Hence

$$\text{Holding value} = e^{-rT}[K - e^{(r-\delta)h} \times S_0] = Ke^{-rT} - S_0 e^{-\delta h} \stackrel{(\delta=0)}{=} Ke^{-rT} - S_0,$$

which is always less than the exercise value of $K - S_0$ for any positive r.

4.4   Options on Other Assets

The fundamental ideas underlying the binomial option pricing model have been sketched in the previous three sections. In this section, we extend our analysis to options on currencies and options on futures, and explore these options' similarities and peculiarities compared with stock options. It is shown that the term "stock" is essentially a label and that as soon as the correct "dividend yields" (in a general sense) applicable to currencies and futures are identified, options on these assets can be priced in technically the same way as stock options.

4.4.1 Case Study 1: Currency Options

Currency options are options whose underlying assets are currencies. For concreteness, consider two given currencies, say dollars ($) and euros (€). We are interested in options on euros (underlying asset).

Suppose that

the current dollar-euro exchange rate is $\$X(0)/$€,

the exchange rate at any future time t is $\$X(t)/$€ (random),

the continuously compounded risk-free interest rate on euros is $r_€$,

the continuously compounded risk-free interest rate on dollars is $r_\$$.

Notice that although the numbers of the two currencies will grow without uncertainty in the future (e.g., $1 will grow to $\$e^{r_\$ t}$ and €1 will grow to €$e^{r_€ t}$ at time t), their relative values (i.e., the exchange rate) in the future are random. Such randomness provides the fundamental motivation for trading currency options, which allow one to hedge against currency risk, also known as exchange rate risk.

Consider a dollar-denominated $\$K$-strike T-year call option on (1 unit of) euro. By "dollar-denominated," we mean that the strike price and premium of the option are both expressed in dollars. This call option gives you the right to give up $\$K$ in return for €1 at time T, with a payoff of

$$\text{Currency call payoff} = (€1 - \$K)_+ = (\ \underbrace{\$X(T) - \$K}_{\text{same denomination}}\ )_+.$$

It allows its holder to benefit from an appreciation of euros against dollars. An otherwise identical put option allows you to give up €1 in return for $\$K$, with a time-$T$ payoff of

$$\text{Currency put payoff} = (\$K - €1)_+ = (\$K - \$X(T))_+,$$

and offers protection against the depreciation of euros against dollars.

Valuing currency options in the binomial tree framework is in principle no different than valuing options on stocks, as soon as you recognize the same roles that $r_€$ and $r_\$$ play as the dividend yield of the underlying asset and the continuously compounded risk-free interest rate, respectively. With this identification, the risk-neutral probability of an increase in the *dollar-euro exchange rate* at the end of a binomial period is

$$\boxed{p^* = \frac{e^{(r_\$ - r_€)h} - d}{u - d}.}$$

The value of a currency option can then be recursively calculated at each node of a binomial tree using risk-neutral pricing as usual.

Example 4.4.1. (SOA Exam IFM Advanced Derivatives Sample Question 5: Currency option) Consider a 9-month dollar-denominated American put option on British pounds. You are given that:

(i) The current exchange rate is 1.43 US dollars per pound.

(ii) The strike price of the put is 1.56 US dollars per pound.

(iii) The volatility of the exchange rate is $\sigma = 0.3$.

(iv) The US dollar continuously compounded risk-free interest rate is 8%.

(v) The British pound continuously compounded risk-free interest rate is 9%.

Using a three-period binomial model, calculate the price of the put.

(A) 0.23

(B) 0.25

(C) 0.27

(D) 0.29

(E) 0.31

Solution. The method of constructing the binomial tree is not specified, so we use a forward tree, which has parameters

$$u = \exp[(r_\$ - r_£)h + \sigma\sqrt{h}] = \exp[(0.08 - 0.09)(0.25) + 0.3\sqrt{0.25}] = 1.158933,$$
$$d = \exp[(r_\$ - r_£)h - \sigma\sqrt{h}] = \exp[(0.08 - 0.09)(0.25) - 0.3\sqrt{0.25}] = 0.858559.$$

The risk-neutral probability of an up move is

$$p^* = \frac{1}{1 + e^{\sigma\sqrt{h}}} = \frac{1}{1 + e^{0.3\sqrt{0.25}}} = 0.462570.$$

The three-period binomial tree for the exchange rate (not stock price) is shown in Figure 4.4.1, along with the values of the put option. Note that early exercise is optimal at the dd node. **(Answer: (A))**   □

4.4.2   Case Study 2: Options on Futures

Options on futures are options that easily cause confusion and deserve special attention. One of the reasons is that there are two derivatives, the concerned option maturing at time T, and the underlying futures maturing at a later time T_f with $T \leq T_f$. If we repeat the one-period replicating procedure discussed in Subsection 4.1.1, with Δ now being the number of futures to buy, then the two replicating equations become

$$\begin{cases} \Delta \times (F_u - F_0) + Be^{rh} = V_u, \\ \Delta \times (F_d - F_0) + Be^{rh} = V_d, \end{cases}$$

where $F_u - F_0$ and $F_d - F_0$ are the mark-to-market payments (recall what you learnt in Section 2.4) of the futures in the u node and d node, respectively. The solutions for Δ and B are

$$\Delta = \frac{V_u - V_d}{F_u - F_d} \quad \text{and} \quad B = e^{-rh}\left[V_u\left(\frac{1-d}{u-d}\right) + V_d\left(\frac{u-1}{u-d}\right)\right].$$

Because no investment is required to enter a futures contract, the price of the option on the futures is simply

$$V_0 = \Delta \times 0 + B = e^{-rh}\left[V_u\left(\frac{1-d}{u-d}\right) + V_d\left(\frac{u-1}{u-d}\right)\right]. \tag{4.4.1}$$

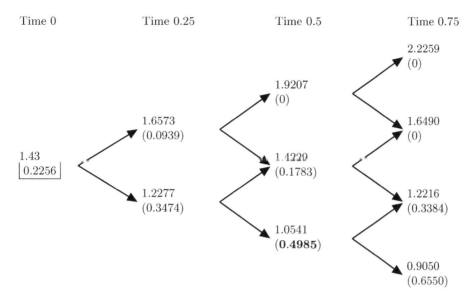

FIGURE 4.4.1
The exchange rate evolution in Example 4.4.1.

Viewed from the risk-neutral pricing perspective, this formula implies that the risk-neutral probability of an up move *in the futures price* in the current setting is

$$p^* = \frac{1-d}{u-d}. \tag{4.4.2}$$

The one-period pricing formula (4.4.1) extends easily to a multi-period setting, with (4.4.2) serving as the risk-neutral probability.

Parenthetically, the risk-neutral probability given in equation (4.4.2) can be obtained by equating the dividend yield δ of the futures with the continuously compounded risk-free interest rate r. Heuristic justification of this somewhat mysterious fact can be found on pages 324, 327, and 328 of Sundaram and Das (2016).

Example 4.4.2. (SOA Exam IFM Advanced Derivatives Sample Question 46: Futures option) You are to price options on a futures contract. The movements of the futures price are modeled by a binomial tree. You are given:

(i) Each period is 6 months.

(ii) $u/d = 4/3$, where u is one plus the rate of gain on the futures price if it goes up, and d is one plus the rate of loss if it goes down.

(iii) The risk-neutral probability of an up move is $1/3$.

(iv) The initial futures price is 80.

(v) The continuously compounded risk-free interest rate is 5%.

Let C_I be the price of a 1-year 85-strike European call option on the futures contract, and C_{II} be the price of an otherwise identical American call option.
    Determine $C_{II} - C_I$.

(A) 0

(B) 0.022

(C) 0.044

(D) 0.066

(E) 0.088

Solution. To find the individual values of u and d, consider

$$p^* = \frac{1-d}{u-d} = \frac{1/d-1}{u/d-1} = \frac{1/d-1}{4/3-1} = \frac{1}{3},$$

which gives $d = 0.9$, and so $u = 1.2$. Now we can construct the binomial futures price tree in Figure 4.4.2.

To calculate the difference $C_{II} - C_I$, we can compute C_I and C_{II} separately. Alternatively, one can observe that $C_{II} - C_I$ returns the current value of the early exercise right carried by the American futures call. Prior to maturity, the only node at which the call is in-the-money is the u node, where the holding value is

$$C_u = e^{-0.05(0.5)} \left[\frac{1}{3}(30.2) + \frac{2}{3}(1.4) \right] = 10.7284,$$

whereas the exercise value is $(96 - 85)_+ = 11$, which is higher. Thus the right to early exercise brings us an extra value of $11 - 10.7284 = 0.2716$ realized at the u node. Then $C_{II} - C_I$ equals this extra value weighed by the risk-neutral probability of entering the u node, discounted for half a year, or

$$C_{II} - C_I = e^{-0.05(0.5)} \times (1/3) \times 0.2716 = \boxed{0.0883}. \quad \textbf{(Answer: (E))}$$

$\square$

4.5 Epilogue: Pricing by Real Probabilities of Stock Price Movements

The risk-neutral pricing formula in the form of (4.2.2) is simple, amazing yet somehow perplexing. One of its far-reaching implications is that the true probability distribution of the stock price plays no role in the pricing problem – nowhere in the pricing formula do we have the true probability of the stock price going up. The only important ingredients are the risk-neutral probability of an increase in the stock price and the risk-free interest rate.

One would wonder:

Shouldn't the true stock price probability distribution be an important input for determining option prices? If the stock price has a higher probability of moving up, shouldn't a call option be more liable to end up in-the-money and hence become more expensive?

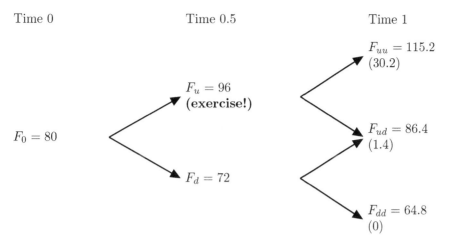

Time 0   Time 0.5   Time 1

$F_{uu} = 115.2$
(30.2)

$F_u = 96$
(exercise!)

$F_0 = 80$

$F_{ud} = 86.4$
(1.4)

$F_d = 72$

$F_{dd} = 64.8$
(0)

FIGURE 4.4.2
The binomial futures price tree in Example 4.4.2.

More generally, two interrelated questions arise naturally:

Question 1: Is risk-neutral pricing, albeit technically correct, consistent with standard discounted cash flow calculations, for which we discount the *genuine* expected values of cash flows at a rate of return that reflects the expected rate of return on an asset of equivalent risk?

Question 2: How can pricing be performed using the true probability distribution of stock prices?

We shall address these two theoretically and practically important questions in this section.

True probability of an up move in stock price.

Let α be the continuously compounded expected return on the stock that pays continuous proportional dividends at a rate of δ and p be the *true* probability (also known as the *real* or *physical* probability) of the stock price going up. By definition,

$$S_0 e^{\alpha h} = \mathbb{E}[\text{Stock price}] = p(e^{\delta h} S_u) + (1-p)(e^{\delta h} S_d) = e^{\delta h}[p(u \times S_0) + (1-p)(d \times S_0)].$$

Canceling S_0 on both sides and solving for p, we have

$$p = \frac{e^{(\alpha - \delta)h} - d}{u - d}. \tag{4.5.1}$$

Comparing (4.1.3) and (4.5.1), one notices that the risk-free rate r goes with the risk-neutral probability p^*, whereas the true rate of return of the stock α stays in company with the true probability p.

Example 4.5.1. (SOA Exam MFE Spring 2007 Question 2: True probability of an up move) For a one-period binomial model for the price of a stock, you are given:

(i) The period is one year.

(ii) The stock pays no dividends.

(iii) $u = 1.433$, where u is one plus the rate of capital gain on the stock if the price goes up.

(iv) $d = 0.756$, where d is one plus the rate of capital loss on the stock if the price goes down.

(v) The continuously compounded annual expected return on the stock is 10%.

Calculate the true probability of the stock price going up.

(A) 0.52

(B) 0.57

(C) 0.62

(D) 0.67

(E) 0.72

Solution. Directly applying equation (4.5.1) yields

$$p = \frac{e^{(\alpha - \delta)h} - d}{u - d} = \frac{e^{(0.1-0)(1)} - 0.756}{1.433 - 0.756} = \boxed{0.5158}. \quad \textbf{(Answer: (A))}$$

$\square$

Discounted rate for the option.

Using the true probability p, we can compute the actual expected payoff of the option as

$$pV_u + (1 - p)V_d.$$

Denote by $\boxed{\gamma}$ the *discount rate* for (or the *expected rate of return* on) the option. To determine γ in terms of α and r, we use the following familiar fact from corporate finance:

The rate of return on a portfolio is the weighted average of the constituent rates of return.

With this fact, we can relate γ, α and r via

$$e^{\gamma h} = \underbrace{\frac{S_0 \Delta}{S_0 \Delta + B}}_{\text{weight on stock}} e^{\alpha h} + \underbrace{\frac{B}{S_0 \Delta + B}}_{\substack{\text{weight on} \\ \text{risk-free bond}}} e^{rh}. \qquad (4.5.2)$$

(Note: It is *not* true that $e^{-\gamma h} = \frac{S_0 \Delta}{S_0 \Delta + B} e^{-\alpha h} + \frac{B}{S_0 \Delta + B} e^{-rh}$.) The price of the option can then be evaluated as

$$\boxed{V_0^{\text{real}} = e^{-\gamma h}[pV_u + (1 - p)V_d].} \qquad (4.5.3)$$

At this point, we have two "prices": One computed by risk-neutral valuation given in equation (4.1.4), and another computed by actual probabilities given in (4.5.3). If these two prices are not the same, then risk-neutral pricing is defective and there is no reason for us to dwell on it (and this book should not have been published!). Fortunately, we are reassured that these two prices are indeed identical with each other. Here is an algebraic proof.

Proof. We are to show

$$e^{-\gamma h}[pV_u + (1-p)V_d] = e^{-rh}[p^*V_u + (1-p^*)V_d]. \qquad (4.5.4)$$

Due to (4.1.1) and (4.5.1), the left-hand side equals

$$\frac{\Delta S_0 + B}{\Delta S_0 e^{\alpha h} + B e^{rh}}[pV_u + (1-p)V_d]$$

$$= \frac{\Delta S + B}{e^{(\alpha-\delta)h}\left(\frac{V_u - V_d}{u-d}\right) + \frac{uV_d - dV_u}{u-d}}\left[\frac{e^{(\alpha-\delta)h} - d}{u-d} \times V_u + \frac{u - e^{(\alpha-\delta)h}}{u-d} \times V_d\right]$$

$$= \frac{\Delta S + B}{\frac{e^{(\alpha-\delta)h} - d}{u-d} \times V_u + \frac{u - e^{(\alpha-\delta)h}}{u-d} \times V_d}\left[\frac{e^{(\alpha-\delta)h} - d}{u-d} \times V_u + \frac{u - e^{(\alpha-\delta)h}}{u-d} \times V_d\right]$$

$$= \Delta S + B$$

$$\overset{(4.1.2)}{=} e^{-rh}[p^*V_u + (1-p^*)V_d]$$

$$= V_0.$$

$\square$

We are now in a position to give complete answers to the two questions posted at the beginning of this section.

Question 1: Risk-neutral pricing is compatible with traditional discounted cash flow calculations. More precisely, these two pricing methods give identical results.

Question 2: To perform standard discounted cash flow calculations, if your course instructor stubbornly insists, then the following three-step procedure motivated from equation (4.5.3) can be adopted:

Step 1.   Find the *true* probability p of an increase in the stock price using (4.5.1). The expected rate of return on the stock α is needed.

Step 2.   Compute the appropriate discount rate γ using (4.5.2).

Step 3.   Use (4.5.3) to calculate the derivative price.

In comparison, risk-neutral pricing is considerably simpler because it makes Step 2 unnecessary. There is no need to identify the correct discount rate at all—the discount rate is simply the risk-free interest rate, which is common among all derivatives.

Incidentally, a by-product of the equality (4.5.4) of risk-neutral valuation and physical valuation is the following alternative and easier equation for γ:

$$\underbrace{e^{-\gamma h}[pV_u + (1-p)V_d]}_{\text{involving }(p,\gamma)} = \underbrace{e^{-rh}[p^*V_u + (1-p^*)V_d]}_{\text{involving }(p^*,r)}.$$

The use of this equation obviates the need for determining the replicating portfolio (Δ, B). More importantly, determining the rate of return on the derivative this way is valid even for *multi-period* binomial trees.

Example 4.5.2. (CAS Exam 8 Spring 2003 Question 36: Rate of return in a one-period tree) A price of a nondividend-paying stock is currently $40.

It is known that at the end of one month the stock's price will be either $42 or $38. The risk-free interest rate is 8% per annum with continuous compounding.

(a) Determine the value of a one-month European call option with a strike price of $39.

(b) Assume that the expected return on the stock is 10% as opposed to the risk-free rate.

What is the correct discount rate to be applied to the payoff in the real world?

Show all work.

Solution. (a) With $u = 42/40 = 1.05$ and $d = 38/40 = 0.95$, the risk-neutral probability of an up move is

$$p^* = \frac{e^{0.08/12} - 0.95}{1.05 - 0.95} = 0.566889.$$

Hence

$$C_0 = e^{-0.08/12}[0.566889(42 - 39)_+ + (1 - 0.566889)(38 - 39)_+] = \boxed{1.6894}.$$

(b) The true probability of an up move is

$$p = \frac{e^{0.1/12} - 0.95}{1.05 - 0.95} = 0.583682.$$

Setting

$$e^{-\gamma/12}[p(42 - 39)_+ + (1 - p)(38 - 39)_+] = C_0 = 1.6894$$

yields $\gamma = \boxed{0.4301}$.   □

Example 4.5.3. (SOA Exam MFE Spring 2009 Question 7: Modification to a binomial tree and its implications) The following one-period binomial stock price model was used to calculate the price of a one-year 10-strike call option on the stock.

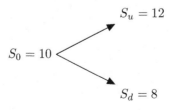

$$S_0 = 10 \qquad \begin{matrix} S_u = 12 \\ \\ S_d = 8 \end{matrix}$$

You are given:

(i) The period is one year.

(ii) The true probability of an up move is 0.75.

(iii) The stock pays no dividends.

(iv) The price of the one-year 10-strike call is $1.13.

Upon review, the analyst realizes that there was an error in the model construction and that S_d, the value of the stock on a down-move, should have been 6 rather than 8. The true probability of an up move does not change in the new model, and all other assumptions were correct.
    Recalculate the price of the call option.

(A) $1.13

(B) $1.20

(C) $1.33

(D) $1.40

(E) $1.53

Ambrose's comments:

This is one of my favorite MFE questions. It shows a counter-intuitive but interesting result about the use of risk-neutral pricing and physical valuation.

Solution. We can perform risk-neutral valuation to deduce the value of e^{-r}, which remains the same after the model review:

$$e^{-r}[p^*(2) + (1 - p^*)(0)] = 1.13$$
$$e^{-r}\left[2\left(\frac{10e^r - 8}{12 - 8}\right)\right] = 1.13$$
$$e^{-r} = 0.9675.$$

After the model review, the risk-neutral probability becomes

$$(p^*)' = \frac{10e^r - 6}{12 - 6} = 0.7226.$$

The new price of the call option is then $e^{-r}[2(p^*)'] = \boxed{1.3983}$. **(Answer: (D))**   □

Remark. (i) Because the call option pays off only at the u node, and S_u and p are both unchanged after the model review, it is true that the expected payoff of the call is unchanged as well after the model review. This seems to imply no change in the call price, as Answer A suggests! However, we have no reason to believe that the expected rate of return on the call, γ, remains the same. In fact, a by-product of this question is that γ has indeed altered. (Exercise: Try to calculate the change.)

(ii) The true probability of an up move given in (ii) of the question is designed by the SOA to distract you!

Example 4.5.4. (Expected rate of return as a function of the scaled strike price) Consider a one-period binomial tree. The length of the period is h years. The stock price moves from S to $S \times u$ or to $S \times d$, $0 < d < u$.

The stock pays no dividends.

Let α be the continuously compounded expected rate of return on the stock.

Consider a one-period put option on the stock with strike price $S \times k$, for some $k > 0$.

Let $\gamma(k)$ denote the continuously compounded expected rate of return on the put option, considered as a function of k.

Determine $\gamma(k)$. Your answer should involve k, r, α, h, u and d. Note that k ranges from 0 to ∞.

Solution. The key equation is

$$\mathrm{e}^{-\gamma(k)h}[pP_u + (1-p)P_d] = \mathrm{e}^{-rh}[p^*P_u + (1-p^*)P_d],$$

or, upon canceling the initial stock price S_0,

$$\mathrm{e}^{-\gamma(k)h}[p(k-u)_+ + (1-p)(k-d)_+] = \mathrm{e}^{-rh}[p^*(k-u)_+ + (1-p^*)(k-d)_+]. \quad (4.5.5)$$

There are three cases:

Case 1. $0 < k < d$

Here, (4.5.5) becomes $\mathrm{e}^{-\gamma(k)h}(0) = \mathrm{e}^{-rh}(0)$. Hence $\gamma(k)$ is not well defined; it can be any real number.

Case 2. $d < k < u$

Here, (4.5.5) becomes $\mathrm{e}^{-\gamma(k)h}[(1-p)(k-d)] = \mathrm{e}^{-rh}[(1-p^*)(k-d)]$. Thus

$$\gamma(k) = r - \frac{1}{h}\ln\left(\frac{1-p^*}{1-p}\right) = r - \frac{1}{h}\ln\left(\frac{u-\mathrm{e}^{rh}}{u-\mathrm{e}^{\alpha h}}\right).$$

Case 3. $u < k < \infty$

Here, (4.5.5) becomes

$$\mathrm{e}^{-\gamma(k)h}[p(k-u) + (1-p)(k-d)] = \mathrm{e}^{-rh}[p^*(k-u) + (1-p^*)(k-d)].$$

As $p(k-u) + (1-p)(k-d) = k - [pu + (1-p)d] = k - \mathrm{e}^{\alpha h}$ and $p^*(k-u) + (1-p^*)(k-d) = k - \mathrm{e}^{rh}$,

$$\gamma(k) = r - \frac{1}{h}\ln\left(\frac{k-\mathrm{e}^{rh}}{k-\mathrm{e}^{\alpha h}}\right).$$

$\square$

Example 4.5.5. (Rate of return in a two-period tree) You use the following information to construct a binomial forward tree for modeling the price movements of a stock.

(i) The length of each period is one year.

(ii) The current stock price is 82.

(iii) The stock's volatility is 30%.

(iv) The stock pays no dividends.

(v) The continuously compounded risk-free interest rate is 5%.

(vi) The continuously compounded expected return on the stock is 10%.

Calculate the continuously compounded expected rate of return on a two-year 80-strike European call option on the stock.

Solution. The forward tree parameters are

$$u = e^{0.05(1)+0.3\sqrt{1}} = e^{0.35} = 1.419068 \quad \text{and} \quad d = e^{0.05(1)-0.3\sqrt{1}} = e^{-0.25} = 0.778801.$$

The risk-neutral probability of an up move is

$$p^* = \frac{1}{1 + e^{0.3\sqrt{1}}} = 0.425557.$$

With $S_{uu} = 165.1277$, $S_{ud} = 90.6240$ and $S_{dd} = 49.7355$, we have $C_{uu} = 85.1277$, $C_{ud} = 10.6240$ and $C_{dd} = 0$. The price of the call option is

$$
\begin{aligned}
& e^{-2r}[(p^*)^2 C_{uu} + 2p^*(1 - p^*)C_{ud} + (1 - p^*)^2 C_{dd}] \\
=~ & e^{-0.05(2)}[(0.425557)^2(85.1277) + 2(0.425557)(1 - 0.425557)(10.6240)] \\
=~ & 18.6494.
\end{aligned}
$$

With the true probability of an up move being

$$p = \frac{e^{0.1} - 0.778801}{1.419068 - 0.778801} = 0.509740,$$

we solve

$$e^{-2\gamma}[\underbrace{p^2 C_{uu} + 2p(1 - p)C_{ud} + (1 - p)^2 C_{dd}}_{(0.509740)^2(85.1277)+2(0.509740)(1-0.509740)(10.6240)}] = 18.6494,$$

resulting in $\gamma = \boxed{0.1929}$. $\qquad\square$

4.6   Problems

One-period binomial trees

Problem 4.6.1. (Valuing a strangle) Consider a 50-65 1-year strangle strategy. You are given:

(i) The stock currently sells for $55.

(ii) In one year, the stock will either sell for $70 or $45.

(iii) The *effective annual* risk-free interest rate is 10%.

Calculate the price you now pay for the strangle.

Problem 4.6.2. (Valuing a derivative that pays the square of terminal stock price) The current price of a nondividend-paying stock is $S(0) = 100$. The price of the stock at the end of one year, $S(1)$, is either 90 or 120. The continuously compounded risk-free interest rate is 10%.

Consider a derivative security that pays $[S(1)]^2$ at the end of one year.

(a) Determine the replicating portfolio for the derivative, i.e., find B and Δ.

(b) Calculate the current price of the derivative.

Problem 4.6.3. (Valuing a strange derivative) Suppose that the current stock price is $30 per share. At the end of 6 months, the stock price will be either $25 or $38. The 6-month effective risk-free interest rate is 10%.

A European-type derivative written on this stock has its payoff in 6 months equal to

$$[S(T) - 32]_+^2 + [28 - S(T)]_+^3,$$

where $S(T)$ is the stock price at maturity.

Calculate the current price of this derivative.

Problem 4.6.4. (Arbitraging a mispriced bear spread in a one-period binomial model) You are given:

(i) The current price of a stock is $65.

(ii) One year from now the stock will sell for either $60 or $70.

(iii) The stock pays dividends continuously at a rate proportional to its price. The dividend yield is 4%.

(iv) The continuously compounded risk-free interest rate is 6%.

(v) The current price of a one-year 65-75 European put bear spread on the above stock is $6.50.

Describe transactions (i.e., what to buy/sell/borrow/lend) that one should enter into to exploit an arbitrage opportunity (if one exists). Show your work.

Problem 4.6.5. [HARDER!] (A market consisting only of risky securities – I)
For a one-period arbitrage-free binomial model with two nondividend-paying securities, you
are given:

(i) The following price evolution of the two securities:

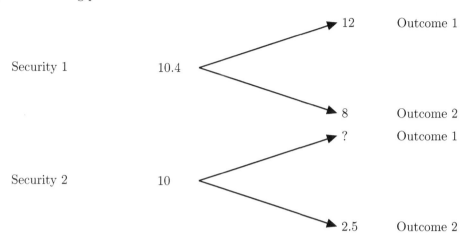

	12	Outcome 1
Security 1	10.4	
	8	Outcome 2
	?	Outcome 1
Security 2	10	
	2.5	Outcome 2

(ii) The following information about two European call options:

Call Option	Underlying Asset	Strike Price	Current Price
A	Security 1	9	1.8
B	Security 2	11	??

Calculate the current price of call option B.
   (Hints:

(i) This market does not have a risk-free bond, at least in its current form. Without the risk-
   free rate, the risk-neutral probability cannot be defined. If you use risk-neutral valuation,
   you need to explain in your solution how the risk-free bond can be constructed and what
   the risk-free interest rate is. If these prove too hard, go back to basics and use the method
   of replication instead.

(ii) In any case, you may find it useful to realize that a replicating portfolio can consist of any
   securities at your choosing. Why not replicate using the given securities and/or call options
   with known prices and/or payoffs!?)

Problem 4.6.6. (A market consisting only of risky securities – II) In an arbitrage-
free securities market, there are two nondividend-paying stocks, A and B, both with current
price $90. There are two possible outcomes for the prices of A and B one year from now:

Outcome	A	B
1	$100	$80
2	$60	$x

The current price of a one-year 100-strike European put option on B is $15.
Determine all possible values of x.

Problem 4.6.7. (Calculation of p^* given volatility information) You are to use a
binomial forward tree model to model the price movements of a stock that pays dividends
continuously at a rate proportional to its price. The length of each period is three months.
   If $\mathrm{Var}[\ln[S(t)]] = 0.25t$ for $t > 0$, find p^*, the risk-neutral probability of an up move.

Multi-period binomial trees (European options)

Problem 4.6.8. [HARDER!] (A two-period binomial model with varying u, d, and i) You are given the following with respect to a public company:

- The common shares of the company were trading at 100 as of December 31, 2015.

- No dividends are paid.

- An industry analyst has projected the possible stock prices over the next two years as a function of the performance of the US economy:

US Economy		Share Price of Company	
2016	2017	December 31, 2016	December 31, 2017
Expansion	Expansion	110	120
Expansion	Recession	110	100
Recession	Expansion	95	110
Recession	Recession	95	90

- The effective annual risk-free interest rate that prevails in Year 2016 (i.e., from January 1, 2016 to December 31, 2016) is 5%.

- The effective annual risk-free interest rate that prevails in Year 2017 is 6% if 2016 has seen an expansion, and 4% if 2016 has seen a recession.

Using the analyst's projections, determine the price, as of December 31, 2015, of a two-year European strangle constructed by buying a 100-strike European put option and a 110-strike European call option.

(Remember that a multi-period binomial tree is essentially a series of one-period binomial trees. In the context of this problem, these one-period trees have different u, d, i, and p^*.)

Problem 4.6.9. (Pricing a European strangle in a three-period binomial forward tree) You use the following information to construct a binomial forward tree for modeling the price movements of a stock:

(i) The length of each period is 1 year.

(ii) The current stock price is 190.

(iii) The stock's volatility is 30%.

(iv) The stock pays dividends continuously at a rate proportional to its price. The dividend yield is 5%.

(v) The continuously compounded risk-free interest rate is 3%.

Calculate the price of a 3-year 150-250 European strangle.

Problem 4.6.10. [HARDER!] (Valuing a straddle in a 10-period binomial tree) For a 10-period binomial stock price model, you are given:

(i) The length of each period is one year.

(ii) The current stock price is 1,000.

(iii) At the end of every year, the stock price will either increase by 5% or decrease by 5% in proportion.

(iv) The stock pays dividends continuously at a rate proportional to its price. The dividend yield is 1%.

(v) The continuously compounded risk-free interest rate is 2%.

Calculate the current price of a 10-year 1,400-strike European straddle on the stock.

(Hint: Think twice before you discount the eleven possible 10-year payoffs of the straddle back to time 0! That could take you half an hour!)

Problem 4.6.11. (A derivative with payoff dependent on interim stock price) Let $S(t)$ be the time-t price of a nondividend-paying stock. For a three-period binomial stock price model, you are given:

(i) The length of each period is one year.

(ii) $S(0) = 100$.

(iii) $u = 1.1$, where u is one plus the percentage change in the stock price per period if the price goes up.

(iv) $d = 1/1.1$, where d is one plus the percentage change in the stock price per period if the price goes down.

(v) The continuously compounded risk-free interest rate is 5%.

Consider a special derivative which pays, at the end of three years,

$$\max\{S(2) - 100, 0\} + \max\{S(3) - 100, 0\}.$$

Calculate the current price of this derivative.

(Note: Note that the time-2 stock price $S(2)$ that constitutes the payoff formula is paid at time 3. This derivative, whose payoff depends on intermediate stock prices, is an example of a *path-dependent* derivative, more examples of which will be given in Sections 8.4 to 8.7.)

Problem 4.6.12. [HARDER!] (Choosing between a call or a put) For a two-period binomial model for stock prices, you are given:

(i) The length of each period is one year.

(ii) The current price of a nondividend-paying stock is $150.

(iii) $u = 1.25$, where u is one plus the percentage change in the stock price per period if the price goes up.

(iv) $d = 0.80$, where d is one plus the percentage change in the stock price per period if the price goes down.

(v) The continuously compounded risk-free interest rate is 6%.

Consider a *chooser option* (also known as an as-you-like-it option) on the stock. At the end of the first year, its holder will choose, to his/her advantage, whether it becomes a European call option or a European put option, both of which will expire at the end of the second year with a strike price of $150.

Calculate the current price of the chooser option.

(Hint: If you were the holder of the chooser option and you were rational, which option, the call or the put, would you choose at the end of one year?)

Problem 4.6.13. [HARDER!] (Valuing an option on another option in the binomial setting) For a two-period binomial model for stock prices, you are given:

(i) The length of each period is one year.

(ii) The current price of a nondividend-paying stock is $40.

(iii) $u = 1.05$, where u is one plus the percentage change in the stock price per period if the price goes up.

(iv) $d = 0.9$, where d is one plus the percentage change in the stock price per period if the price goes down.

(v) The continuously compounded risk-free interest rate is 3%.

Consider Derivative X, which gives its holder the right, but not the obligation, to buy a $38-strike European put option at the end of the first year for $0.5. This put option is written on the stock and will mature at the end of the second year.

(a) Calculate the current price of Derivative X.

(Hint: What is the underlying asset of Derivative X? The stock, or something else? Does the price of this asset evolve according to a binomial tree model?)

(b) Using the result of part (a), calculate the current price of Derivative Y, which is identical to Derivative X, except that it gives its holder the right to *sell* the same put option for $0.5 at the end of the first year.

Problem 4.6.14. [HARDER!] (Valuing a forward on a call in the binomial setting) For a binomial forward tree modeling the price movements of a stock, you are given:

(i) The length of each period is 6 months.

(ii) The current price of a nondividend-paying stock is $9,000.

(iii) The stock's volatility is 32%.

(iv) The continuously compounded risk-free interest rate is 20%.

Consider the following offer:

By receiving this offer, 6 months from now you are *obligated* to buy a 6-month $9,000-strike European call option on the stock for $1,500. This call option expires one year from now.

Calculate the current fair price of this offer.

Problem 4.6.15. [HARDER!] (Arbitraging a mispriced option in a two-period binomial tree) You are given the following regarding the stock of Iowa Actuarial Association (IAA):

(i) The stock is currently selling for $100.

(ii) $u = 1.1$, where u is one plus the percentage change in the stock price per period if the price goes up.

(iii) $d = 0.9$, where d is one plus the percentage change in the stock price per period if the price goes down.

(iv) The stock pays no dividends.

The effective annual risk-free interest rate is 2%.

While reading the *Well Street Journal*, Peter notices that a two-year at-the-money European call written on the stock of IAA is selling for $7.5. Peter wonders whether this call is fairly priced. He uses the binomial option pricing model to determine if an arbitrage opportunity exists.

Construct the trading strategies for Peter to exploit the arbitrage opportunity (if one exists).

(Hint:

- This problem is a two-period version of Example 4.1.5. Now "Michael" becomes "Peter" and "Financial Post" becomes "*Well Street Journal*"!

- Peter may not just sit on the sofa and rest leisurely as soon as the strategies are set up at time 0. He may need to "do something" in response to the stock market environment. In your solution, describe clearly what Peter needs to do at what time and at what stock price level. Check, at least mentally, that your strategies indeed lead to an arbitrage.

- This problem is another illustration of the fundamental value of the method of replication.)

Problem 4.6.16. (Swapping parameters: Binomial tree setting) Two actuaries, A and B, use a two-period binomial forward tree to compute the prices of a European call and a European put using different parameters.

You are given:

Actuary	Option	Underlying Stock Price	Strike Price	Dividend Yield	Risk-free Interest Rate	Stock Volatility	Option Maturity
A	Call	190	200	5%	3%	30%	1
B	Put	200	190	3%	5%	30%	1

Describe the relationship between the call price computed by Actuary A and the put price computed by Actuary B.

Multi-period binomial trees (American options)

Problem 4.6.17. (European vs American prices: Two-period version) You use the following information to construct a binomial forward tree for modeling the price movements of a nondividend-paying stock:

(i) The length of each period is 6 months.

(ii) The current stock price is 100.

(iii) The stock's volatility is 20%.

(iv) The continuously compounded risk-free interest rate is 6%.

Let P_I be the price of a 120-strike 1-year European put option on the stock, and P_{II} be the price of an otherwise identical American put option.
　　Calculate $P_{II} - P_I$.
　　(Hint: Be sure to check whether early exercise is optimal at ALL possible nodes, including...)

Problem 4.6.18. (European vs American prices: Three-period version) You use the following information to construct a binomial forward tree for modeling the price movements of a stock.

(i) The length of each period is 4 months.

(ii) The current stock price is 100.

(iii) The stock's volatility is 30%.

(iv) The stock pays no dividends.

(v) The continuously compounded risk-free interest rate is 8%.

Let P_I be the price of a 95-strike 1-year European put option on the stock, and P_{II} be the price of an otherwise identical American put option.
　　Calculate $P_{II} - P_I$.

Problem 4.6.19. (Three-period binomial tree for an American put) You use the following information to construct a binomial forward tree for modeling the price movements of a stock:

(i) The length of each period is 4 months.

(ii) The current stock price is 55.

(iii) The stock's volatility is 30%.

(iv) The stock pays dividends continuously at a rate proportional to its price. The dividend yield is 3.5%.

(v) The continuously compounded risk-free interest rate is 5%.

Calculate the price of a 1-year 50-strike American put option on the stock.

Problem 4.6.20. (SOA Exam FETE Fall 2011 Question 7 (c): Pricing a warrant as an American call with varying strike prices) The Ashwaubenon Company (Ash Co) needs to raise capital to support its rapidly growing business. One proposal is to publicly issue a certain number of equity units, each of which consists of one share of stock and a warrant to purchase one share of stock.

Assume that the price of the underlying asset follows a binomial tree with 1-year time steps as follows:

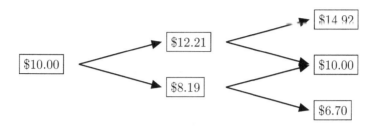

The warrant provides the right to purchase one share of the stock for $9 at the first anniversary or, if not exercised, for $10 at the second anniversary.

Assume further that:

- The stock pays no dividend.

- The risk-free interest rate is 4% per annum.

Calculate the value of the warrant using the binomial tree.

(Note: There are many reasons for structuring the warrant to have the exercise price change with the exercise date. One of them is to encourage earlier exercise of the warrant allowing for better alignment of cash flow needs for a rapidly growing company.)

Options on other assets

Problem 4.6.21. (Pricing an American currency put using a binomial forward tree) You use the following information to construct a two-period binomial forward tree for modeling the movements of the dollar-euro exchange rate:

(i) The current dollar-euro exchange rate is $1.50/€.

(ii) The volatility of the exchange rate is 20%.

(iii) The continuously compounded risk-free interest rate on dollars is 4%.

(iv) The continuously compounded risk-free interest rate on euros is 5%.

Calculate the price of a 6-month $1.80-strike dollar-denominated American put option on euros.

Problem 4.6.22. [HARDER!] (Four-period binomial tree for a Bermudan currency call) For a four-period binomial tree model for the dollar/pound exchange rate, you are given:

(i) The length of each period is 3 months.

(ii) The current dollar/pound exchange rate is 1.4.

(iii) $u = 1.1$ and $d = 0.9$, where u and d are one plus the percentage change in the dollar/pound exchange rate per period if the exchange rate goes up and if the exchange rate goes down, respectively.

(iv) The continuously compounded risk-free interest rate on dollars is 8%.

(v) The continuously compounded risk-free interest rate on pounds is 7%.

Calculate the price of a 1-year at-the-money dollar-denominated Bermudan call option on pounds, where exercise is allowed at any time following an initial 9-month period of call protection (i.e., exercise is allowed in 9 months and thereafter).

(Hint: There is no need to construct the whole 4-period binomial tree. Generate only as many exchange rates as needed.)

Problem 4.6.23. (Replicating portfolio for a futures call option) You use the following information to construct a binomial tree for modeling the price movements of a futures contract on the S&V 150:

(i) The length of each period is 6 months.

(ii) The initial futures price is 500.

(iii) $u = 1.2363$, where u is one plus the percentage change in the futures price per period if the price goes up.

(iv) $d = 0.8089$, where d is one plus the percentage change in the futures price per period if the price goes down.

(v) S&V 150 pays dividends continuously at a rate proportional to its price. The dividend yield is 2%.

(vi) The continuously compounded risk-free interest rate is 6%.

Consider a 1-year at-the-money *American* call option on the above S&V 150 futures contract.

Determine the replicating portfolio of the put option *at the initial node*.

Pricing by true probabilities

Problem 4.6.24. (Expected rate of return on an option in a one-period binomial model) You use the following information to construct a one-period binomial forward tree for modeling the price movements of a nondividend-paying stock:

(i) The current stock price is 82.

(ii) The stock's volatility is 30%.

(iii) The continuously compounded risk-free interest rate is 8%.

(iv) The continuously compounded expected return on the stock is 15%.

Calculate the expected rate of return on a 80-strike 1-year European call option.

Problem 4.6.25. (First-period expected rate of return on an option in a two-period binomial model) You use the following information to construct a binomial forward tree for modeling the price movements of a nondividend-paying stock:

(i) The length of each period is 4 months.

(ii) The current stock price is 50.

(iii) The stock's volatility is 30%.

(iv) The continuously compounded risk-free interest rate is 8%.

(v) The continuously compounded expected return on the stock is 15%.

Calculate the continuously compounded true discount rate for an 8-month 40-strike European call option during the first time period.

Problem 4.6.26. (Expected rate of return on an option in a three-period binomial model) You use the following information to construct a binomial forward tree for modeling the price movements of a nondividend-paying stock:

(i) The length of each period is 4 months.

(ii) The current stock price is 123.

(iii) The stock's volatility is 30%.

(iv) The continuously compounded risk-free interest rate is 8%.

(v) The continuously compounded expected return on the stock is 12%.

Calculate the continuously compounded expected rate of return on a 120-strike 1-year European call option *over its 1-year life*.

5

Mathematical Foundations of the Black-Scholes Framework

Chapter overview: In this rudimentary chapter, we migrate from the discrete-time binomial tree model in the preceding chapter to the most celebrated continuous-time option pricing model commonly referred to as the *Black-Scholes option pricing model*, rightfully due to the two prominent mathematicians Fischer Black and Myron Scholes. In contrast to the binomial tree model, which postulates that future stock prices unfold in a binomial manner, the principal assumption of the Black-Scholes model is that future stock prices evolve continuously in accordance with a lognormal distribution. The primary objective of this chapter is to make this statement mathematically precise and set up the lognormal stock price model so as to pave the way for performing option pricing in Chapter 6. In Section 5.1, we introduce the lognormal distribution, which is the continuous probability distribution that underlies the Black-Scholes model and governs the probabilistic behavior of each future stock price, and formulate the so-called "Black-Scholes framework." In the Black-Scholes context, some simple probabilistic calculations are performed in Section 5.2, including the calculation of the exercise probability of a European option, unconditional and conditional expected stock prices, and the construction of lognormal prediction intervals for stock prices. These probabilistic quantities will be instrumental in deriving the Black-Scholes option pricing formula for European options in the next chapter.

5.1   A Lognormal Model of Stock Prices

The Black-Scholes option pricing model assumes that stock prices in the future obey a lognormal distribution, which can be seen as the limit of the binomial tree model as the number of periods n approaches infinity. Because the stock prices process is a collection of lognormal random variables lumped over time, we will occasionally use the non-standard term "*lognormal process.*"[i] Given the close relevance of the lognormal distribution to continuous-time option pricing, we begin with a description of its key properties.

Definition of the lognormal distribution.

I assume that you are no strangers to the normal distribution, justifiably the most well-known continuous distribution, based on your learning in prior statistics courses. If you need to refresh your memory, you may refer to your favorite probability textbook.

   The normal distribution gives rise to a closely related distribution, called the lognormal distribution, which is useful for modeling stock prices. A random variable Y follows a

---

[i]The mathematically impeccable assumption of the Black-Scholes model is that stock prices follow a geometric Brownian motion, which entails more than just the stock prices being marginally lognormally distributed. The notion of this stochastic process is beyond the scope of this book.

lognormal distribution with parameters m and v^2 if the natural logarithm of Y follows a normal distribution with the same parameters m and v^2 (mean and variance, respectively). The nomenclature can be explained by the fact that if we "log" a lognormal random variable, then it becomes "normal." We shall write $Y \sim \mathrm{LN}(m, v^2)$ to represent such a lognormal random variable. The probability density function of Y is

$$f_Y(y; m, v^2) = \frac{1}{\sqrt{2\pi v^2} y} \exp\left[-\frac{(\ln y - m)^2}{2v^2}\right], \quad y > 0,$$

although we have rather limited use of this expression in this book.

Moments of a lognormal random variable.

It is possible to give a direct but clumsy proof of the expectation of a lognormal random variable by brute-force integration. A more instructive way to obtain lognormal moments is to realize that if Y is lognormal with parameters m and v^2, then Y has the same distribution as e^X, where X is a normal random variable with mean m and variance v^2. Therefore, for any real k,

$$\mathbb{E}[Y^k] = \mathbb{E}[e^{kX}] = M_X(k) = \underbrace{\exp\left(km + \frac{1}{2}k^2 v^2\right)}_{\text{remember?}},$$

which is the moment generating function of X evaluated at k. In particular,

$$\mathbb{E}[Y] = e^{m + v^2/2}, \qquad \mathbb{E}[Y^2] = e^{2(m + v^2)},$$

and the variance of Y is

$$\mathrm{Var}(Y) = \mathbb{E}[Y^2] - \mathbb{E}[Y]^2 = e^{2m + v^2}(e^{v^2} - 1).$$

Lognormal stock price model.

Armed with the basic facts about the lognormal distribution, we are now ready to formulate a lognormal model of stock prices, which lies at the heart of the Black-Scholes framework. For $t \geq 0$, we denote by $S(t)$ the time-t price of the underlying stock paying continuous proportional dividends at a rate of δ. In the Black-Scholes model, we assume that stock prices are configured by

$$\boxed{S(t) = S(0) \exp\left[\left(\alpha - \delta - \frac{1}{2}\sigma^2\right) t + \sigma\sqrt{t}Z\right],} \qquad (5.1.1)$$

for some non-negative parameters α and σ, and some standard normal random variable Z[ii] governing the evolution of the stock prices. This is indeed a lognormal model because the logarithm of each stock price, given by $\ln S(t) = \ln S(0) + (\alpha - \delta - \sigma^2/2)t + \sigma\sqrt{t}Z$, is normally distributed as

$$\boxed{\ln S(t) \sim \mathrm{N}\left[\ln S(0) + \left(\alpha - \delta - \frac{1}{2}\sigma^2\right) t, \; \sigma^2 t\right],} \qquad (5.1.2)$$

[ii]Rigorously speaking, the random variable Z varies with the time t. That is, for different choices of time, the stock prices will involve different Z's, which are dependent random variables. This subtle fact will not concern us because in this book we will mostly look at stock prices in isolation.

or equivalently,

$$S(t) \sim LN \left[\underbrace{\ln S(0)}_{\text{Don't omit this!}} + \left(\alpha - \delta - \frac{1}{2}\sigma^2 \right) t, \ \sigma^2 t \right].$$

Another way to express (5.1.1) is that the continuously compounded expected rate of return (also known as the capital gain) on the stock from time 0 to any future time t, $\ln[S(t)/S(0)]$, is normally distributed, with the following specification:

$$\ln \left[\frac{S(t)}{S(0)} \right] \sim N \left[\left(\alpha - \delta - \frac{1}{2}\sigma^2 \right) t, \ \sigma^2 t \right] \tag{5.1.3}$$

To put it simply, the *normality* of the continuously compounded rates of return on the stock is identical to the *lognormality* of the stock prices.

It is undeniably true that the configuration of the parameters α and σ given in (5.1.3) is rather difficult to remember and not intuitive at all. In particular, it is not immediately clear why $\sigma^2/2$ needs to be subtracted from the mean of the rate of return. In fact, α and σ enter in such a way that makes α the continuously compounded (annualized) expected *total* rate of return on the stock and σ the (annualized) volatility of the stock. To see this, we take expectation on both sides of (5.1.1) and get

$$\mathbb{E}[S(t)] = S(0)e^{(\alpha-\delta-\sigma^2/2)t} \times \underbrace{e^{\sigma^2 t/2}}_{\text{why?}} = S(0)e^{(\alpha-\delta)t},$$

which is equivalent to

$$\underbrace{S(0)}_{\text{beginning value}} e^{\alpha t} = \underbrace{e^{\delta t}\mathbb{E}[S(t)]}_{\text{expected ending value}}.$$

This confirms that α is the continuously compounded expected rate of return on the stock. Moreover, it follows from (5.1.2) that

$$\text{Var}[\ln S(t)] = \text{Var} \left[\ln \frac{S(t)}{S(0)} \right] = \sigma^2 t,$$

showing that σ is the volatility of the stock price (recall the general definition of volatility on page 126). In brief, the seemingly complex arrangement of terms in (5.1.1) is meant to make the model parameters easily interpretable.

Figure 5.1.1 exhibits a typical path of the stock prices in the Black-Scholes framework over a 1-year horizon with the parameters $S(0) = 100$, $\alpha = 0.08$, and $\sigma = 0.3$. The path can be viewed as the superposition of infinitely many binomial trees to produce continuous evolution.

Example 5.1.1. (Specifying the distribution of $S(T)$) Assume the Black-Scholes framework. Let $S(t)$ denote the time-t price of a nondividend-paying stock. You are given:

(i) The current stock price is 38.

(ii) The stock's volatility is 35%.

(iii) The continuously compounded expected return on the stock is 16%.

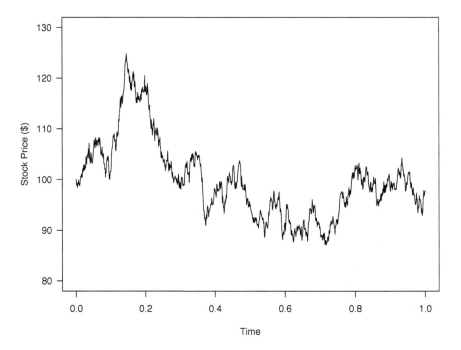

FIGURE 5.1.1
A stock price path in the Black-Scholes stock price model with $S(0) = 100$, $\alpha = 0.08$, and $\sigma = 0.3$.

(iv) The continuously compounded risk-free interest rate is 5%.

Determine the probability distribution of $S(0.5)$.

Solution. The 6-month stock price $S(0.5)$ is lognormal with parameters

$$m = \ln S(0) + \left(\alpha - \delta - \frac{1}{2}\sigma^2\right)t = \ln 38 + (0.16 - \frac{1}{2} \times 0.35^2)(0.5) = \boxed{3.6951}$$

and

$$v^2 = 0.35^2(0.5) = \boxed{0.06125}.$$

$\square$

5.2   Lognormal-Based Probabilistic Quantities

The fact that each future stock price is lognormally distributed according to (5.1.1) allows us to easily calculate a number of distributional quantities of theoretical and practical interest. In this section, we concern ourselves with the exercise probability of a European option, unconditional and conditional expected stock prices, and lognormal prediction intervals for stock prices. Some of these quantities will provide us with the necessary ingredients to form the Black-Scholes option pricing formula for European options in Chapter 6.

Quantity 1: Exercise probabilities.

The first probabilistic quantity of interest is the (physical) probability that a T-year K-strike European call will be exercised at expiration:

$$\mathbb{P}\left(S(T) > K\right).$$

Using (5.1.1), we have

$$
\begin{aligned}
\mathbb{P}\left(S(T) > K\right) &= \mathbb{P}\left(S(0)\exp\left[\left(\alpha - \delta - \frac{1}{2}\sigma^2\right)T + \sigma\sqrt{T}Z\right] > \ln K\right) \\
&= \mathbb{P}\left(Z > \frac{\ln[K/S(0)] - (\alpha - \delta - \sigma^2/2)T}{\sigma\sqrt{T}}\right) \\
&= 1 - N\left[\frac{\ln[K/S(0)] - (\alpha - \delta - \sigma^2/2)T}{\sigma\sqrt{T}}\right] \\
&= N\left[\frac{\ln[S(0)/K] + (\alpha - \delta - \sigma^2/2)T}{\sigma\sqrt{T}}\right],
\end{aligned}
$$

where $N(\cdot)$ is the distribution function of the standard normal distribution (you may have used $\Phi(\cdot)$ in your prior statistics courses) and the last equality follows from the symmetry of $N(\cdot)$ about zero, i.e., $N(x) + N(-x) = 1$ for all $x \in \mathbb{R}$. If we let

$$\hat{d}_2 = \frac{\ln[S(0)/K] + (\alpha - \delta - \sigma^2/2)T}{\sigma\sqrt{T}}, \tag{5.2.1}$$

then the above exercise probability can be compactly written as

$$\boxed{\mathbb{P}\left(S(T) > K\right) = N(\hat{d}_2).} \tag{5.2.2}$$

The reason why we decorate the symbol d by the subscript "2" and with a hat will become apparent in Chapter 6. Likewise, the exercise probability for an otherwise identical European put is

$$\boxed{P\left(S(T) < K\right) = 1 - \mathbb{P}\left(S(T) > K\right) = N(-\hat{d}_2).}$$

As you work out problems in this book, you will frequently need values of the standard normal distribution function $N(x)$ and standard normal quantiles $N^{-1}(p)$ for various $x \in \mathbb{R}$ and $p \in (0, 1)$. For this purpose, it is strongly suggested that you have access to a standard normal distribution function calculator (e.g., actuarial students who will take Exam IFM, which is computerized, can use `https://www.prometric.com/en-us/clients/soa/pages/mfe3f_calculator.aspx`, which is capable of computing $N(x)$ and $N^{-1}(p)$ to five decimal places). For your convenience, a standard normal distribution table is provided in Appendix A of this book. In what follows we will follow the IFM standard normal calculator and display intermediate results to five decimal places.

Example 5.2.1. (CAS Exam 8 Spring 2000 Question 30: Probability that a call will be exercised − I) You are given the following information about a European call option.

- The current stock price is \$35.

- The exercise price is \$40.

- The option matures in 6 months.

- The expected return on the stock is 18% per annum.

- The volatility, σ, of the stock price is 24% per annum.

- The stock's price at each future time T, given its current price, is lognormally distributed.

- The stock pays no dividends.

What is the probability that the call option will be exercised?

Solution. Since

$$\hat{d}_2 = \frac{\ln(35/40) + (0.18 - 0 - 0.24^2/2)(0.5)}{0.24\sqrt{0.5}} = -0.34136,$$

the exercise probability for the call option is

$$\mathbb{P}\left(S(0.5) > 40\right) = N(\hat{d}_2) = 1 - \underbrace{N(0.34136)}_{0.63358} = \boxed{0.36642}.$$

□

Example 5.2.2. (SOA Exam MFE Spring 2009 Question 16: Probability that a call will be exercised II) You are given the following information about a nondividend-paying stock:

(i) The current stock price is 100.

(ii) Stock prices are lognormally distributed.

(iii) The continuously compounded expected return on the stock is 10%.

(iv) The stock's volatility is 30%.

Consider a nine-month 125-strike European call option on the stock.
   Calculate the probability that the call will be exercised.

(A) 24.2%

(B) 25.1%

(C) 28.4%

(D) 30.6%

(E) 33.0%

Solution. With

$$\hat{d}_2 = \frac{\ln(100/125) + (0.1 - 0 - 0.3^2/2)(0.75)}{0.3\sqrt{0.75}} = -0.70011,$$

the exercise probability for the call option is

$$\mathbb{P}\left(S(0.75) > 125\right) = N(\hat{d}_2) = N(-0.70011) = \boxed{0.24193}. \quad \textbf{(Answer: (A))}$$

$\square$

Quantity 2: Lognormal prediction intervals for stock prices.

The next probabilistic quantity we are interested in is a range of values in which the stock price has a high probability of lying. To be precise, given any $t \geq 0$ and $p \in (0, 1)$, we are to find a pair of constants (S^L, S^U) (depending on t and p) such that

$$\mathbb{P}(S^L < S(t) < S^U) = 1 - p.$$

We then call (S^L, S^U) a $100(1 - p)\%$ *lognormal prediction interval* for $S(t)$. Quite often, we require the prediction interval to be equal-tailed in the sense that

$$\mathbb{P}(S(t) < S^L) = \mathbb{P}(S(t) > S^U) = \frac{p}{2}.$$

As you will see soon, using $1 - p$ as the probability level instead of p makes the construction of the prediction interval easier.

A note about the term "prediction interval" is in order. Here we are forming two deterministic constants, S^L and S^U, which will bound the *random* variable $S(t)$ with the prespecified probability of $1 - p$. This is in contradistinction to the usual setting in mathematical statistics in which we use a pair of random variables, called a *confidence interval*, to estimate a *constant* parameter (e.g., the mean or variance of a normal distribution). The differences between prediction intervals and confidence intervals are summarized in the following table:

Type of interval	Nature of bounds of interval	Nature of target
<u>Confidence</u> interval	Random variables (i.e., estimators/statistics)	Constant parameter
<u>Prediction</u> interval	Can be random variables or constants	Random variable

To determine a $100(1 - p)\%$ prediction interval for $S(t)$ for any $t \geq 0$, we first construct a $100(1 - p)\%$ prediction interval for Z, the standard normal random variable "driving" $S(t)$. In terms of the inverse of the standard normal distribution function, we have

$$\mathbb{P}\left(N^{-1}(p/2) < Z < N^{-1}(1 - p/2)\right) = 1 - p,$$

so the $100(1-p)\%$ equal-tailed prediction interval for Z is $\left(N^{-1}(p/2), N^{-1}(1 - p/2)\right)$. Then we go from Z to $S(t)$ by

$$\mathbb{P}\left(S(0)e^{(\alpha-\delta-\sigma^2/2)t+\sigma\sqrt{t}N^{-1}(p/2)} < S(t) < S(0)e^{(\alpha-\delta-\sigma^2/2)t+\sigma\sqrt{t}N^{-1}(1-p/2)}\right) = 1 - p,$$

meaning that

$$\boxed{(S^L, S^U) := \left(S(0)e^{(\alpha-\delta-\sigma^2/2)t+\sigma\sqrt{t}N^{-1}(p/2)}, \; S(0)e^{(\alpha-\delta-\sigma^2/2)t+\sigma\sqrt{t}N^{-1}(1-p/2)}\right)} \quad (5.2.3)$$

is the desired $100(1-p)\%$ equal-tailed prediction interval for $S(t)$. An easy way to remember (5.2.3) is that the two end-points in (5.2.3) are obtained by replacing Z in the stock price equation (5.1.1) by the two end-points of the corresponding prediction interval for Z, i.e., $N^{-1}(p/2)$ and $N^{-1}(1-p/2)$.

As a matter of fact, prediction intervals are not unique. For example,

$$\left(0, S(0)\exp\left[\left(\alpha-\delta-\frac{1}{2}\sigma^2\right)t+\sigma\sqrt{t}N^{-1}(1-p)\right]\right)$$

and

$$\left(S(0)\exp\left[\left(\alpha-\delta-\frac{1}{2}\sigma^2\right)t+\sigma\sqrt{t}N^{-1}(p)\right],\infty\right)$$

are also $100(1-p)\%$ prediction intervals for $S(t)$ (why do they work?). They are sometimes referred to as one-sided or one-tailed prediction intervals. However, unless otherwise stated, in this book we always use (5.2.3), which is a two-sided prediction interval based on a *symmetric* prediction interval for the corresponding normal random variable $\ln S(t)$ (although (5.2.3) is not symmetric about the mean of $S(t)$).

Example 5.2.3. (SOA Exam IFM Advanced Derivatives Sample Question 50: Upper limit of a lognormal prediction interval) Assume the Black-Scholes framework.

You are given the following information for a stock that pays dividends continuously at a rate proportional to its price.

(i) The current stock price is 0.25.

(ii) The stock's volatility is 0.35.

(iii) The continuously compounded expected rate of stock-price appreciation is 15%.

Calculate the upper limit of the 90% lognormal prediction interval for the price of the stock in 6 months.

(A) 0.393

(B) 0.425

(C) 0.451

(D) 0.486

(E) 0.529

Solution. We are given from (i), (ii), and (iii) that

$$S(0) = 0.25, \quad \sigma = 0.35, \quad \alpha - \delta = 0.15.$$

With $t = 0.5$ and $p = 1 - 0.9 = 0.1$, the upper limit of the 90% lognormal prediction interval for the 6-month stock price is

$$
\begin{aligned}
S^U(0.5) &= S(0.5)\exp\left[\left(\alpha-\delta-\frac{1}{2}\sigma^2\right)t+N^{-1}\left(1-\frac{p}{2}\right)\sigma\sqrt{t}\right] \\
&= 0.25\exp\left\{\left[0.15-\frac{1}{2}(0.35)^2\right]\times 0.5 + \underbrace{N^{-1}(0.95)}_{1.64485}\,0.35\sqrt{0.5}\right\} \\
&= \boxed{0.3926}. \quad \textbf{(Answer: (A))}
\end{aligned}
$$

□

Example 5.2.4. (SOA Exam FETE Spring 2012 Question 7 (a)–(b): Distributional quantities of a lognormal stock price) You are given the following information on a stock:

Initial Price	$25
Expected Annual Return per annum	8%
Estimated Annual Volatility	20%

Assume that the log returns are normally distributed.

(a) Calculate the 95% prediction interval for the stock price in one year.

(b) Calculate the expected stock price and the standard deviation of the stock price in one year.

Solution. (a) As

$$\ln S(1) \sim \mathrm{N}(\ln 25 + (0.08 - \frac{1}{2} \times 0.2^2), 0.2^2) \equiv \mathrm{N}(3.278876, 0.04),$$

a 95% prediction interval for $S(1)$ is

$$e^{3.278876 \pm 1.96\sqrt{0.04}} = \boxed{(17.94, 39.29)}.$$

(b) • $\mathbb{E}[S(1)] = S(0)e^{\alpha T} = 25e^{0.08(1)} = \boxed{27.08}$

  • $\mathrm{Var}[S(1)] = e^{2(3.278876)+0.04}(e^{0.04} - 1) = 29.93$, so $\sqrt{\mathrm{Var}[S(1)]} = \sqrt{29.93} = \boxed{5.47}$. □

Quantity 3: Conditional expected stock prices.

Apart from exercise probabilities for European options and lognormal prediction intervals, we are also interested in the expected value of the terminal stock price, given that the European option in question finishes in-the-money. Mathematically, we aim to determine the two conditional expectations

$$\mathbb{E}[S(T)| \underbrace{S(T) > K}_{\substack{\text{call expires} \\ \text{in-the-money}}}] \quad \text{and} \quad \mathbb{E}[S(T)| \underbrace{S(T) < K}_{\substack{\text{put expires} \\ \text{in-the-money}}}],$$

where T and K are the maturity time and strike price of the European option, respectively.

To evaluate these conditional expectations, we rewrite them in terms of unconditional ones, recalling the definition for any random variable X and event A with a non-zero probability:

$$\mathbb{E}[X|A] := \frac{\mathbb{E}[X1_A]}{\mathbb{P}(A)},$$

where 1_A is the indicator function of A, i.e., $1_A = 1$ if A is true, and $1_A = 0$ otherwise. Applying this definition to $X = S(T)$ and $A = \{S(T) > K\}$, we have

$$\mathbb{E}[S(T)|S(T) > K] = \frac{\mathbb{E}[S(T)1_{\{S(T)>K\}}]}{\mathbb{P}(S(T) > K)}.$$

The denominator is precisely the exercise probability of the K-strike T-year European call studied earlier, i.e., $\mathbb{P}(S(T) > K) = N(\hat{d}_2)$. For the numerator, recall from page 175 that

$$S(T) > K \quad \Leftrightarrow \quad Z > \frac{\ln[K/S(0)] - (\alpha - \delta - \sigma^2/2)T}{\sigma\sqrt{T}} = -\hat{d}_2.$$

Regarding $S(T) = S(0)\exp[(\alpha - \delta - \sigma^2/2)t + \sigma\sqrt{t}Z]$ as a function of the standard normal random variable Z, we can evaluate

$$
\begin{aligned}
\mathbb{E}[S(T)1_{\{S(T)>K\}}] &= \int_{-\hat{d}_2}^{\infty} S(0)e^{(\alpha-\delta-\sigma^2/2)T+\sigma\sqrt{T}z} \times \underbrace{\frac{1}{\sqrt{2\pi}}e^{-z^2/2}}_{\text{standard normal p.d.f. } \phi(z)} \, dz \\
&= S(0)e^{(\alpha-\delta)T} \int_{-\hat{d}_2}^{\infty} \frac{1}{\sqrt{2\pi}}e^{-(z-\sigma\sqrt{T})^2/2} \, dz \\
&= S(0)e^{(\alpha-\delta)T} \int_{-\hat{d}_2-\sigma\sqrt{T}}^{\infty} \frac{1}{\sqrt{2\pi}}e^{-z^2/2} \, dz \quad \text{(change of variable)} \\
&= S(0)e^{(\alpha-\delta)T} \int_{-\infty}^{\hat{d}_2+\sigma\sqrt{T}} \frac{1}{\sqrt{2\pi}}e^{-z^2/2} \, dz \quad \text{(symmetry of } \phi(z)\text{)} \\
&= S(0)e^{(\alpha-\delta)T} N(\hat{d}_1),
\end{aligned}
$$

where

$$\hat{d}_1 = \hat{d}_2 + \sigma\sqrt{T} = \frac{\ln[S(0)/K] + \left(\alpha - \delta \boxed{+} \sigma^2/2\right)T}{\sigma\sqrt{T}}. \tag{5.2.4}$$

Combining all pieces, we obtain

$$\boxed{\mathbb{E}[S(T)|S(T) > K] = S(0)e^{(\alpha-\delta)T}\frac{N(\hat{d}_1)}{N(\hat{d}_2)}.} \tag{5.2.5}$$

Observe that such a conditional expected stock price equals the unconditional expected T-year stock price $\mathbb{E}[S(T)] = S(0)e^{(\alpha-\delta)T}$ scaled by a factor of $N(\hat{d}_1)/N(\hat{d}_2)$, which, because $\hat{d}_1 > \hat{d}_2$, is always greater than 1. When $S(T)$ is *a priori* assumed to be greater than K, the conditional expected stock price exceeds the unconditional expected stock price, as expected.

To determine the expected stock price given that the put expires in-the-money, one can adapt the above derivations to the case of $S(T) < K$. Alternatively, we can appeal to the law of total probability, which says that

$$\mathbb{E}[S(T)] = \mathbb{E}[S(T)|S(T) > K]\mathbb{P}(S(T) > K) + \mathbb{E}[S(T)|S(T) < K]\mathbb{P}(S(T) < K),$$

or

$$S(0)e^{(\alpha-\delta)T} = S(0)e^{(\alpha-\delta)T}N(\hat{d}_1) + \mathbb{E}[S(T)|S(T) < K]N(-\hat{d}_2).$$

A rearrangement of terms readily gives

$$\boxed{\mathbb{E}[S(T)|S(T) < K] = S(0)e^{(\alpha-\delta)T}\frac{N(-\hat{d}_1)}{N(-\hat{d}_2)},} \tag{5.2.6}$$

which is again in the form of the unconditional expected T-year stock price multiplied by an adjusted factor, this time being less than unity.

Example 5.2.5. (A "sandwiched" conditional stock price) Assume the Black-Scholes framework.

Determine an expression for $\mathbb{E}[S(T)|a < S(T) < b]$, where a and b are positive constants with $a < b$.

Solution. Mimicking the derivations above, we have

$$\mathbb{E}[S(T)|a < S(T) < b] = \frac{\mathbb{E}[S(T)1_{\{S(T)>a\}}] - \mathbb{E}[S(T)1_{\{S(T)>b\}}]}{\mathbb{P}(a < S(T) < b)},$$

where

$$\mathbb{P}(a < S(T) < b) = \mathbb{P}(S(T) > a) - \mathbb{P}(S(T) > b) = N(\hat{d}_2^{@K=a}) - N(\hat{d}_2^{@K=b})$$

and

$$\mathbb{E}[S(T)1_{\{S(T)>a\}}] - \mathbb{E}[S(T)1_{\{S(T)>b\}}] = S(0)e^{(\alpha-\delta)T}[N(\hat{d}_1^{@K=a}) - N(\hat{d}_1^{@K=b})]$$

In conclusion,

$$\mathbb{E}[S(T)|a < S(T) < b] = \boxed{\frac{S(0)e^{(\alpha-\delta)T}[N(\hat{d}_1^{@K=a}) - N(\hat{d}_1^{@K=b})]}{N(\hat{d}_2^{@K=a}) - N(\hat{d}_2^{@K=b})}}.$$

$\square$

Remark. Letting $b \to \infty$ retrieves (5.2.5) with $K = a$ (because $\hat{d}_1^{@K=b} = \hat{d}_2^{@K=b} \to -\infty$ and $N(\hat{d}_1^{@K=b}) = N(\hat{d}_2^{@K=b}) \to 0$), while setting $a = 0$ recovers (5.2.6) with $K = b$ (because $\hat{d}_1^{@K=a} = \hat{d}_2^{@K=a} \to +\infty$ and $N(\hat{d}_1^{@K=a}) = N(\hat{d}_2^{@K=a}) \to 1$, and note that $N(x) = 1 - N(-x)$ for any $x \in \mathbb{R}$).

5.3   Problems

Lognormal probabilistic quantities

Problem 5.3.1. (Exercise probability of a call) Consider a nondividend-paying stock whose current price is 100.

The stock-price process is a lognormal process with volatility 30%.

The continuously compounded expected return on the stock is 10%.

The continuously compounded risk-free interest rate is 7%.

Calculate the probability that a nine-month 75-strike European call option on the stock will be exercised.

Problem 5.3.2. (Calculations of miscellaneous probabilistic quantities) Assume the Black-Scholes framework. You are given:

(i) The current stock price is 100.

(ii) The stock pays dividends continuously at a rate proportional to its price. The dividend yield is 2.5%.

(iii) The continuously compounded expected rate of return on the stock is 6%.

(iv) The stock's volatility is 30%.

Calculate:

(a) The probability that a 3-year at-the-money European put option on the stock is exercised

(b) The 95% percentile of the 3-year stock price

(c) The expected 3-year stock price

(d) The expected 3-year stock price, given that the put option in (a) pays off at maturity

(e) The variance of the 3-year stock price

Problem 5.3.3. (Lognormal prediction interval) Assume the Black-Scholes framework. You are given:

(i) The stock, whose current price is 100, pays dividends continuously at a rate proportional to its price.

(ii) The stock's volatility is 0.35.

(iii) The continuously compounded expected rate of stock-price appreciation is 15%.

(iv) The continuously compounded risk-free interest rate is 12%.

Construct a 95% lognormal prediction interval for the price of the stock in 3 months.

Problem 5.3.4. (Given one prediction interval, find another) Assume the Black-Scholes framework. Let $S(t)$ denote the time-t price of a stock, which pays dividends continuously at a rate proportional to its price.

You are given:

(i) $S(0) = 8$

(ii) The 90% lognormal prediction interval for $S(2)$ is $(13.10, 41.93)$.

Calculate the width of the 95% lognormal prediction interval for $S(4)$

Problem 5.3.5. [HARDER!] (Given two prediction intervals, find one more) Assume the Black-Scholes framework. For $t \geq 0$, let $S(t)$ be the time-t price of a stock. You are given:

(i) The stock pays dividends continuously at a rate proportional to its price.

(ii) The 90% lognormal prediction interval for $S(2)$ is $(13.1072, 41.9448)$.

(iii) The 95% lognormal prediction interval for $S(4)$ is $(25.7923, 183.1083)$.

Calculate the width of the 99% lognormal prediction interval for $S(6)$.

Problem 5.3.6. (What can you deduce from an exercise probability and a lognormal prediction interval?) Assume the Black-Scholes framework. For a stock which pays dividends continuously at a rate proportional to its price, you are given:

(i) The probability that a 2-year at-the-money European put option on the stock will be exercised is 0.5279.

(ii) The 99% lognormal prediction interval for the 3-year stock price is $(9.4493, 524.0208)$.

Calculate the expected 4-year stock price.

Problem 5.3.7. (Going from unconditional to conditional expected stock price) Assume the Black-Scholes framework. You are given:

(i) The current price of a stock is 80.

(ii) The stock's volatility is 25%.

(iii) The stock pays dividends continuously at a rate proportional to its price.

(iv) The continuously compounded risk-free interest rate is 6%.

(v) The expected 1-year stock price is 84.2069.

Calculate the expected 1-year stock price, given that a 1-year at-the-money (when issued) European put option on the stock expires in-the-money.

Problem 5.3.8. (Given the unconditional and conditional expected stock prices – I) Assume the Black-Scholes framework. You are given:

(i) The stock pays dividends continuously at a rate proportional to its price. The dividend yield is identical to the continuously compounded (total) expected rate of return on the stock.

(ii) The expected 6-month stock price is 150.

(iii) The expected 6-month stock price, given that an at-the-money (when issued) 6-month European put option expires in-the-money, is 125.21.

Determine the 95% lognormal prediction interval for the 1-year stock price.

Problem 5.3.9. (Given the unconditional and conditional expected stock prices – II) Assume the Black-Scholes framework. For a stock which pays dividends continuously at a rate proportional to its price, you are given:

(i) The probability that an 8-month European put option on the stock will be exercised is 0.512.

(ii) The expected 8-month stock price is 10.134.

(iii) The expected 8-month stock price, given that the put option in (i) expires in-the-money, is 8.483.

Determine the 98% lognormal prediction interval for the 8-month stock price.
    (Note: The strike price of the put option is not given.)

6

The Black-Scholes Formula

Chapter overview: Now that the Black-Scholes framework and its technical underpinnings have been set up, in this chapter we are ready to derive, interpret, and apply the all-important Black-Scholes option pricing formula—one of the highlights of this book—for European call and put options. To begin with, Section 6.1, our first encounter with the Black-Scholes formula, revolves around the simple case of stocks paying continuous proportional dividends. Section 6.2 extends the analysis in Section 6.1 and presents the Black-Scholes formula for options on other underlying assets, including stocks that pay discrete dividends, currencies, and futures contracts. It is shown that options on these seemingly unrelated assets can all be priced by suitably imposing the Black-Scholes framework and cosmetically modifying the basic Black-Scholes formula in Section 6.1. With the Black-Scholes formula at our disposal, we proceed to investigate in Section 6.3 the sensitivity of the option price to various model parameters as quantified by option Greeks. They are the partial derivatives of the option price with respect to the option input under consideration and are widely used in practice to assess the risk of an option position. Further aspects and applications of the Black-Scholes formula and option Greeks will be studied in Chapters 7 and 8.

6.1 The Black-Scholes Formula for Stocks Paying Continuous Proportional Dividends

What is the Black-Scholes formula?

The *Black-Scholes option pricing formula*, or *Black-Scholes formula* in short, is an analytic (i.e., exact, closed-form) formula for the prices of *European*[i] call and put options in terms of parameters in the Black-Scholes framework. It is justifiably the best known option pricing formula—so famous that "Black-Scholes" has become a household item in modern finance. Its validity is predicated upon several assumptions, which can be broadly categorized into two groups:

1. *Behavior of future stock prices:* As you can expect, the Black-Scholes formula lives in the Black-Scholes framework introduced in Chapter 5. In the simplest setting, it is assumed that future stock prices are lognormally distributed over the life of the option in question.[ii] The stock pays continuous proportional dividends at the dividend yield of δ and its volatility is constant over time[iii] at σ.

2. *Economic environment:* It is also assumed that the (continuously compounded) risk-free

---

[i]The prices of American options in the Black-Scholes framework are beyond the scope of this book.

[ii]This assumption will be weakened in Section 6.2.

[iii]It is possible to generalize the Black-Scholes formula to the case with deterministic time-dependent risk-free interest rate and volatility. See, e.g., Exercise 5.4 on page 253 of Shreve (2004).

interest rate is known and constant at r and the market is frictionless, i.e., there are no taxes, transaction costs, bid/ask spreads or restrictions on short sales.

Example 6.1.1. (CAS Exam 3 Fall 2007 Question 20: Black-Scholes assumptions) Which of the following is an assumption of the Black-Scholes option pricing model?

(A) Stock prices are normally distributed.

(B) Stock price volatility is a constant.

(C) Changes in stock price are lognormally distributed.

(D) All transaction costs are included in stock returns.

(E) The risk-free interest rate is a random variable.

Solution. Only (B) is correct. For (A) and (C), the correct statements are that stock prices are *lognormally* distributed whereas changes in the stock price are *normally* distributed. There are no transaction costs and the risk-free interest rate is a constant, rendering (D) and (E) incorrect. **(Answer: (B))**   □

Sidebar: (Very) Brief history of the Black-Scholes formula.

The Black-Scholes formula was first published in the *Journal of Political Economy* in 1973 by Fischer *Black* and Myron *Scholes* (hence the names "Black" and "Scholes"). Their paper, Black and Scholes (1973), along with related work by Robert Merton(the Black-Scholes formula is also known as the Black-Scholes-Merton formula), was said to revolutionize the theory and practice of finance. In recognition of their groundbreaking work on option pricing, Robert Merton and Myron Scholes won the Nobel Prize in Economics in 1997. Regrettably, Fischer Black was ineligible for the prize because he had already passed away.

Black-Scholes pricing formula for European calls and puts.

Without further ado, we now formally state the Black-Scholes formula for the price of a T-year K-strike European call option on a stock that pays continuous proportional dividends at a rate of δ. The formula reads

$$C = \text{BS}(S(0), \delta; K, r; \sigma, T) = S(0)e^{-\delta T} N(d_1) - Ke^{-rT} N(d_2), \qquad (6.1.1)$$

where C is the (time-0) price of the call, $N(\cdot)$ is the standard normal distribution function, and the two extra parameters d_1 and d_2 are defined by

$$d_1 = \frac{\ln[S(0)/K] + (r - \delta + \sigma^2/2)T}{\sigma\sqrt{T}},$$

$$d_2 = d_1 - \sigma\sqrt{T} = \frac{\ln[S(0)/K] + (r - \delta - \sigma^2/2)T}{\sigma\sqrt{T}}.$$

Note that d_1 and d_2 are identical to $\hat{d}_1$ and $\hat{d}_2$ in (5.2.4) and (5.2.1), respectively, except that the stock's expected rate of return α is replaced by the risk-free interest rate r.

By put-call parity (which holds for any no-arbitrage option pricing model), the accompanying Black-Scholes European put price formula is

$$P = \text{BS}(K, r; S(0), \delta; \sigma, T) = Ke^{-rT} N(-d_2) - S(0)e^{-\delta T} N(-d_1),$$ (6.1.2)

with the same definitions of d_1 and d_2 above. When you apply (6.1.2), make sure that you *do not omit the negative signs in front of d_1 and d_2!*

At first sight, the Black-Scholes pricing formulas, (6.1.1) and (6.1.2), look intimidating and not particularly informative. The monstrous forms of (6.1.1), (6.1.2), and the mysterious quantities d_1 and d_2 seem to make it a daunting task to remember the formulas by heart. To your dismay, however, two principal and inter-related learning objectives of this chapter (indeed, of this book) are:

(1) Remember the Black-Scholes pricing formula impeccably and keep it ingrained in your mind; every piece of the formula must appear in the right place.[iv] You cannot afford making any mistakes in writing the pricing formula!

(2) Calculate Black-Scholes option prices proficiently.

With respect to (2), the best way to gain proficiency with the use of the Black-Scholes formula is to work out numerous computational problems. This will be deferred to page 190 after we have learned how to interpret the pricing formula. With respect to (1), two useful pedagogical vehicles are to view the Black-Scholes formula from a unifying function and to understand the economic meaning of the two terms that constitute the formula from a cost-benefit perspective.

General Black-Scholes pricing function.

To emphasize the symmetry and common structure obeyed by the Black-Scholes call price and put price in (6.1.1), (6.1.2), as well as many other Black-Scholes pricing formulas in the later part of this book, we have devised the generic Black-Scholes-type pricing function defined by

$$\text{BS}(s_1, \delta_1; s_2, \delta_2; \sigma, T) := s_1 e^{-\delta_1 T} N(d_1) - s_2 e^{-\delta_2 T} N(d_2),$$ (6.1.3)

where

$$d_1 = \frac{\ln(s_1/s_2) + (\delta_2 - \delta_1 + \sigma^2/2)T}{\sigma\sqrt{T}} \quad \text{and} \quad d_2 = d_1 - \sigma\sqrt{T}.$$

All of the Black-Scholes-type pricing formulas[v] we encounter in this book can be expressed in the form of the "BS" function for some appropriate choices of the arguments specific to a given context. Further discussions on the meaning of the variables inside the "BS" function will be provided when the notion of exchange options is introduced in Section 8.2. For the time being, it suffices to say that s_1 (resp. s_2) is the current price of the asset you acquire (resp. lose) if you exercise the option, δ_1 (resp. δ_2) is the "dividend yield" (in a general sense) of this asset, σ is the volatility of the asset which is assumed to follow the Black-Scholes framework, and T is the time to maturity of the option.

If you exercise a call, you lose the cash of \$$K$ and receive the stock. The dividend yield of the stock is δ while the "dividend yield" of cash is the risk-free interest rate r—\$1 at time 0 grows at the rate of r to \$$e^{rt}$ at any future time t. This suggests setting $s_1 = S(0)$, $\delta_1 = \delta$, $s_2 = K$, and $\delta_2 = r$ in (6.1.3) to obtain the Black-Scholes call price

$$C = \text{BS}(S(0), \delta; K, r; \sigma, T),$$

[iv]If you misstate the BS (Black-Scholes) formula, it will become the BS (Bxllsxit) formula...

[v]The only exception is the pricing formula for gap options in Section 8.1.

with

$$d_1^C = \frac{\ln[S(0)/K] + (r - \delta + \sigma^2/2)T}{\sigma\sqrt{T}} \quad \text{and} \quad d_2^C = \frac{\ln[S(0)/K] + (r - \delta - \sigma^2/2)T}{\sigma\sqrt{T}}.$$

(The superscript "C" on d_1 and d_2 suggests "call") A put is the opposite—you lose the stock and receive the cash of $\$K$ upon exercising a put. To get the Black-Scholes put price, we set $s_1 = K$, $\delta_1 = r$, $s_2 = S(0)$, and $\delta_2 = \delta$ in (6.1.3), leading to

$$P = \mathrm{BS}(K, r; S(0), \delta; \sigma, T),$$

with

$$d_1^P = \frac{\ln[K/S(0)] + (\delta - r + \sigma^2/2)T}{\sigma\sqrt{T}} = -d_2^C \quad \text{and} \quad d_2^P = d_1^P - \sigma\sqrt{T} = -d_1^C.$$

However, it is customary to use the d_1 and d_2 for calls and omit the superscripts "C" and "P" altogether.

Cost-benefit interpretation of the Black-Scholes formula.

There are different ways to make sense of the abstruse Black-Scholes formula. One instructive way is to view the cost and benefit (in an accounting sense) of exercising the option.

Call.   The first term of the pricing formula, $S(0)\mathrm{e}^{-\delta T}N(d_1) = F_{0,T}^P(S)N(d_1)$, represents the (risk-neutral) expected present value of what you receive at maturity upon exercising the call, namely the stock with a random time-T value of $S(T)$. The appearance of the term $N(d_1)$, which is between zero and one, stems from the fact that the call will be exercised only when $S(T) > K$, an even with a probability of between zero and one. The second term, $Ke^{-rT}N(d_2)$, is the (risk-neutral) expected present value of what you *pay* at maturity if you exercise the call, namely the strike price of K, provided that $S(T) > K$. Here $N(d_2)$ is the risk-neutral probability that the call will be exercised (in contrast, $N(\hat{d}_2)$ is the *real* exercise probability).

Put.   Analogously, the put price formula can be thought of as the difference between the (risk-neutral) expected present value of what you *receive* (the strike price of K, provided that $K < S(T)$) and that of what you *give up* (the random amount of $S(T)$, provided that $K < S(T)$) by exercising the put.

Regardless of whether it is a call or a put, the Black-Scholes formula expresses the price of the option as the (risk-neutral) expected present value of the benefit of exercising the option less the (risk-neutral) expected present value of the corresponding cost.

Deriving the Black-Scholes formula.

There are many possible derivations of the Black-Scholes formula in the option pricing literature, including but not limited to:

1. The method of risk-neutral pricing

2. The method of dynamic replication

3. The method of solving the Black-Scholes partial differential equation (see page 249)

4. The method of Esscher transform (see the celebrated paper by Gerber and Shiu (1994))

5. By the Capital Asset Pricing Model (CAPM) (see the original paper of Black and Scholes (1973))

6. Taking the limit of the binomial option pricing formula as the number of periods approaches infinity and the length of each period approaches zero (the binomial formula will converge to the Black-Scholes formula; see the Appendix of Chapter 13 of Hull (2015))

In this book, we will only discuss Method 1 as it is the least technically demanding but most germane to our treatment.

To kick-start the risk-neutral valuation, recall that under the risk-neutral probability distribution (whose existence is posited)[vi], the stock (with dividends) earns the risk-free interest rate. Upon the substitution of the expected rate of return α by the risk-free interest rate r, we have the following important distributional representation result:

$$S(T) \stackrel{d}{=} S(0) \exp\left[\left(r - \delta - \frac{\sigma^2}{2}\right)T + \sigma\sqrt{T}Z\right],$$

where Z denotes a standard normal random variable *under the risk-neutral probability*. By risk-neutral pricing, the time-0 price of the European call option takes the discounted expectation in the form of

$$C = \mathbb{E}^*\left[e^{-rT}(S(T) - K)_+\right].$$

Because $x_+ = x$ if $x \geq 0$ and $x_+ = 0$ if $x < 0$, we can appeal to double expectation (also known as iterated expectation) and simplify the preceding expectation as

$$
\begin{aligned}
C &= \mathbb{E}^*\left[e^{-rT}(S(T) - K)_+\middle|\, S(T) > K\right]\mathbb{P}^*(S(T) > K) \\[2ex]
&\quad + \mathbb{E}^*\left[e^{-rT}\underbrace{(S(T) - K)_+}_{0}\middle|\, S(T) \leq K\right]\mathbb{P}^*(S(T) \leq K) \\[2ex]
&= e^{-rT}\mathbb{E}^*\left[S(T) - K\,\middle|\, S(T) > K\right]\mathbb{P}^*(S(T) > K).
\end{aligned}
$$

Now $\mathbb{P}^*(S(T) > K)$ is the *risk-neutral* exercise probability of the call, which, by (5.2.2) with $\hat{d}_2$ replaced by d_2, equals $N(d_2)$. Moreover, the linearity of conditional expectations yields

$$
\begin{aligned}
\mathbb{E}^*\left[S(T) - K\,\middle|\, S(T) > K\right] &= \mathbb{E}^*\left[S(T)\middle|\, S(T) > K\right] - K \\[1ex]
&= S(0)e^{(r-\delta)T}\frac{N(d_1)}{N(d_2)} - K,
\end{aligned}
$$

where the last equality follows from (5.2.5) with $\alpha, \hat{d}_1, \hat{d}_2$ changed to r, d_1, d_2, respectively. Combining all pieces, we have

$$C = e^{-rT}\left[S(0)e^{(r-\delta)T}\frac{N(d_1)}{N(d_2)} - K\right]N(d_2) = S(0)e^{-\delta T}N(d_1) - Ke^{-rT}N(d_2),$$

which is the same as (6.1.1). We are done with deriving the Black-Scholes formula (and ready to clinch Scholes and Merton's Nobel Prize)!!

[vi]Indeed, a lot of work has to be done to justify the existence of and to define the risk-neutral probability distribution.

The Black-Scholes formula in action.

We end this introductory section with several concrete numerical examples that are designed to sharpen your computational proficiency with the use of the Black-Scholes formula. It is strongly suggested that you try these calculations out on a piece of paper and see how (un)interesting they are. It is time to get our hands dirty!

Example 6.1.2. (SOA Exam IFM Advanced Derivatives Sample Question 6: Black-Scholes call price) You are considering the purchase of 100 units of a 3-month 25-strike European call option on stock.
  You are given:

(i) The Black-Scholes framework holds.

(ii) The stock is currently selling for 20.

(iii) The stock's volatility is 24%.

(iv) The stock pays dividends continuously at a rate proportional to its price. The dividend yield is 3%.

(v) The continuously compounded risk-free interest rate is 5%.

Calculate the price of the block of 100 options.

(A) 0.04

(B) 1.93

(C) 3.63

(D) 4.22

(E) 5.09

Solution. With $S = 20$, $K = 25$, $\sigma = 0.24$, $r = 0.05$, $\delta = 0.03$ and $T = 0.25$, we have

$$d_1 = \frac{\ln(20/25) + (0.05 - 0.03 + 0.24^2/2)(0.25)}{0.24\sqrt{0.25}} = -1.75786,$$

$$d_2 = d_1 - 0.24\sqrt{0.25} = -1.87786,$$

$$N(d_1) = 0.03939,$$

$$N(d_2) = 0.03020.$$

The Black-Scholes price of each call is

$$\begin{aligned}
C &= S(0)e^{-\delta T}N(d_1) - Ke^{-rT}N(d_2) \\
&= 20e^{-0.03(0.25)}(0.03939) - 25e^{-0.05(0.25)}(0.03020) \\
&= 0.036292.
\end{aligned}$$

Hence the price of the block of 100 options is $100C = \boxed{3.6292}$. (**Answer: (C)**)   □

Example 6.1.3. (SOA Exam MFE Spring 2007 Question 3: Black-Scholes put price) You are asked to determine the price of a European put option on a stock. Assuming the Black-Scholes framework holds, you are given:

(i) The stock price is $100.

(ii) The put option will expire in 6 months.

(iii) The strike price is $98.

(iv) The continuously compounded risk-free interest rate is $r = 0.055$.

(v) $\delta = 0.01$.

(vi) $\sigma = 0.50$.

Calculate the price of this put option.

(A) $3.50

(B) $8.60

(C) $11.90

(D) $16.00

(E) $20.40

Solution. With

$$d_1 = \frac{\ln(100/98) + (0.055 - 0.01 + 0.5^2/2)(0.5)}{0.5\sqrt{0.5}} = 0.29756,$$

$$d_2 = d_1 - 0.5\sqrt{0.5} = -0.05600,$$

$$N(-d_1) = 0.38302,$$

$$N(-d_2) = 0.52233,$$

the price of the put is

$$\begin{aligned}
P &= Ke^{-rT} N(-d_2) - S(0)e^{-\delta T} N(-d_1) \\
&= 98e^{-0.055(0.5)}(0.52233) - 100e^{-0.01(0.5)}(0.38302) \\
&= \boxed{11.6889}. \quad \textbf{(Answer: (C))}
\end{aligned}$$

$\square$

6.2 Applying the Black-Scholes Formula to Other Underlying Assets

The Black-Scholes option pricing formula is by no means confined to European options on stocks that pay continuous proportional dividends. In this section, we broaden the applicability of the Black-Scholes formula by extending its scope to European options on more diverse underlying assets, including stocks that pay non-random, discrete dividends at known times, foreign currencies, and futures contracts. Despite the varied nature of these assets, it will be shown that with the right perspective, European options on each of these assets can be priced by a mostly cosmetic modification of the basic Black-Scholes formulas (6.1.1) and (6.1.2) with appropriate choices of the initial stock price $S(0)$, continuous dividend yield δ, and with σ being the volatility of an appropriate asset on which the Black-Scholes framework is imposed. The mathematical structure of the pricing formula is preserved.

6.2.1   Case study 1: Stocks paying non-random, discrete dividends.

The discrete-dividend case is an intriguing item that is often not given due attention in the literature and in the classroom. Pricing European options in this setting requires a subtle application of the standard Black-Scholes pricing formula for nondividend-paying stocks presented in the last section.

Pricing assumption.

To set the stage, consider a stock that, at each *known* time t_i, will make a dividend payment with a *known* amount of $D(t_i)$, where $0 < t_1 < t_2 < \cdots < t_n \leq T$ and T is the maturity time of a European option written on such a stock. This is the same discrete-dividend setting in Subsection 2.2.2. In this case, the stock price process $\{S(t)\}$ cannot be a lognormal process, because there must be a downward jump in the stock price immediately after each dividend is paid, making the stock price path discontinuous. So, what process is lognormal here? It turns out that the Black-Scholes pricing formula in this discrete-dividend setting can be derived from the assumption that the stochastic process of the *prepaid forward prices* for the time-T delivery of one share of the stock, that is,

$$\{F_{t,T}^P(S)\}_{t \in [0,T]},$$

is a lognormal process, with σ now being the standard deviation per unit time of its natural logarithm:

$$\text{Var}[\ln F_{t,T}^P(S)] = \sigma^2 t, \quad 0 \leq t \leq T.$$

Why does applying the Black-Scholes framework to $\{F_{t,T}^P(S)\}_{t \in [0,T]}$ make (practical or technical) sense? To see this, recall from Subsection 2.2.2 that the time-t price of a prepaid forward maturing at time T is the current stock price less the present value of the dividends payable over the remaining life of the option, that is,

$$F_{t,T}^P(S) = S(t) - \text{PV}_{t,T}(\text{Div}) = S(t) - \sum_{\{i:t_i \geq t\}} D(t_i) e^{-r(t_i - t)}.$$

See (2.2.1) on page 32 with the valuation date changed to time t. Although the stock price decreases abruptly by the amount of each discrete dividend as the date of each dividend payment is approached from the left, so does the present value factor $\text{PV}_{t,T}(\text{Div})$—that particular dividend need not be discounted. This preserves the continuity of the prepaid forward price $F_{t,T}^P(S)$ as a function of time t.

Pricing formula and its proof.

Under the lognormality assumption on the prepaid forward prices on the above stock that pays discrete dividends, the Black-Scholes formula for the price of a T-year K-strike European call option on the stock is given by

$$\boxed{C = \text{BS}\left(F_{0,T}^P(S), 0; K, r; \sigma, T\right) = F_{0,T}^P(S)\, N(d_1) - K e^{-rT}\, N(d_2),} \qquad (6.2.1)$$

where

$$d_1 = \frac{\ln[F_{0,T}^P(S)/K] + (r + \sigma^2/2)T}{\sigma\sqrt{T}},$$

$$d_2 = d_1 - \sigma\sqrt{T}.$$

The corresponding European put price is

$$P = \text{BS}\left(K, r; F_{0,T}^P(S), 0; \sigma, T\right) = Ke^{-rT}N(-d_2) - F_{0,T}^P(S)N(-d_1).$$

The proof of (6.2.1) is interesting in its own right because it introduces an unconventional but useful way to view options—any option can be considered an option on a prepaid forward written on the underlying asset. To derive (6.2.1), we appeal to risk-neutral valuation again, whence

$$C = e^{-rT}\mathbb{E}^*\left[(S(T) - K)_+\right].$$

Because $F_{T,T}^P(S) = S(T)$, the preceding expectation can be written as

$$C = e^{-rT}\mathbb{E}^*\left[\left(\boxed{F_{T,T}^P(S)} - K\right)_+\right],$$

which suggests that such a call on the stock can also be viewed as a K-strike T-year European call on a *prepaid forward contract* on the stock maturing at time T. Because $\{F_{t,T}^P(S)\}_{t\in[0,T]}$ is a lognormal process and the prepaid forward pays no dividends, we can apply the standard Black-Scholes formula (6.1.1) with

the time-0 price of the stock, $S(0)$, changed to $F_{0,T}^P(S)$ (i.e., the time-0 price of the T-year prepaid forward, which is the underlying asset),

the dividend yield on the underlying asset, δ, set to 0, and

the parameter, σ, being the volatility of the prepaid forward.

This yields the call price formula

$$C = \text{BS}\left(F_{0,T}^P(S), 0; K, r; \sigma, T\right) = F_{0,T}^P(S)N(d_1) - Ke^{-rT}N(d_2),$$

where

$$d_1 = \frac{\ln[F_{0,T}^P(S)/K] + (r + \sigma^2/2)T}{\sigma\sqrt{T}} \quad \text{and} \quad d_2 = d_1 - \sigma\sqrt{T}.$$

Example 6.2.1. (SOA Exam MFE Spring 2007 Question 15: Discrete dividend − I) For a six-month European put option on a stock, you are given:

(i) The strike price is \$50.00.

(ii) The current stock price is \$50.00.

(iii) The only dividend during this time period is \$1.50 to be paid in four months.

(iv) $\sigma = 0.30$.

(v) The continuously compounded risk-free interest rate is 5%.

Under the Black-Scholes framework, calculate the price of the put option.

(A) $3.50

(B) $3.95

(C) $4.19

(D) $4.73

(E) $4.93

Solution. The 6-month prepaid forward price of the stock is

$$F^P_{0,1/2}(S) = S(0) - \text{PV}_{0,1/2}(\text{Div}) = 50 - 1.5e^{-0.05/3} = 48.52479282.$$

Then

$$
\begin{aligned}
d_1 &= \frac{\ln(48.52479282/50) + (0.05 + 0.3^2/2)(0.5)}{0.3\sqrt{0.5}} = 0.08274, \\
d_2 &= d_1 - 0.3\sqrt{0.5} = -0.12939, \\
N(-d_1) &= 0.46703, \\
N(-d_2) &= 0.55148.
\end{aligned}
$$

The put option price is

$$
\begin{aligned}
P &= Ke^{-rT}N(-d_2) - F^P_{0,1/2}(S)N(-d_1) \\
&= 50e^{-0.05(0.5)}(0.55148) - 48.52479282(0.46703) \\
&= \boxed{4.2307}. \quad \textbf{(Answer: (C))}
\end{aligned}
$$

□

Remark. (i) The difference between the computed put price and the price in Answer Choice C is due to the way the normal c.d.f. $N(\cdot)$ is computed. Using the rounding rules in a standard normal distribution table, we have

$$N(-d_1) = 0.4681, \quad \text{and} \quad N(-d_2) = 0.5517,$$

and the final answer becomes

$$50e^{-0.05(0.5)}(0.5517) - 48.52479282(0.4681) = \boxed{4.19}. \quad \textbf{(Answer: (C))}$$

(ii) Make sure you know the precise meaning of σ in this context. It is the volatility of the prepaid forward that calls for the delivery of the stock in six months.

Example 6.2.2. (SOA Exam MFE Spring 2009 Question 19: Discrete dividend − II) Consider a one-year 45-strike European put option on a stock S. You are given:

(i) The current stock price, $S(0)$, is 50.00.

(ii) The only dividend is 5.00 to be paid in nine months.

(iii) $\mathrm{Var}[\ln F_{t,1}^P(S)] = 0.01 \times t, \quad 0 \le t \le 1$.

(iv) The continuously compounded risk-free interest rate is 12%.

Under the Black-Scholes framework, calculate the price of 100 units of the put option.

(A) 1.87

(B) 18.39

(C) 18.69

(D) 19.41

(E) 23.76

Solution. This question is similar to the preceding one in terms of content and wording, except that we are not directly told the value of σ—we need to identify it from Point (iii): $\sigma = \sqrt{0.01} = 0.1$. The time-0 prepaid forward price for time-1 delivery of the stock is

$$F_{0,1}^P(S) = S(0) - \mathrm{PV}_{0,1}(\mathrm{Div}) = 50 - 5e^{-0.12(0.75)} = 45.4303.$$

Thus

$$d_1 = \frac{\ln[F_{0,1}^P(S)/K] + (r + \sigma^2/2)(1)}{\sigma\sqrt{1}} = \frac{\ln(45.4303/45) + (0.12 + 0.1^2/2)(1)}{0.1\sqrt{1}} = 1.34518$$

and

$$d_2 = d_1 - \sigma\sqrt{1} = 1.24518.$$

The price of the one unit of the put option is

$$
\begin{aligned}
P &= Ke^{-rT}N(-d_2) - F_{0,1}^P(S)N(-d_1) \\
&= 45e^{-0.12(1)}(0.10653) - 45.4303(0.08928) \\
&= 0.1957.
\end{aligned}
$$

For 100 units, the price is $100(0.1957) = \boxed{19.57}$. (**Answer: (D)**)   □

Remark. Using the rounding rules in a standard normal distribution table, we have

$$N(-d_1) = 1 - 0.9115 = 0.0885, \quad \text{and} \quad N(-d_2) = 1 - 0.8944 = 0.1056,$$

and the final answer becomes

$$P = 100[39.9114(0.1056) - 45.4303(0.0885)] = \boxed{19.41}. \quad (\textbf{Answer: (D)})$$

Remark.

The proof of (6.2.1) and the symbol $\mathrm{BS}\left(F_{0,T}^P(S), 0\,;\,\cdots\right)$ suggest that one can always treat the underlying asset of an option as a prepaid forward on the asset with the same time to maturity as the option. In fact, the lognormality of the prepaid forward prices $F_{t,T}^P(S)$ for $t \in [0, T]$ is a more general assumption for the Black-Scholes formula for dividend-paying stocks to hold true (see Subsection 8.2.2 for a further generalization), and (6.2.1) accommodates not only stocks paying discrete dividends of known amounts at known times, as discussed in this subsection, but also stocks paying dividends continuously at a rate proportional to its price, as in Section 6.1. If the stock pays continuous proportional dividends, then it can be shown that $\{F_{t,T}^P(S)\}_{0 \le t \le T}$ is a lognormal process if and only if the stock price process $\{S(t)\}$ is a lognormal process with both stochastic processes sharing the same volatility σ. Furthermore, in this continuous-dividend case $F_{0,T}^P(S) = S(0)e^{-\delta T}$, so (6.2.1) reduces to

$$C = S(0)e^{-\delta T} N(d_1) - Ke^{-rT} N(d_2),$$

with

$$d_1 = \frac{\ln[S(0)e^{-\delta T}/K] + (r + \sigma^2/2)T}{\sigma\sqrt{T}} = \frac{\ln[S(0)/K] + (r - \delta + \sigma^2/2)T}{\sigma\sqrt{T}}$$

and $d_2 = d_1 - \sigma\sqrt{T}$. This call price formula is nothing but (6.1.1) on page 186.

6.2.2 Case Study 2: Currency options.

In addition to stocks paying discrete dividends, the Black-Scholes methodology can also be applied to price options on currencies, perhaps more easily than one would have imagined.

For specificity, let $X(t)$ be the time-t exchange rate, which is the value of one unit of the foreign currency in question measured in terms of the domestic currency at time t. Valuing a currency option in the Black-Scholes framework hinges upon the realization that the underlying asset of the currency option is the foreign risk-free asset, which plays the same role as the underlying for stock options. Just as the number of shares of a stock paying continuous proportional dividends grows exponentially via the reinvestment of dividends at the dividend yield δ, an investment in the foreign currency grows exponentially via the reinvestment of interest at the foreign risk-free interest rate r_f. With this resemblance (see Table 6.1) in mind, we can price currency options by assuming that future exchange rates (as opposed to stock prices) are lognormally distributed with constant volatility σ and applying the Black-Scholes formula in the form of (6.1.1) and (6.1.2), with

the current stock price $S(0)$ replaced by the current exchange rate $X(0)$,

the dividend yield δ by r_f, and

the (continuously compounded) risk-free interest rate r by the risk-free interest rate of the domestic currency, r_d.

This yields the following prices, expressed in the domestic currency, for a K-strike T-year European currency call and an otherwise identical currency put on the foreign currency:

$$\boxed{\begin{aligned} C &= \mathrm{BS}\,(X(0), r_f; K, r_d; \sigma, T) = X(0)e^{-r_f T} N(d_1) - Ke^{-r_d T} N(d_2), \\ P &= \mathrm{BS}\,(K, r_d; X(0), r_f; \sigma, T) = Ke^{-r_d T} N(-d_2) - X(0)e^{-r_f T} N(-d_1), \end{aligned}} \tag{6.2.2}$$

where

$$d_1 = \frac{\ln[X(0)/K] + (r_d - r_f + \sigma^2/2)T}{\sigma\sqrt{T}} \quad \text{and} \quad d_2 = d_1 - \sigma\sqrt{T}.$$

These pricing formulas for currency options are sometimes known as the *Garman-Kohlhagen* formula, in recognition of Garman and Kohlhagen (1983).

Example 6.2.3. (CAS Exam 3 Fall 2007 Question 21: Straightforward application of (6.2.2)) On January 1st, 2007, the following currency information is given:

	Stock Options	Currency Options
Underlying asset	Stock	Foreign currency
Time-t value of underlying	$S(t)$	$X(t)$
Dividend yield	δ	r_f
Pricing assumption	$S(t)$'s are lognormally distributed	$X(t)$'s are lognormally distributed
Meaning of σ	Volatility of the stock	Volatility of the exchange rate

TABLE 6.1
Comparing stock options and currency options in the Black-Scholes framework.

- Spot exchange rate = \$0.82/euro

- Dollar interest rate = 5.0% compounded continuously

- Euro interest rate = 2.5% compounded continuously

- Exchange rate volatility = 0.10

What is the price of 850 dollar-denominated euro call options with a strike exchange rate of \$0.80/euro that expire on January 1st, 2008?

(A) Less than \$10

(B) At least \$10, but less than \$20

(C) At least \$20, but less than \$30

(D) At least \$30, but less than \$40

(E) At least \$40

Solution. As

$$d_1 = \frac{\ln(0.82/0.8) + (0.05 - 0.025 + 0.1^2/2)(1)}{0.1\sqrt{1}} = 0.54693,$$

$$d_2 = d_1 - 0.1\sqrt{1} = 0.44693,$$

$$N(d_1) = 0.70779,$$

$$N(d_2) = 0.67254,$$

the price of the euro call is

$$C = 0.82e^{-0.025}(0.70779) - 0.8e^{-0.05}(0.67254) = 0.05427,$$

so the total price is $850(0.05427) = \boxed{46.1262}$. (**Answer: (E)**) □

Currency options bear particular relevance to multinational corporations, which export and import goods to and from international markets as part of their regular operating cycle. Because their cash inflows and outflows depend on the future exchange rates of foreign currencies, these companies are susceptible to exchange rate fluctuations. The following example illustrates how currency options can be leveraged to hedge against currency risk.

Example 6.2.4. (SOA Exam IFM Advanced Derivatives Sample Question 7: How to use currency options to hedge against currency risk?) Company A is a US international company, and Company B is a Japanese local company. Company A is negotiating with Company B to sell its operation in Tokyo to Company B. The deal will be settled in Japanese yen. To avoid a loss at the time when the deal is closed due to a sudden devaluation of yen relative to dollar, Company A has decided to buy at-the-money dollar-denominated yen puts of the European type to hedge this risk.

You are given the following information:

(i) The deal will be closed 3 months from now.

(ii) The sale price of the Tokyo operation has been settled at 120 billion Japanese yen.

(iii) The continuously compounded risk-free interest rate in the United States is 3.5%.

(iv) The continuously compounded risk-free interest rate in Japan is 1.5%.

(v) The current exchange rate is 1 US dollar = 120 Japanese yen.

(vi) The daily volatility of the yen per dollar exchange rate is 0.261712%.

(vii) 1 year = 365 days; 3 months = 1/4 year.

Calculate Company A's option cost.

(A) 7.32 million

(B) 7.42 million

(C) 7.52 million

(D) 7.62 million

(E) 7.72 million

Solution. Prelude: Before computing the required option cost, we explain why currency puts are of value to Company A. In this example, Company A is a US-based company which has a business in Japan. Let $X(t)$ be the exchange rate of US dollar per Japanese yen at time t. That is, at time t, $¥1 = \$X(t)$.

Time 0. According to (v), we have $X(0) = 1/120$.

Time 0.25. According to (ii), the sale price of the business received at time $t = 0.25$ is fixed at ¥120 billion. Being US-based, Company A naturally wants to convert the sale price back to US dollars. Then the sale price of the business in US dollars will be $\$120X(0.25)$ billion, which is a random amount because $X(0.25)$ is not known at time 0. If $X(0.25)$ turns out to be very low (i.e., Japanese yen depreciates substantially against the US dollars), then Company A will only receive a disappointingly low revenue in US dollars from selling its business. This is a great cause for concern to Company A.

To hedge such exchange rate risk, dollar-denominated currency puts on yen come to Company A's rescue. If at-the-money ones are used and the downside risk is to be completely eliminated, then Company A shall need to buy 120 billion yen puts, and its overall payoff in 3 months, in billion dollars, becomes

$$120 \left[X(0.25) + \underbrace{\left(120^{-1} - X(0.25)\right)_{+}}_{\text{payoff of each yen put}} \right]$$

$$= \left[120X(0.25) + (1 - 120X(0.25))_{+}\right] = \max\{1, 120X(0.25)\},$$

which is bounded from below by \$1 billion. In the language of Section 3.1, the 120 billion yen puts place a floor on the revenue Company A will receive in US dollars in 3 months. The cash flows between different parties are depicted in Figure 6.2.1.

Calculations: Using the Black-Scholes formula with $S(0) = K = \$1/120$, $r = 0.035$ (interest rate in the US), $\delta = 0.015$ (interest rate in Japan), $\sigma/\sqrt{365} = 0.261712\%$, or $\sigma = 0.05$, we have

$$
\begin{aligned}
d_1 &= \frac{(r - \delta + \sigma^2/2)T}{\sigma\sqrt{T}} = \frac{(0.035 - 0.015 + 0.05^2/2)(0.25)}{0.05\sqrt{0.25}} = 0.2125, \\
d_2 &= d_1 - 0.05\sqrt{0.25} = 0.1875, \\
N(-d_1) &= 0.41586, \\
N(-d_2) &= 0.42563,
\end{aligned}
$$

and the total cost of the put options in dollars is

$$\$120 \text{ billion} \times \left[\frac{1}{120}e^{-0.035(0.25)}(0.42563) - \frac{1}{120}e^{-0.015(0.25)}(0.41586) \right]$$

$$= \boxed{\$0.00761854 \text{ billions}},$$

or \$7.61854 millions. **(Answer: (D))**   □

Remark. (i) As an alternative to currency puts, Company A could have locked in the sale price of its Tokyo operation via selling currency yen forwards with forward price $X(0)e^{(r_\$ - r_¥)/4}$ (analog of $S(0)e^{(r-\delta)T}$ for forwards on stocks) from the market or synthetically creating short currency forwards. The latter can be accomplished by borrowing $¥120e^{-r_¥/4}$ billion at time 0, immediately converting this amount to $\$e^{-r_¥/4}$, and depositing it in the risk-free account in the US. After repaying the yen loan with interest (i.e., ¥120), the overall 3-month payoff of Company A, in billion US dollars, becomes constant at

$$120X(0.25) + [e^{(r_\$ - r_¥)/4} - 120X(0.25)] = e^{(r_\$ - r_¥)/4}.$$

The downside currency risk is eliminated. On the other hand, with the currency options, Company A has the opportunity to benefit from the appreciation of the yen against the dollar.

(ii) A more challenging version of this example is Problem 6.4.11, where you need to figure out whether currency calls or puts should be purchased. You cannot just calculate blindly!

Company A's overall payoff
$= \max(120X(0.25), 1)$ in billion

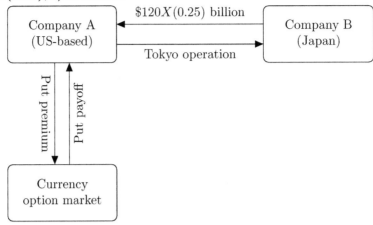

FIGURE 6.2.1
The cash flows between different parties in Example 6.2.4.

6.2.3   Case Study 3: Futures options.

As with currency options, futures options can be embedded in the standard Black-Scholes pricing framework by assuming that future futures prices (not a typo!) are lognormally distributed with constant volatility σ and treating the underlying futures as a stock whose "current price" is the current futures price and the "dividend yield" is the continuously compounded risk-free interest rate r (see the discussions in Subsection 4.4.2).[vii] This yields the following Black-Scholes pricing formulas for K-strike T-year European futures options:

$$
\begin{aligned}
C &= \mathrm{BS}\left(F_{0,T_f}, r; K, r; \sigma, T\right) = F_{0,T_f}\mathrm{e}^{-rT}N(d_1) - K\mathrm{e}^{-rT}N(d_2), \\
P &= \mathrm{BS}\left(K, r; F_{0,T_f}, r; \sigma, T\right) = K\mathrm{e}^{-rT}N(-d_2) - F_{0,T_f}\mathrm{e}^{-rT}N(-d_1),
\end{aligned}
\tag{6.2.3}
$$

where

T is the time to maturity of the futures call and put,

T_f is the time to maturity of the underlying futures with $T \le T_f$,

F_{0,T_f} is the current price of the T_f-year futures, and

$$
d_1 = \frac{\ln(F_{0,T_f}/K) + \sigma^2 T/2}{\sigma\sqrt{T}}, \qquad d_2 = d_1 - \sigma\sqrt{T}.
$$

Example 6.2.5. (SOA Exam IFM Advanced Derivatives Sample Question 55: Put on futures) Assume the Black-Scholes framework. Consider a 9-month at-the-money European put option on a futures contract. You are given:

(i) The continuously compounded risk-free interest rate is 10%.

(ii) The strike price of the option is 20.

[vii]This can also be seen by the Itô's Lemma, which is beyond the scope of this book.

(iii) The price of the put option is 1.625.

If three months later the futures price is 17.7, what is the price of the put option at that time?

(A) 2.09

(B) 2.25

(C) 2.45

(D) 2.66

(E) 2.83

Solution. Since the put option is at-the-money, we have

$$d_1 = \frac{\overbrace{\ln(F_{0,T_f}/K)}^{0} + \sigma^2 T/2}{\sigma\sqrt{T}} = \frac{\sigma\sqrt{T}}{2} \quad \text{and} \quad d_2 = d_1 - \sigma\sqrt{T} = -d_1.$$

With $N(-d_2) = N(d_1) = 1 - N(-d_1)$, it follows from Point (iii) that

$$P = e^{-rT}[K \times N(-d_2) - K \times N(-d_1)] = 20e^{-0.1(0.75)}[1 - 2N(-d_1)] = 1.625,$$

whence $N(-d_1) = 0.45621$ and $d_1 = -N^{-1}(0.45621) = 0.10999$. Then $\sigma = 2d_1/\sqrt{0.75} = 0.25401$.

Three months later, the put option has 6 months to go before it expires. With the Black-Scholes parameters updated to

$$d_1 = \frac{\ln(17.7/20) + (0.25401^2/2)(0.5)}{0.25401\sqrt{0.5}} = -0.59037,$$
$$d_2 = d_1 - 0.25401\sqrt{0.5} = -0.76998,$$
$$N(-d_1) = 0.72253,$$
$$N(-d_2) = 0.77934,$$

the price of the put option at that time is

$$P = 20e^{-0.1(0.5)}(0.77934) - 17.7e^{-0.1(0.5)}(0.72253) = \boxed{2.66156}. \quad \textbf{(Answer: (D))}$$

□

Remark. As this example shows, at-the-money futures options provide a rare instance in which observed option prices can be used to infer volatility analytically, given other model parameters (note that the Black-Scholes pricing formula is a highly non-linear function of the volatility; σ appears in both d_1 and d_2). Such a volatility is aptly named as the *implied volatility*.

Option Greek	Greek Symbol	Partial Derivative of the Option Price With Respect To
Delta	Δ	Stock price (first derivative)
Gamma	Γ	Stock price (second derivative)
Vega	N.A.	Volatility
Theta	θ	Passage of time (not time to maturity)
Rho	ρ	Interest rate
Psi	Ψ	Dividend yield

TABLE 6.2
The definitions and symbols of the six most common option Greeks.

6.3 Option Greeks

The explicit pricing formulas given in Sections 6.1 and 6.2 take the model parameters (e.g., stock price, time to maturity, risk-free interest rate, dividend yield, volatility) as input and provide the value of a European option at a single point of time. As time elapses, these model parameters, especially the stock price, are likely to change, which in turn alters the value of the option. An appeal of the Black-Scholes formula is that its explicitness easily allows us to quantify the sensitivity of the option price to changes in these model parameters. Such *sensitivity analysis* can be formally performed by investigating vehicles known as option Greeks. These metrics give a great deal of information about the risk exposure of an option trader and can be carefully exploited to optimize one's risk position. In this section, we study the definitions, computations, interpretation, and qualitative properties of option Greeks. Their practical uses in hedging and risk management will be examined in Chapter 7.

Definition.

Mathematically, option Greeks are partial derivatives of the option price with respect to the option parameter in question, holding other inputs fixed. The six most common option Greeks are given in Table 6.2.

Option Greeks inherit the "approximation" interpretation that partial derivatives carry in multi-variable calculus. For instance, if the delta of an option equals 0.8, then a unit increase in the *current* price of the underlying stock causes the price of the option to increase by *approximately* 0.8, provided that the values of other option parameters are held constant. With this way of interpretation, option Greeks are natural indicators of the sensitivity of option prices to changes in the option parameters. The larger the magnitude of an option Greek, the more sensitive the option price is to the underlying risk factor.

Questions of interest.

For each option Greek, the following questions are often of both theoretical and practical interest:

- Question 1: Is there a convenient computing formula for the Greek?

  The definitions of the six option Greeks, in terms of partial derivatives, are suitable mostly for making interpretation, but do not lend themselves to practical calculations.[viii]
  To obtain explicit expressions of the option Greeks, one can partially differentiate the

---

[viii]Actuarial students taking Exam IFM should know that one of the learning outcomes of Exam IFM is to *compute* and interpret Option Greeks

option price with respect to the relevant option parameters. The manipulations, as will be shown shortly, are not as straightforward as you may have thought.

- Question 2: How do the Greeks of otherwise identical European calls and puts compare?

  The key to this question is partially differentiating both sides of put-call parity. The difference between the call Greek and the put Greek is then related to the right-hand side of put-call parity differentiated with respect to the option parameter in question. Such a relation between the two Greeks makes it easy to translate our analysis for the call Greek to the put Greek, or the other way round.

- Question 3: Is the Greek always positive, always negative, or sometimes positive or negative? How can this behavior be explained?

  In most cases, an option Greek takes a definite sign, which can be justified algebraically using its closed-form expression or more easily and instructively by verbal reasoning.

- Question 4: How does the Greek vary with the moneyness and time to expiration of the option? Why does it exhibit this behavior?

  In addition to its sign, a Greek typically varies regularly with the price of the underlying asset (or equivalently, its moneyness) and the time to expiration of the option. These regular patterns, again, admit intuitive explanations.

To avoid repetition, the answers to these four questions will be epitomized by delta and gamma, which are arguably the Greeks of predominant interest. For concreteness, throughout this section we concentrate on European options on stocks that pay continuous proportional dividends, although option Greeks can be evaluated and studied for options on many other kinds of underlying asset (examples can be found in Section 6.2). We will also simplify notation and write S for $S(0)$ when no confusion arises.

6.3.1 Option Delta

Definition.

The *delta* of an option is defined as the partial derivative of the option price with respect to the current price of the underlying asset:

$$\Delta := \frac{\partial V}{\partial S}.$$

It serves to measure the sensitivity of the option price to changes in the current price of the underlying stock and is arguably the most important option Greek in theory as well as in practice.

Question 1: Closed-form expression for delta.

In the Black-Scholes model, we have a closed-form formula for the prices of European call and put options. Partially differentiating the call and put prices with respect to the current stock price yields

$$\boxed{\Delta_C := \frac{\partial C}{\partial S} = e^{-\delta T} N(d_1) \quad \text{and} \quad \Delta_P := \frac{\partial P}{\partial S} = -e^{-\delta T} N(-d_1).} \tag{6.3.1}$$

Proof of (6.3.1). First consider the call delta. At first glance, it seems that we directly have

$$\Delta_C = \frac{\partial}{\partial S}[Se^{-\delta T} N(d_1) - Ke^{-rT} N(d_2)] = e^{-\delta T} N(d_1).$$

However, this completely ignores the fact that d_1 and d_2 themselves are also functions of S! This should be firmly kept in mind in using the product rule of differentiation. It turns out, however, that the above naive differentiation is coincidentally true because of the following simple but useful auxiliary result:

$$F_{0,T}^P(S)N'(d_1) = F_{0,T}^P(K)N'(d_2). \tag{6.3.2}$$

This result can be remembered as the strikingly simple fact that if you "accidentally" mistake the standard normal distribution function $N(\cdot)$ as the standard normal probability density function $N'(\cdot)$ given by

$$N'(x) = \frac{1}{\sqrt{2\pi}}\exp(-x^2/2), \tag{6.3.3}$$

your Black-Scholes call price must be zero (don't make this fatal mistake in the exam!). To check (6.3.2), we use the identity

$$e^{-(x+y)^2/2} = e^{-(x-y)^2/2} \times e^{-2xy}$$

for any real x and y and the prepaid forward versions of d_1 and d_2 to yield

$$F_{0,T}^P(S)N'(d_1)$$

$$= F_{0,T}^P(S) \times \frac{1}{\sqrt{2\pi}}\exp\left[-\frac{\left(\overbrace{\ln[F_{0,T}^P(S)/F_{0,T}^P(K)]/\sigma\sqrt{T}}^{x} + \overbrace{\sigma\sqrt{T}/2}^{y}\right)^2}{2}\right]$$

$$= F_{0,T}^P(S) \times \frac{1}{\sqrt{2\pi}}\exp\left[-\frac{\left(\ln[F_{0,T}^P(S)/F_{0,T}^P(K)]/\sigma\sqrt{T}\boxed{-}\sigma\sqrt{T}/2\right)^2}{2} - \ln\frac{F_{0,T}^P(S)}{F_{0,T}^P(K)}\right]$$

$$= F_{0,T}^P(S) \times \frac{1}{\sqrt{2\pi}}\exp(-d_2^2/2) \times \frac{F_{0,T}^P(K)}{F_{0,T}^P(S)}$$

$$= F_{0,T}^P(K)N'(d_2).$$

Using the chain rule, we have

$$\begin{aligned}
\Delta_C &= \frac{\partial}{\partial S}[Se^{-\delta T}N(d_1) - Ke^{-rT}N(d_2)] \\
&= e^{-\delta T}N(d_1) + Se^{-\delta T}N'(d_1)\frac{\partial d_1}{\partial S} - Ke^{-rT}N'(d_2)\frac{\partial d_2}{\partial S}. \\
&= e^{-\delta T}N(d_1) + \underbrace{[Se^{-\delta T}N'(d_1) - Ke^{-rT}N'(d_2)]}_{0} \times \frac{\partial d_1}{\partial S} \\
&= e^{-\delta T}N(d_1),
\end{aligned}$$

where the third equality is due to (6.3.2) and $\partial d_1/\partial S = \partial d_2/\partial S$ (because $d_2 = d_1 - \sigma\sqrt{T}$).

To derive the put delta, the easiest way is to use put-call parity, which says that

$$C - P = Se^{-\delta T} - Ke^{-rT}.$$

Partially differentiating both sides with respect to S yields

$$\Delta_C - \Delta_P = e^{-\delta T},$$

whence

$$\Delta_P = \Delta_C - e^{-\delta T} = -e^{-\delta T} N(-d_1).$$

This addresses Question 2.   □

Example 6.3.1. (Direct calculation of delta) Assume the Black-Scholes framework. You are given:

(i) The stock price is 100.

(ii) The stock pays dividends continuously at a rate proportional to its price. The dividend yield is 2%.

(iii) The continuously compounded risk-free interest rate is 6%.

(iv) The stock's volatility is 40%.

Calculate the delta of a one-year 105-strike European call option.

Solution. As

$$d_1 = \frac{\ln(100/105) + (0.06 - 0.02 + 0.4^2/2)(1)}{0.4\sqrt{1}} = 0.17802,$$

the delta of the call option is

$$\Delta_C = \underbrace{e^{-0.02(1)}}_{\text{Don't omit this!}} \times \underbrace{N(d_1)}_{0.57065} = \boxed{0.55935}.$$

□

Example 6.3.2. (SOA Exam IFM Advanced Derivatives Sample Question 8) You are considering the purchase of a three-month 41.5-strike American call option on a nondividend-paying stock.

You are given:

(i) The Black-Scholes framework holds.

(ii) The stock is currently selling for 40.

(iii) The stock's volatility is 30%.

(iv) The current call option delta is 0.5.

Determine the current price of the option.

(A) $20 - 20.453 \int_{-\infty}^{0.15} e^{-x^2/2} \, dx$

(B) $20 - 16.138 \int_{-\infty}^{0.15} e^{-x^2/2} \, dx$

(C) $20 - 40.453 \int_{-\infty}^{0.15} e^{-x^2/2} \, dx$

(D) $16.138 \int_{-\infty}^{0.15} e^{-x^2/2} \, dx - 20.453$

(E) $40.453 \int_{-\infty}^{0.15} e^{-x^2/2} \, dx - 20.453$

Solution. It can be shown that it is never optimal to exercise an American call option on a nondividend-paying stock before maturity (see Chapter 9), so the current price of the American call is the same as that of the otherwise identical European call.

Because $N(d_1) = 0.5$, we have $d_1 = 0$. It follows from

$$d_1 = \frac{\ln(40/41.5) + (r + 0.3^2/2)(0.25)}{0.3\sqrt{0.25}} = 0$$

that $r = 0.102256$. Then $d_2 = d_1 - 0.3\sqrt{0.25} = -0.15$. Finally, by (6.1.1), the price of the call option is

$$
\begin{aligned}
C &= S(0)\Delta - Ke^{-rT}N(d_2) \\
&= 40(0.5) - 41.5e^{-(0.102256)(0.25)}N(-0.15) \\
&= 20 - 40.452540[1 - N(0.15)] \\
&= 40.452540 \int_{-\infty}^{0.15} \frac{1}{\sqrt{2\pi}} e^{-x^2/2} \, dx - 20.4525 \\
&= \boxed{16.138 \int_{-\infty}^{0.15} e^{-x^2/2} \, dx - 20.453}. \quad \textbf{(Answer: (D))}
\end{aligned}
$$

$\square$

Sidebar: Delta in the Black-Scholes framework vs delta in the binomial framework.

We have seen in the (one-period) binomial tree setting of Chapter 4 that the fair price of a derivative, V, equals the cost of setting up the replicating portfolio (Δ, B). Mathematically,

$$V = S\Delta + B, \tag{6.3.4}$$

where Δ is the number of stocks to buy and B is the amount of risk-free investment made initially so as to mimic the payoff of the derivative at the end of the period. Surprisingly, the Black-Scholes option pricing formula, when viewed in the correct light, shares the same structure as (6.3.4). In the case of a European call, for instance, we have

$$C = S\boxed{e^{-\delta T}N(d_1)} - Ke^{-rT}N(d_2) = S\Delta_C + B,$$

with $\Delta_C := e^{-\delta T}N(d_1)$, which is the delta of the call (see (6.3.1)), and $B := -Ke^{-rT}N(d_2)$. It can be shown that $(\Delta_C, B) = (e^{-\delta T}N(d_1), -Ke^{-rT}N(d_2))$ provides the ingredients for the replicating portfolio of the call *at time 0*. To replicate the payoff of the call at maturity, it is necessary to adjust the values of Δ and B *continuously* in response to the emergence of information in the market (the technical details are beyond the scope of this book). The bottom line is that the Black-Scholes formula provides not only the fair price of an option, but also the *initial* replicating portfolio of the option.

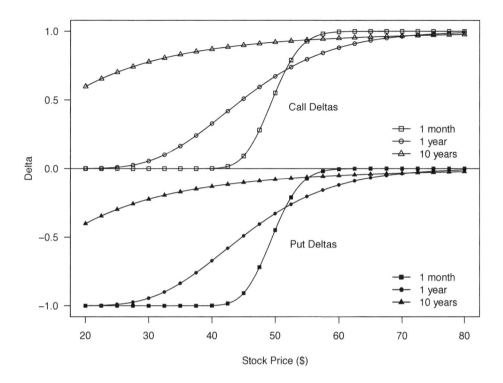

FIGURE 6.3.1
The variation of call and put deltas with the current stock price for different times to
expiration.

Question 3: Signs of delta.

Intuition suggests and (6.3.1) confirms that call prices (resp. put prices) increase (resp.
decrease) with the price of the underlying asset, because $\Delta_C > 0$ and $\Delta_P < 0$. As the
stock price increases, the call (resp. put), which is a right to acquire (resp. give up) the
stock, becomes more (resp. less) valuable. Moreover, the delta of a call is always less than
1 because the call price cannot increase by more than one dollar for every dollar increase
in the price of the underlying asset. Similarly, the delta of a put is always greater than -1
because a unit decrease in the price of the underlying asset cannot lead to more than a
dollar increase in the put price.

Question 4: Graphs of delta as a function of S.

Figure 6.3.1 shows how the deltas of a call option and a put option on a nondividend-paying
stock varies with the current stock price S for different times to expiration. These curves
(and other option curves in this section) are generated using the following common set of
baseline parameters:

$$K = 40, \qquad \sigma = 30\%, \qquad r = 8\%, \qquad \delta = 0.$$

Some salient features of the figure deserve special attention:

1. The delta curve of a put is the delta curve of an otherwise identical call translated
   downward by 1.

Explanations: This can be easily seen by differentiating both sides of put-call parity, yielding

$$\Delta_C - \Delta_P = e^{-\delta T} \quad \Rightarrow \quad \Delta_P = \Delta_C - e^{-\delta T} \overset{(\text{if } \delta = 0)}{=} \Delta_C - 1.$$

The fact that the put delta is simply a downward translation of the call delta means that we can easily understand the put delta if we can analyze the behavior of the call delta.

2. The *magnitude* of the delta depends closely on the option's *moneyness*. More specifically:

   - Deep in-the-money options have deltas close to 1 in absolute value (i.e., $\Delta_C \approx 1$ for very high S and $\Delta_P \approx -1$ for very low S).

     Explanations (for call): A deep in-the-money call is very likely to be exercised at expiration, with a terminal payoff of approximately $S(T) - K$. The current price is then approximately $S(0) - Ke^{-rT}$, leading to an approximate delta of 1. Such a call is very sensitive to changes in the stock price, as a unit increase in S will lead to an approximate unit increase in the call price.

   - Deep out-of-the-money options have deltas close to 0 (i.e., $\Delta_C \approx 0$ for very low S and $\Delta_P \approx 0$ for very high S).

     Explanations: A deep out-of-the-money call is very unlikely to be exercised at expiration and is very insensitive to changes in the stock price – even if S increases by one unit, the deep out-of-the-money call remains deep out-of-the-money, with a price close to zero.

   You can also see these phenomena mathematically by noticing that

   $$d_1 = \frac{\ln(S/K) + (r + \sigma^2/2)T}{\sigma\sqrt{T}} = \begin{cases} \to \infty, & \text{as } S \to \infty, \\ \to -\infty, & \text{as } S \to 0, \end{cases}$$

   so

   $$\Delta_C = N(d_1) = \begin{cases} \to 1, & \text{as } S \to \infty, \\ \to 0, & \text{as } S \to 0. \end{cases}$$

3. Comparing the three delta curves for each of the call and put options shows how the delta varies with the *time to expiration*, when the stock price is fixed. Typically, the delta of a deep out-of-the-money call becomes greater and the delta of a deep in-the-money call becomes lower as the time to expiration increases.

   Explanations: With greater time to expiration, it is more probable that a deep out-of-the-money (resp. in-the-money) option will expire in-the-money (out-of-the-money), thereby becoming more (resp. less) sensitive to the stock price.

6.3.2 Option Gamma

Definition.

The *gamma* of an option is defined as the second partial derivative of the option price with respect to the current price of the underlying asset. Equivalently, it is the partial derivative of the option delta with respect to the current price of the underlying asset:

$$\Gamma = \frac{\partial \Delta}{\partial S} = \frac{\partial^2 V}{\partial S^2}.$$

Corresponding to these two equivalent definitions, gamma can be thought of as a measure of the sensitivity of the delta to changes in the current stock price, and a measure of the curvature of the option price as a function of S.

Question 1: Closed-form expression for gamma.

Differentiating the expression for delta, by virtue of the chain rule again, shows that gamma is given by

$$\Gamma_C = \Gamma_P = e^{-\delta T} N'(d_1) \frac{\partial d_1}{\partial S} = e^{-\delta T} \times \frac{1}{\sqrt{2\pi}} e^{-d_1^2/2} \times \frac{1}{S\sigma\sqrt{T}},$$

where $N'(\cdot)$ is the probability density function of the standard normal distribution given in (6.3.3). Because every term that constitutes Γ is positive, so is Γ, answering Question 3. This comes as no surprise to us because delta as a function of S is increasing, as we have seen in the previous subsection. Mathematically, the positivity of gamma for both call and put options means that they are *convex* functions of the stock price. The convexity holds generally, even outside the Black-Scholes framework.

Example 6.3.3. (Direct calculation of gamma) Assume the Black-Scholes framework. You are given:

(i) The stock price is 100.

(ii) The stock pays dividends continuously at a rate proportional to its price. The dividend yield is 2%.

(iii) The continuously compounded risk-free interest rate is 6%.

(iv) The stock's volatility is 40%.

Calculate the gamma of a 1-year 105-strike European call option.

Solution. As

$$d_1 = \frac{\ln(100/105) + (0.06 - 0.02 + 0.4^2/2)(1)}{0.4\sqrt{1}} = 0.17802,$$

the gamma of the call option is

$$\Gamma = e^{-0.02} \times \frac{1}{\sqrt{2\pi}} e^{-0.17802^2/2} \times \frac{1}{100(0.4)\sqrt{1}} = \boxed{0.009622}.$$

□

Graphs of gamma.

Figure 6.3.2 shows how the gamma of a call and put on a nondividend-paying stock varies with the current stock price S for different times to expiration, using the same set of baseline parameters on page 207. Again, many features of the figure can be justified intuitively:

1. The gamma curve of a call and that of an otherwise identical put coincide. This can be seen by differentiating both sides of put-call parity with respect to S, yielding

$$\Gamma_C - \Gamma_P = 0 \quad \Rightarrow \quad \Gamma_C = \Gamma_P.$$

This settles Question 2 regarding the relationship between the gamma of otherwise identical calls and puts.

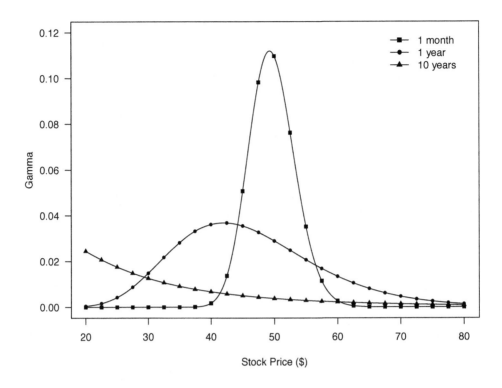

FIGURE 6.3.2
The variation of gamma with the current stock price for different times to expiration.

2. Gamma is almost zero when the option is deep in-the-money or deep out-of-the-money.

   Explanations: Without loss of generality, consider a call (since a put has the same gamma). When the call is deep in-the-money (resp. deep out-of-the-money), delta is very close to one (resp. zero) and does not change much as the stock price moves. Gamma, being the rate of change of delta, is then almost zero.

3. For short-lived options, gamma peaks prominently near the strike price ($40). For longer-lived options, gamma peaks further to the left of the strike price and less steeply.

   Explanations: For short-lived options, delta increases very abruptly from 0 or 1 around the strike price. Defined as the rate of change in delta, gamma takes very positive values in the same region. The increase in delta is more gradual for longer-lived options, so they have a lower gamma.

Example 6.3.4. (Highest gamma) Which of the following otherwise identical European options has the highest gamma?

(A) 1-day deep out-of-the-money call option

(B) 10-day deep in-the-money call option

(C) 1-year at-the-money put option

(D) 10-year at-the-money put option

(E) There is not enough information to determine the answer.

(Answer: (C))

6.3.3 Option Greeks of a Portfolio

Option Greeks are by no means confined to call or put options; they can be defined even for a portfolio of derivatives and evaluated easily. Since partial differentiation is a linear operator, the Greek of a portfolio equals the *sum* of the constituent Greeks. Symbolically, we have:

$$V_{\text{portfolio}} = \sum V_{\text{component}} \overset{\text{(by partial differentiation)}}{\Rightarrow} \text{Greek}_{\text{portfolio}} = \sum \text{Greek}_{\text{component}}.$$

(6.3.5)

The additivity of option Greeks is practically important because it means not only that the risk of a portfolio as reflected by an option Greek can be easily calculated from the component risks, but also that risk management and hedging at the portfolio level can be readily implemented. This will prove useful in the next chapter when we study delta-hedging and delta-gamma-hedging.

Example 6.3.5. (SOA Exam IFM Advanced Derivatives Sample Question 31: Delta of a bull spread) You compute the current delta for a 50-60 bull spread with the following information:

(i) The continuously compounded risk-free rate is 5%.

(ii) The underlying stock pays no dividends.

(iii) The current stock price is $50 per share.

(iv) The stock's volatility is 20%.

(v) The time to expiration is 3 months.

How much does delta change after 1 month, if the stock price does not change?

(A) increases by 0.04

(B) increases by 0.02

(C) does not change, within rounding to 0.01

(D) decreases by 0.02

(E) decreases by 0.04

Ambrose's comments:

This question involves the calculations of four d_1's! Be patient!

Solution. Assume without loss of generality that the bull spread is a call bull spread. The same answers would be obtained by assuming put options instead of call options. Then the delta of the bull spread equals

$$\Delta_{\text{bull spread}} = \Delta_C^{\text{50-strike}} - \Delta_C^{\text{60-strike}}.$$

Initially:

- For the 50-strike call,

$$d_1 = \frac{\ln(50/50) + (0.05 + 0.2^2/2)(0.25)}{0.2\sqrt{0.25}} = 0.175,$$

$$\Delta_C^{\text{50-strike}} = N(d_1) = 0.56946.$$

- For the 60-strike call,

$$d_1 = \frac{\ln(50/60) + (0.05 + 0.2^2/2)(0.25)}{0.2\sqrt{0.25}} = -1.64822,$$

$$\Delta_C^{\text{60-strike}} = N(d_1) = 0.04965.$$

The original delta is thus $0.56946 - 0.04965 = 0.51981$.

After 1 month (the remaining time to expiration is 2 months):

- For the 50-strike call,

$$d_1 = \frac{\ln(50/50) + (0.05 + 0.2^2/2)/6}{0.2\sqrt{1/6}} = 0.14289,$$

$$\Delta_C^{\text{50-strike}} = N(d_1) = 0.55681.$$

- For the 60-strike call,

$$d_1 = \frac{\ln(50/60) + (0.05 + 0.2^2/2)/6}{0.2\sqrt{1/6}} = -2.09009,$$

$$\Delta_C^{\text{60-strike}} = N(d_1) = 0.01830.$$

The new delta is $0.55681 - 0.01830 = 0.53851$. Therefore, the change in delta is $0.53851 - 0.51981 = \boxed{0.0187}$. (**Answer: (B)**) $\qquad \square$

Remark. The calculation of the change in the price of the bull spread by hand is much more cumbersome.

6.3.4 Option Elasticity

Motivation.

Whereas the delta of an option quantifies the approximate change in the option price for a dollar increase in the underlying stock, such a sensitivity measure can be criticized since *absolute* changes in the option price and *absolute* changes in the stock price are, in fact, not directly comparable. Call prices, for example, are constrained by the price of the stock, so that a $1 change in the call price can mean a lot compared to a $1 change in the stock price. A fairer and dimensionless sensitivity measure can be developed by weighing the *percentage* changes in the option price against the *percentage* changes in the stock price. This leads to the notion of option elasticity, which emanates from economics.

Computing formula.

Mathematically, let $V(S)$ be the price of a generic option (in fact, any derivative) when the current price of the underlying stock is S (other arguments, e.g., strike price, time to maturity, are suppressed and assumed to be fixed here). The *elasticity* of the option, denoted by Ω, measures the *percentage* change in the option price relative to the *percentage* change in the stock price. It provides answers to the question

"If the price of the stock increases by 1%, by how much does the price of the option change in proportion?"

To derive the expression for the elasticity of an option, consider

$$\Omega(S) = \lim_{\epsilon \to 0} \overbrace{\frac{[V(S + \epsilon) - V(S)]/V(S)}{\underbrace{\epsilon/S}_{\% \text{ change in stock price}}}}^{\% \text{ change in option price}} = \frac{S}{V(S)} \lim_{\epsilon \to 0} \frac{V(S + \epsilon) - V(S)}{\epsilon} = \frac{SV'(S)}{V(S)},$$

or, more compactly,

$$\boxed{\Omega = \frac{S\Delta_V}{V}.} \tag{6.3.6}$$

This provides a convenient formula for computing the elasticity of an option. When viewed in the form

$$\Omega = \frac{\Delta_V/V}{1/S},$$

the formula confirms that the elasticity indeed captures the proportional changes in the option price and the stock price.

Example 6.3.6. (CAS Exam 3 Fall 2007 Question 22: Warm-up calculation of Ω) A call option is modeled using the Black-Scholes formula with the following parameters.

- $S = 25$

- $K = 24$

- $r = 4\%$

- $\delta = 0\%$

- $\sigma = 20\%$

- $T = 1$

Calculate the call option elasticity, Ω.

(A) Less than 5

(B) At least 5, but less than 6

(C) At least 6, but less than 7

(D) At least 7, but less than 8

(E) At least 8

Solution. With

$$d_1 = \frac{\ln(25/24) + (0.04 + 0.2^2/2)(1)}{0.2\sqrt{1}} = 0.50411,$$

$$d_2 = d_1 - 0.2\sqrt{1} = 0.30411,$$

$$N(d_1) = 0.69291, \quad (= \Delta_C)$$

$$N(d_2) = 0.61948,$$

the Black-Scholes call price is

$$C = 25(0.69291) - 24e^{-0.04(1)}(0.61948) = 3.03819.$$

The call option elasticity is

$$\Omega_C = \frac{S\Delta_C}{C} = \frac{25(0.69291)}{3.03819} = \boxed{5.7017}. \quad (\textbf{Answer: (B)})$$

□

Example 6.3.7. (SOA Exam IFM Advanced Derivatives Sample Question 41: Elasticity of a capped stock) Assume the Black-Scholes framework. Consider a 1-year European contingent claim on a stock.

You are given:

(i) The time-0 stock price is 45.

(ii) The stock's volatility is 25%.

(iii) The stock pays dividends continuously at a rate proportional to its price. The dividend yield is 3%.

(iv) The continuously compounded risk-free interest rate is 7%.

(v) The time-1 payoff of the contingent claim is as follows:

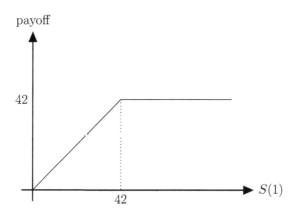

Calculate the time-0 contingent-claim elasticity.

(A) 0.24

(B) 0.29

(C) 0.34

(D) 0.39

(E) 0.44

Solution. There are two ways to view the contingent claim. We discuss both below. As with option Greeks, we decorate Ω by the subscripts "C" and "P" to denote the elasticity of a call and that of a put, respectively.

- *Short put plus bond:* The first way is to consider the contingent claim as a short 42-strike 1-year put coupled with a long 1-year zero-coupon bond with a face value of 42. As

$$d_1 = \frac{\ln(45/42) + (0.07 - 0.03 + 0.25^2/2)(1)}{0.25\sqrt{1}} = 0.56097,$$

$$d_2 = d_1 - 0.25\sqrt{1} = 0.31097,$$

$$N(-d_1) = 0.28741,$$

$$N(-d_2) = 0.37791,$$

the price of the put option is

$$P = 42e^{-0.07}(0.37791) - 45e^{-0.03}(0.28741) = 2.24795,$$

so the time-0 price of the contingent claim is

$$V = \text{PV}_{0,1}(42) - P = 42e^{-0.07} - 2.24795 = 36.91259.$$

Because the zero-coupon bond has zero delta, the delta of the contingent claim is

$$\Delta_V = -\Delta_P = -[-e^{-0.03}(0.28741)] = 0.27892.$$

Thus the contingent-claim elasticity is

$$\Omega_V = \frac{S\Delta_V}{V} = \frac{45(0.27892)}{36.91259} = \boxed{0.34003}. \quad \textbf{(Answer: (C))}$$

- *Long stock plus short call:* Another way to view the contingent claim is to regard it as one unit of the stock *at time 1* (i.e., a 1-year prepaid forward on the stock) and a short 42-strike 1-year call. In the language of Section 3.1, the contingent claim is a written covered call.

As $N(d_1) = 0.71259$ and $N(d_2) = 0.62209$, the price of the call is

$$C = 45e^{-0.03}(0.71259) - 42e^{-0.07}(0.62209) = 6.75746,$$

so the time-0 price of the contingent claim is

$$V = F_{0,1}^P(S) - C = 45e^{-0.03} - 6.75746 = 36.91259.$$

The delta of the contingent claim is

$$\Delta_V = e^{-\delta T} - \Delta_C = e^{-0.03} - e^{-0.03}(0.71259) = 0.27892.$$

Therefore, the contingent-claim elasticity is

$$\Omega_V = \frac{S\Delta_V}{V} = \frac{45(0.27892)}{36.91259} = \boxed{0.34003}. \quad (\textbf{Answer: (C)})$$

□

Elasticity for call and put options.

Like option delta, the elasticity of call and put options enjoys bounds that can be justified either algebraically or verbally, although there are no straightforward connections between the elasticity of a call and the elasticity of a put having the same strike price and time to expiration.

Call.   For a call, $\Omega > 1$ always, because

$$\Omega_C = \frac{S\Delta_C}{C} = \frac{Se^{-\delta T}N(d_1)}{Se^{-\delta T}N(d_1) - Ke^{-rT}N(d_2)} > \frac{Se^{-\delta T}N(d_1)}{Se^{-\delta T}N(d_1)} = 1.$$

As a result, a call can be said to be riskier than the underlying stock in the sense that the call price increases more vigorously in proportion than the stock price.

Put.   For a put, $\Omega \leq 0$, because the stock price and the put price exhibit changes in different directions, as reflected by the negativity of the delta of a put. It is possible, however, that $-1 \leq \Omega_P \leq 0$, which occurs when the put is deep in-the-money, i.e., when the current stock price is substantially below the strike price. In this case, a 1% percent increase in the stock price means a very small absolute increase in the stock price. As $\Delta_P > -1$, the absolute decrease in the put price is minuscule as well. This, together with the high put price, in turn causes a negligible percentage drop in the put price and explains the small magnitude of the put elasticity.

Elasticity of a portfolio.

Unlike the portfolio Greek, the elasticity of a portfolio is not the sum, but a *weighted average* of the constituent elasticities. To see this and identify the weights, consider a portfolio of

n_1 option 1,

n_2 option 2,

$\vdots$

and n_N option N,

all of which are on the same underlying stock. Denote by V_i and Δ_i the value and delta of the ith option, for $i = 1, 2, \ldots, N$. Applying (6.3.6) to this portfolio treated as another asset and using the additivity of delta, we have

$$\Omega_{\text{portfolio}} = \frac{S\Delta_{\text{portfolio}}}{V_{\text{portfolio}}} = \frac{S\sum_{i=1}^{N} n_i\Delta_i}{\sum_{i=1}^{N} n_i V_i}, \tag{6.3.7}$$

which can be further rearranged as

$$\Omega_{\text{portfolio}} = \sum_{i=1}^{N} \left(\frac{n_i V_i}{\sum_{j=1}^{N} n_j V_j} \right) \left(\frac{S\Delta_i}{V_i} \right) = \sum_{i=1}^{N} \omega_i \Omega_i, \tag{6.3.8}$$

where $\omega_i = n_i V_i / \sum_{j=1}^{N} n_j V_j$ is the portion of the portfolio invested in option i.

As a matter of fact, (6.3.7) is what you usually use in a computational problem, while (6.3.8) is useful only when you are given minimal information, e.g., only constituent elasticities and prices are given, but not S, r, δ, T, σ, etc.

Example 6.3.8. (SOA Exam IFM Advanced Derivatives Sample Question 20: Elasticity of portfolio) Assume the Black-Scholes framework. Consider a stock, and a European call option and a European put option on the stock. The current stock price, call price, and put price are 45.00, 4.45, and 1.90, respectively.

Investor A purchases two calls and one put. Investor B purchases two calls and writes three puts.

The current elasticity of Investor A's portfolio is 5.0. The current delta of Investor B's portfolio is 3.4.

Calculate the current put-option elasticity.

(A) -0.55

(B) -1.15

(C) -8.64

(D) -13.03

(E) -27.24

Solution. Considering the elasticity of Investor A's portfolio, we get

$$\frac{S}{2C + P}(2\Delta_C + \Delta_P) = 5.0 \quad \Rightarrow \quad \frac{45}{2(4.45) + 1.90}(2\Delta_C + \Delta_P) = 5.0$$

$$\Rightarrow \quad 2\Delta_C + \Delta_P = 1.2.$$

Moreover, the current delta of Investor B's portfolio is

$$2\Delta_C - 3\Delta_P = 3.4.$$

Solving these two equations in Δ_C and Δ_P, we have $\Delta_P = -0.55$ (and $\Delta_C = 0.875$). It follows that the current put-option elasticity is

$$\Omega_P = \frac{S\Delta_P}{P} = \frac{45(-0.55)}{1.90} = \boxed{-13.03}. \quad \textbf{(Answer: (D))}$$

□

Example 6.3.9. (Elasticity of a bull spread) You are given the following information about 50-strike and 60-strike European put options with the same stock and time to expiration:

Strike price	Elasticity	Put premium
50	−4.9953	3.7295
60	−3.4267	9.5865

Calculate the elasticity of a 50-60 European put bull spread.

Solution. The 50-60 put bull spread is set up by buying the 50-strike put and selling the 60-strike put. By (6.3.8), the elasticity of the bull spread as a portfolio is

$$
\begin{aligned}
\Omega_{\text{bull spread}} &= \frac{3.7295}{3.7295 - 9.5865}\Omega_P^{\text{50-strike}} + \left(\frac{-9.5865}{3.7295 - 9.5865}\right)\Omega_P^{\text{60-strike}} \\
&= \frac{3.7295}{3.7295 - 9.5865}(-4.9953) + \left(\frac{-9.5865}{3.7295 - 9.5865}\right)(-3.4267) \\
&= \boxed{-2.4279}.
\end{aligned}
$$

Alternatively, from the two put elasticities, we deduce that

$$
\begin{cases}
\Omega_P^{\text{50-strike}} = \dfrac{S\Delta_P^{\text{50-strike}}}{P^{\text{50-strike}}} = \dfrac{S\Delta_P^{\text{50-strike}}}{3.7295} = -4.9953 \\[2mm]
\Omega_P^{\text{60-strike}} = \dfrac{S\Delta_P^{\text{60-strike}}}{P^{\text{60-strike}}} = \dfrac{S\Delta_P^{\text{60-strike}}}{9.5865} = -3.4267
\end{cases}
$$

$$
\Rightarrow \begin{cases}
S\Delta_P^{\text{50-strike}} = -18.6300 \\
S\Delta_P^{\text{60-strike}} = -32.8501
\end{cases}.
$$

An application of (6.3.7) results in

$$
\Omega_{\text{bull spread}} = \frac{S\Delta_{\text{bull spread}}}{P^{\text{50-strike}} - P^{\text{60-strike}}} = \frac{-18.6300 - (-32.8501)}{3.7295 - 9.5865} = \boxed{-2.4279}.
$$

□

Remark. While a put bull spread has the same delta as a call bull spread (see Example 6.3.5), they do not share the same price. Accordingly, their elasticities also differ.

Volatility of an option in terms of its elasticity

There are other practically useful and theoretically appealing results surrounding the elasticity of an option. For example, in the Black-Scholes framework the volatility of the *option* and the volatility of the underlying stock are related via[ix]

$$\boxed{\sigma_{\text{option}} = |\Omega|\sigma_{\text{stock}}.}$$

(6.3.9)

In this respect, the elasticity serves as an amplifier that is applied to the stock volatility to form the option volatility. However, in contrast to the stock volatility, which is a constant, the volatility of the option, as with elasticity, generally varies with the current stock price and the remaining time to expiration. To be mathematically precise, we should state (6.3.9) as

$$\sigma_{\text{option}}(s, t) = |\Omega(s, t)|\sigma_{\text{stock}}$$

to emphasize the dependence of $\Omega(\cdot, \cdot)$ and $\sigma_{\text{option}}(\cdot, \cdot)$ on the current stock price s and current time t.

[ix]This can be shown by the Itô's Lemma, which is beyond the scope of this book.

6.4   Problems

The Black-Scholes formula for stocks paying continuous proportional dividends

Problem 6.4.1. (Based on SOA Exam MFE Spring 2009 Question 4: Direct calculation of put prices) Your company has just written a one-year European put option on an equity index fund.

The equity index fund is currently trading at 1000. It pays dividends continuously at a rate proportional to its price; the dividend yield is 2%. It has a volatility of 20%.

The strike price of the put option is set in order to insure against a reduction of more than 40% in the value of the equity index fund at the end of one year.

The continuously compounded risk-free interest rate is 2.5%.

Using the Black-Scholes model, determine the price of the put option.

Problem 6.4.2. (Given the price of an ATM option, deduce the current stock price) Assume the Black-Scholes framework. For an at-the-money, 8-month European put option on a stock, you are given:

(i) The stock pays dividends continuously at a rate proportional to its price. The dividend yield is 2%.

(ii) The continuously compounded risk-free interest rate is 5%.

(iii) $\text{Var}[\ln S(t)] = 0.16t$ for all $t \geq 0$, where $S(t)$ is the time-t price of the stock.

(iv) The current price of this put option is 7.

Calculate the current stock price.

Problem 6.4.3. (Black-Scholes price of a strange claim) Assume the Black-Scholes framework. Consider a 3-year European contingent claim on a stock. For $t \geq 0$, let $S(t)$ be the time-t price of the stock.

You are given:

(i) $S(0) = 45$.

(ii) The stock's volatility is 20%.

(iii) The stock pays dividends continuously at a rate proportional to its price. The dividend yield is 3%.

(iv) The continuously compounded risk-free interest rate is 6%.

(v) The 3-year payoff of the contingent claim is

$$\text{Payoff} = \max\{(S(3) - 45)_+, 2(S(3) - 60)\}.$$

Calculate the current price of the contingent claim.

(Hint: You may sketch the claim's payoff diagram, from which you may deduce that the claim equals the sum of two calls with different strike prices.)

Problem 6.4.4. (Given the true exercise probability of a put, find its price)
Assume the Black-Scholes framework. You are given:

(i) The current price of a stock is 80.

(ii) The stock's volatility is 25%.

(iii) The stock pays dividends continuously at a rate proportional to its price.

(iv) The continuously compounded risk-free interest rate is 5%.

(v) The continuously compounded expected rate of return on the stock, α, is 8%.

(vi) The *true* probability that a 1-year at-the-money European put option on the stock will
be exercised is 0.4681.

Calculate the current price of the put option in (vi).

Applying the Black-Scholes formula to other underlying assets

Problem 6.4.5. (Stock paying discrete dividends – I) Consider a stock with current
price $50. You are given:

(i) There will be only one dividend; $2 will be paid in three months.

(ii) $\sigma = 0.30$.

(iii) The continuously compounded risk-free interest rate is 5%.

Use the Black-Scholes methodology to price a nine-month at-the-money European put option on the stock. What is the meaning of σ in this context?

Problem 6.4.6. (Stock paying discrete dividends – II) Assume the Black-Scholes
framework. You are given:

(i) The current stock price is $82.

(ii) The stock's volatility is 30%.

(iii) The stock pays no dividends.

(iv) The continuously compounded risk-free interest rate is 8%.

Using the above information, you calculate the price of a 3-month 80-strike European call.
Immediately after your valuation, it is publicly announced that the stock will pay a
dividend of $6 in 1 month, and no other payouts over the life of the call. Using the Black-
Scholes methodology with the same volatility parameter of 30%, you recalculate the price
of the call.
Calculate the change in the price of the call.

Problem 6.4.7. (Valuing a bear spread in the presence of discrete dividends) Assume the Black-Scholes framework. For a 9-month 45-55 put bear spread on a stock, you are given:

(i) The current stock price is 50.

(ii) The only dividends during this time period are 2.50 to be paid in two months and five months.

(iii) $\mathrm{Var}[\ln F^P_{t,0.75}(S)] = 0.16t$ for $0 \le t \le 0.75$.

(iv) The continuously compounded risk-free interest rate is 8%.

Calculate the current price of the bear spread.

Problem 6.4.8. (Profit on a put in the presence of discrete dividends) Assume the Black-Scholes framework. For a 3-month at-the-money European put option on a stock, you are given:

(i) The stock is currently selling for 50.

(ii) The stock will pay a single dividend of 1.5 in two months.

(iii) $\mathrm{Var}[\ln F^P_{t,0.25}(S)] = 0.09t$, for $0 \le t \le 0.25$.

(iv) The continuously compounded risk-free interest rate is 10%.

The stock price when the put option expires is 45.
    Calculate the 3-month profit on the put option.

Problem 6.4.9. (Profit on a cap on a stock paying discrete dividends) Assume the Black-Scholes framework. Consider a 3-month European contingent claim on a stock. You are given:

(i) The stock is currently selling for 50.

(ii) The stock will pay a single dividend of 1.5 in two months.

(iii) $\mathrm{Var}[\ln F^P_{t,0.25}(S)] = 0.09t$, for $0 \le t \le 0.25$.

(iv) The continuously compounded risk-free interest rate is 10%.

(v) The 3-month payoff of the contingent claim is as follows:

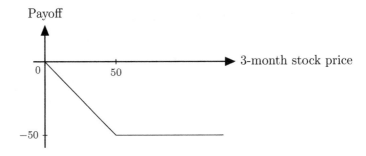

Calculate the profit on the contingent claim if the 3-month stock price is 45.

Problem 6.4.10. (How do a binomial price and a Black-Scholes price compare with each other?) You are given:

(i) The current dollar-euro exchange rate is 1.50$/€.

(ii) The volatility of the exchange rate is 20%.

(iii) The continuously compounded risk-free interest rate on dollars is 3%.

(iv) The continuously compounded risk-free interest rate on euros is 4%.

Consider a 6-month at-the-money dollar-denominated European put option on euros.

Actuary A values the put option using a binomial forward tree, where the length of each period is 3 months, to model the movements of the dollar-euro exchange rate. Actuary B values the same put option assuming the Black-Scholes framework.

Calculate the absolute value of the difference between the prices computed by Actuary A and Actuary B.

Problem 6.4.11. (Use of currency options in a daily life context – "rich" version) You have ordered a Rolls Royce car for the price of 200,000 British pounds, which you will pay when the car is delivered to you in three months. The current exchange rate is 1.60 US dollars per British pound, and your insanely rich mom will give you US $320,000 three months from now. Because the US dollar may lose value, you now buy appropriate 3-month at-the-money currency options of the European type to exactly cover the shortfall in case it occurs.

You are given:

(i) The continuously compounded risk-free interest rate in the United States is 1%.

(ii) The continuously compounded risk-free interest rate in the United Kingdom is 2%.

(iii) The volatility of the dollar/pound exchange rate is 20%.

Under the Black-Scholes framework, determine the total cost of the currency options in US dollars.

(Hint: Are you long or short with respect to the 3-month dollar/pound exchange rate? Are you worried about the exchange rate going up or down in three months? Should you set up a floor or a cap? Hopefully these questions will help you decide whether to buy a pound call or a pound put.)

Problem 6.4.12. (Use of currency options in a daily life context – "poor" version) To settle an urgent debt of US $300,000 payable in three months, you have decided to (reluctantly!) sell your favorite Rolls Royce car for the price of 200,000 British pounds, which you will receive when the car is delivered to the buyer in three months. Because the British pound may lose value relative to the US dollar, you decide to buy appropriate 3-month at-the-money European currency options now to exactly cover the shortfall in case it occurs.

You are given:

(i) The current dollar/pound exchange rate is 1.5.

(ii) The continuously compounded risk-free interest rate in the United States is 4%.

(iii) The continuously compounded risk-free interest rate in the United Kingdom is 8%.

(iv) The future exchange rates of dollar per pound are lognormally distributed with a volatility of 30%.

Calculate the total cost of the currency options in US dollars.

Problem 6.4.13. (Pricing and using a currency put) Assume the Black-Scholes framework. You are given:

(i) The current dollar/euro exchange rate is 1.50$/€.

(ii) The volatility of the exchange rate is 20%.

(iii) The continuously compounded risk-free interest rate on dollars is 4%.

(iv) The continuously compounded risk-free interest rate on euros is 5%.

Consider a 6-month at-the-money dollar-denominated European put option on euros.

(a) Calculate the price of the put option.

(b) Determine whether the euro put above can be used to hedge against exchange rate risk faced by each of the following individuals living in the United States.

(1) Apple (Ambrose's twin brother), a famous chef in Iowa City, regularly imports food raw materials from Europe, with the next order made in 6 months and settled in euros.

(2) Ambrosio (Ambrose's another twin brother), fed up with his newly purchased iPhone XX, has decided to sell it to his aunt living in Europe for €500 in 6 months.

(c) The continuously compounded expected rate of *appreciation* of the dollar/euro exchange rate is 1.5%.

Calculate the true probability that the put option will be exercised.

Problem 6.4.14. (Pricing a futures option – I) Assume the Black-Scholes framework. Consider a 95-strike 9-month European call option on an S&V 150 futures contract which matures one year from now. You are given:

(i) The current price of the S&V 150 index is 100.

(ii) The volatility of the S&V 150 index price is 30%.

(iii) The volatility of the futures price is 30%.

(iv) S&V 150 pays dividends continuously at a rate proportional to its price. The dividend yield is 3%.

(v) The continuously compounded risk-free interest rate is 8%.

Calculate the price of the call option.

(Note: You may assume the fact that when the interest rate is constant, futures prices and forward prices agree.)

Problem 6.4.15. (Pricing a futures option – II) Assume the Black-Scholes framework. You are given:

(i) The current price of the P&K 777 index is 500.

(ii) The P&K 777 index pays dividends continuously at a rate proportional to its price. The dividend yield is 2%.

(iii) The continuously compounded risk-free interest rate is 6%.

(iv) The current prices and volatility of futures contracts on P&K 777 of various maturities:

Maturity (in Years)	1	2	3	4
Current Price	520.41	541.64	563.75	586.76
Volatility	30%			

Calculate the price of a 2-year 550-strike European put option on a 1-year futures contract (i.e., the futures matures at the end of 3 years).

Option Greeks

Problem 6.4.16. (What can you say given the price and delta of a put?) Assume the Black-Scholes framework. The current price of a nondividend-paying stock is $60 and the continuously compounded risk-free interest rate is 5%.

Your boss has asked you to quote a price for Put A, which is a 6-month at-the-money European put option on the stock. Although the market price for Put A is not available, the market price of Put B, which is a 6-month 65-strike European put option on the same stock, is observed to be $6.2514, and its delta is -0.5882.

Calculate the price of Put A.

Problem 6.4.17. (Gamma of a futures option) Assume the Black-Scholes framework. Consider a 6-month 90-strike European put option on a futures contract. You are given:

(i) The price of the underlying futures contract is 95.

(ii) The delta of the put option is -0.3382.

(iii) The continuously compounded risk-free interest rate is 3%.

(iv) The volatility of the futures is less than 50%.

Calculate the gamma of the futures put option.

Problem 6.4.18. (Signs of option Greeks) Assume the Black-Scholes framework. Consider European call and put options on a stock that pays dividends continuously at a rate proportional to its price.

Determine the signs of the following twelve Greeks:

Call delta, call gamma, call theta, call vega, call rho, call psi,

put delta, put gamma, put theta, put vega, put rho, and put psi.

That is, for each of these Greeks, answer "Always positive," "Always negative," or "Sometimes positive and sometimes negative." Explain your answers briefly.

Problem 6.4.19. (Graphical question) Assume the Black-Scholes framework. Which of the following graphs best represents the relationship between the delta of a deep out-of-the-money European call option on a nondividend-paying stock and the time to maturity T?

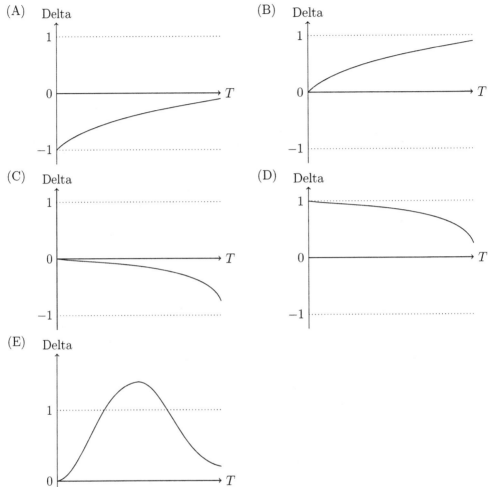

Problem 6.4.20. (Limiting behavior of gamma) Determine the limiting value of gamma as we approach maturity (i.e., find $\lim_{T \to 0} \Gamma$) for each of the following options:

(a) A deep in-the-money European call option

(b) A deep out-of-the-money European call option

(c) An at-the-money European call option

For each option, provide an intuitive explanation and a mathematical explanation.

(Note: It may not be easy to examine the global monotonicity of gamma as a function of the time to maturity. Determining the limiting values of gamma is not as hard.)

Problem 6.4.21. (Relationship between the option price and the strike price)

(a) If the strike price of a European call option decreases, then how will the price of the call change, holding everything else constant?

(b) Verify your conclusion in part (a) by general reasoning.

(c) Verify your conclusion in part (a) by the Black-Scholes formula.

Problem 6.4.22. (Greeks of a straddle) Assume the Black-Scholes framework. Consider a long K-strike European straddle.

For each of delta and gamma, discuss intuitively how the option Greek of the straddle varies with S, the current stock price.

(Hint: Pay special attention to how the characteristics of the call and put Greeks combine to form the overall picture.)

Problem 6.4.23. (Extension of Example 6.3.5: Gamma of a bull spread) You compute the current gamma for a 50-60 bull spread with the following information:

(i) The continuously compounded risk-free interest rate is 5%.

(ii) The underlying stock pays no dividends.

(iii) The current stock price is $50 per share.

(iv) The stock's volatility is 20%.

(v) The time to expiration is 3 months.

Calculate the change in gamma after 1 month, if the stock price does not change.

Problem 6.4.24. (Calculations of Greeks and elasticity for currency options) Assume the Black-Scholes framework. You are given:

(i) The current dollar/euro exchange rate is 1.2.

(ii) The continuously compounded risk-free interest rate in the United States is 2%.

(iii) The continuously compounded risk-free interest rate in Europe is 3%.

(iv) The volatility of the dollar/euro exchange rate is 15%.

For a dollar-denominated nine-month at-the-money European call option on euro, calculate and interpret the values of:

(a) Delta and gamma

(b) Vega, theta, and rho

(c) Elasticity

Problem 6.4.25. (Direct calculation of elasticity) Assume the Black-Scholes framework. You are given:

(i) The current stock price is 82.

(ii) The stock's volatility is 30%.

(iii) The stock pays dividends continuously at a rate proportional to its price. The dividend yield is 3%.

(iv) The continuously compounded risk-free interest rate is 8%.

Calculate the elasticity of a 3-month 80-strike European call option.

Problem 6.4.26. (Given elasticity, what is σ?) Assume the Black-Scholes framework. Consider a one-year at-the-money European put option on a nondividend-paying stock.
    You are given:

(i) The ratio of the current put option price to the current stock price is 0.073445.

(ii) The current put-option elasticity is -5.941861.

(iii) The continuously compounded risk-free interest rate is 1.2%.

Determine the stock's volatility.

Problem 6.4.27. (Elasticity of a straddle) Assume the Black-Scholes framework. For a 3-month 32-strike European straddle on a stock, you are given:

(i) The stock currently sells for $30.

(ii) The stock's volatility is 30%.

(iii) The stock pays dividends continuously at a rate proportional to its price. The dividend yield is 2%.

(iv) The continuously compounded risk-free interest rate is 5%.

Calculate the current elasticity of the straddle.

Problem 6.4.28. (Given the constituent prices, deltas, and elasticities, find the portfolio elasticity) Assume the Black-Scholes framework. Consider a portfolio consisting of three European options, X, Y, and Z, on the same stock. You are given:

	X	Y	Z
Option price	6.8268	?	1.9299
Option delta	?	−0.4269	0.3537
Option elasticity	5.6496	−6.8755	9.1627

Calculate the elasticity of the portfolio.

Problem 6.4.29. (Given the gamma of a derivative, find its elasticity) Assume the Black-Scholes framework. For $t \geq 0$, let $S(t)$ be the time-t price of a stock.

Consider a 9-month European contingent claim on the stock. You are given:

(i) The stock's volatility is 35%.

(ii) The stock pays dividends continuously at a rate proportional to its price. The dividend yield is 2%.

(iii) The continuously compounded risk-free interest rate is 6%.

(iv) The 9-month payoff of the contingent claim is as follows:

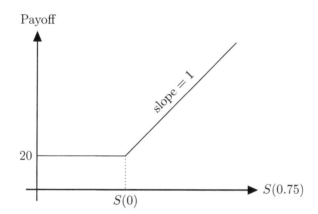

(v) The current gamma of the contingent claim is 0.0314.

Calculate the time-0 contingent-claim elasticity.

Problem 6.4.30. (Option volatility) Assume the Black-Scholes framework. For $t \geq 0$, let $S(t)$ be the time-t price of a nondividend-paying stock. You are given:

(i) $S(0) = 100$.

(ii) $\text{Var}[\ln S(t)] = 0.16t$, for $t \geq 0$.

(iii) The continuously compounded risk-free interest rate is 6%.

Calculate the current volatility of a 105-strike 1-year European call option.

7

Option Greeks and Risk Management

Chapter overview: On the basis of the Black-Scholes pricing formula and option Greeks studied in the previous chapter, we now examine the risk management aspect of the Black-Scholes framework and migrate from the *measurement* to the *management* of risks inherent in options. The central question we address is: How can we implement a hedging strategy using the underlying stock and related options to reduce, if not eliminate, the stock price risk we are exposed to? This problem, which practitioners confront on a day-to-day basis, is addressed in Section 7.1, where the important technique of delta-hedging is formally introduced. Attention is paid to the motivation behind and mechanics of delta-hedging and the calculation of the holding profit of a delta-hedged portfolio. Section 7.2 extends the discussion in Section 7.1 to hedging multiple Greeks, most notably delta and gamma, and demonstrates how delta-gamma-hedging remedies the deficiencies of delta-hedging. Section 7.3 presents a further application of option Greeks to the approximation of option prices, leading naturally to the development of the celebrated Black-Scholes equation (not formula!). Intriguingly, delta-hedging, besides being a widely used hedging strategy in practice, gives rise to new results concerning the pricing of options in theory.

7.1   Delta-hedging

In this chapter, we expand the notation of delta from Δ to $\Delta(S(t), t)$ to indicate its dependence on the current time t and the current stock price $S(t)$. When only one of the two arguments is of interest, the other argument will be suppressed. The same applies to other option Greeks. At minimal technical cost, such a notational expansion is desirable because it emphasizes the dynamic nature of risk management—things change as life moves on!

Why delta-hedge?

To motivate the concept of delta-hedging, let's put ourselves in the shoes of an option trader who has *sold* a European call option on a nondividend-paying stock and has therefore incurred a delta of $-\Delta_C(0)$, where $\Delta_C(0)$ is the time-0 delta of the call option. Because the trader's delta is negative, he is exposed to upside stock price risk, i.e., the risk that the stock price rises in the future. To protect himself against a decline in the option value should the stock price increase, the trader can buy $\Delta_C(0)$ shares[i] of the stock to *delta-hedge* his position. Because the Greek of a portfolio is given by the sum of the Greeks of its components (see (6.3.5) in Subsection 6.3.3), the delta of the trader's delta-hedged portfolio is

$$\underbrace{-\Delta_C(0)}_{\text{short call}} + \underbrace{\Delta_C(0)}_{\text{long stock}} = 0,$$

---

[i]We assume that trading fractional units of the stock is allowed in the market.

meaning that the investor is *locally* protected against stock price risk. The trader's portfolio is said to be *delta-neutral* because its delta is neutralized to exactly zero. In general, *delta-hedging* entails entering into a set of offsetting transactions (such as buying $\Delta_C(0)$ shares in the above example) in such a way that your overall delta is zero.

Example 7.1.1. (SOA Exam FETE Fall 2008 Question 10: Delta-hedging for currency options) Your company has just sold a European put option on 10,000 USD for a premium paid in Japanese yen.

You are given the following:

- The time to maturity $T = 180$ days

- Yen risk-free interest rate $r_j = 0.4988\%$ per annum, continuously compounded

- USD risk-free interest rate $r_u = 4.97\%$ per annum, continuously compounded

- The current exchange rate is $Q = 120$ yen/USD

- The strike price of the option is 117 yen/USD

- The volatility of the exchange rate is 10%

- Assume there are 365 days in a year

(a) Calculate the option premium received, based on the Black-Scholes currency option formula.

Your company plans to delta-hedge the short put position using the option premium received.

A hedge portfolio will be purchased such that its value in Yen will approximately equal the value of the put option at each point in time.

To hedge the put option, you buy Y units of USD and invest the balance of the portfolio at Yen money market rate.

(c) Calculate Y using the Black-Scholes currency option formula.

(Note: Parts (b) and (d) are not included here.)

Solution. (a) Note that the foreign currency in this example is USD while the domestic currency is Japanese Yen. With

$$d_1 = \frac{\ln(120/117) + (0.4988\% - 4.97\% + 0.1^2/2)(180/365)}{0.1\sqrt{180/365}} = 0.08165,$$

$$d_2 = d_1 - 0.1\sqrt{180/365} = 0.01143,$$
$$N(-d_1) = 0.46746,$$
$$N(-d_2) = 0.49544,$$

the price of the currency put is

$$10,000[117e^{-0.4988\%(180/365)}(0.49544) - 120e^{-4.97\%(180/365)}(0.46746)] = \boxed{30,870}.$$

(c) The delta of each currency put option is (Note: Don't omit the negative sign!)

$$\Delta_P = -e^{-4.97\%(180/365)}N(-d_1) = -0.45614.$$

Shorting 10,000 of these put options means that the delta of the company is

$$-10,000(-0.45614) = 4,561,$$

which suggests short selling 4,561 units of USD, i.e., $\boxed{Y = -4,501}$.

□

Calculations of holding profits.

Our next task is to investigate the profit at any future time t before expiration of a trader who has delta-hedged his portfolio. Our analysis holds true not only for a call, but also for a general European derivative V. Recall that we are assuming no dividends are payable over the life of the option.

- *At time 0:* Suppose that the trader sells the derivative for an initial income of $V(0)$ and buys $\Delta(0)$ shares of the stock for the purpose of delta-hedging for a cost of $\Delta(0)S(0)$. The net investment made at time 0 equals

$$\text{Time-0 investment} = \Delta(0)S(0) - V(0).$$

- *At time t:* To close his/her positions, the trader should sell the $\Delta(0)$ shares of the stock he/she is holding for a cash inflow of $\Delta(0)S(t)$ and cover the short position in the derivative by buying it back at the time-t price of $V(t)$. Overall, the time-t payoff is (recall the definition of a payoff in Chapter 1)

$$\text{Time-}t \text{ payoff} = \Delta(0)S(t) - V(t).$$

As the profit is defined simply as the payoff less the *future value* of the previous investments, the holding profit of the trader is given by

$$\text{Holding profit} = \underbrace{[\Delta(0)S(t) - V(t)]}_{\text{time-}t \text{ payoff}} - e^{rt}\underbrace{[\Delta(0)S(0) - V(0)]}_{\text{time-0 investment}}. \quad (7.1.1)$$

When t is 1 day (i.e., $t = 1/365$), the holding profit is referred to as the *overnight profit*—you are computing the profit "overnight."

Note that:

- The same delta of $\Delta(0)$ is used in both the time-t payoff term and the time-0 investment term. Do not use (and there is no need to calculate!) the time-t delta $\Delta(t)$ for the payoff term!

- To calculate the holding profit, you need the option price at time 0 (i.e., $V(0)$), and that at the new time t (i.e., $V(t)$). They require two rounds of Black-Scholes calculations.

Example 7.1.2. (SOA Exam IFM Advanced Derivatives Sample Question 47: Calculation of holding profit) Several months ago, an investor sold 100 units of a one-year European call option on a nondividend-paying stock. She immediately delta-hedged the commitment with shares of the stock, but has not ever re-balanced her portfolio. She now decides to close out all positions.

You are given the following information:

(i) The risk-free interest rate is constant.

(ii)

	Several months ago	Now
Stock price	$40.00	$50.00
Call option price	$8.88	$14.42
Put option price	$1.63	$0.26
Call option delta	0.794	

The put option in the table above is a European option on the same stock and with the same strike price and expiration date as the call option.

Calculate her profit.

(A) $11

(B) $24

(C) $126

(D) $217

(E) $240

Solution. Denote the date several months ago by time 0 and the current date by time t. Through delta-hedging, the net investment of the investor at time 0 was

$$100[\Delta(0)S(0) - C(0)] = 100[0.794(40) - 8.88] = 2,288.$$

After closing out all positions at time t, her profit will be

$$
\begin{aligned}
100[\Delta(0)S(t) - C(t)] - 2,288e^{rt} &= 100[0.794(50) - 14.42] - 2,288e^{rt} \\
&= 2,528 - 2,288e^{rt}.
\end{aligned}
$$

To find the accumulation factor e^{rt}, put-call parity can be applied to the two pairs of call and put prices:

$$
\begin{cases} C(0) - P(0) = S(0) - Ke^{-rT} \\ C(t) - P(t) = S(t) - Ke^{-r(T-t)} \end{cases}
\Rightarrow
\begin{cases} 8.88 - 1.63 = 40 - Ke^{-rT} \\ 14.42 - 0.26 = 50 - Ke^{-r(T-t)} \end{cases}
$$

$$\Rightarrow \quad e^{rt} = 1.094351145.$$

It follows that the profit of the investor is $2,528 - 2,288(1.094351145) = \boxed{24.12}$.
(Answer: (B))   □

Remark. The problem can still be solved if the short-rate is a deterministic function $r(\cdot)$. Then, the accumulation factor e^{rt} is replaced by $\exp\left[\int_0^t r(s)\,ds\right]$, which can be determined using the put-call parity formulas

$$
\begin{cases}
C(0) - P(0) = S(0) - K\exp\left[-\int_0^T r(s)\,ds\right] \\[2ex]
C(t) - P(t) = S(t) - K\exp\left[-\int_t^T r(s)\,ds\right]
\end{cases}.
$$

If interest rates are stochastic, the problem as stated cannot be solved.

A cautionary note is in order. (7.1.1) and (7.1.2) are developed under the assumption that the trader *sells* the derivative and the stock pays no dividends. If you keep the definitions of payoff and profit in mind, you should not have difficulties in developing the holding profit formula when an investor *buys* a derivative and delta-hedges his/her position in the presence of dividends.

Example 7.1.3. (Calculation of the holding profit for a long put position)
Assume the Black-Scholes framework.
   One month ago, Tidy *bought* 1,000 units of a 9-month 60-strike European put option on a stock. He immediately delta-hedged the commitment with shares of the stock, but has not ever re-balanced his portfolio. He now decides to close out all positions.
   You are given:

(i) The stock pays dividends continuously at a rate proportional to its price. The dividend yield is 4%.

(ii) The stock's volatility is 45%.

(iii) The continuously compounded risk-free interest rate is 10%.

(iv) The stock price one month ago was $60.

(v) The current stock price is $65.

Calculate Tidy's profit.

Ambrose's comments:

In this example, we deal with a long position in a put. Also, the stock pays dividends, which are (assumed to be) reinvested, leading to a change in the number of shares in the future.

Solution. We need the put price one month ago and the current put price.
   One month ago, we have

$$
\begin{aligned}
d_1 &= \frac{\ln(60/60) + (0.1 - 0.04 + 0.45^2/2)(0.75)}{0.45\sqrt{0.75}} = 0.31033, \\
d_2 &= d_1 - 0.45\sqrt{0.75} = -0.07939, \\
N(-d_1) &= 0.37816, \\
N(-d_2) &= 0.53164,
\end{aligned}
$$

so the put price was

$$P(0) = 60e^{-0.1(0.75)}(0.53164) - 60e^{-0.04(0.75)}(0.37816) = 7.57451.$$

The put delta was

$$\Delta_P(0) = \underbrace{-e^{-0.04(0.75)}}_{\text{Don't miss this!}}(0.37816) = -0.36698.$$

To delta-hedge his position, the Tidy should *buy* $100(0.36698) = 36.698$ shares to hedge his *long* position in the puts.

Currently, the put has 8 months to live before it expires. With

$$d_1 = \frac{\ln(65/60) + (0.1 - 0.04 + 0.45^2/2)(2/3)}{0.45\sqrt{2/3}} = 0.51043,$$

$$d_2 = d_1 - 0.45\sqrt{2/3} = 0.14300,$$

$$N(-d_1) = 0.30488,$$

$$N(-d_2) = 0.44315,$$

the current put price is

$$P(1/12) = 60e^{-0.1(2/3)}(0.44315) - 65e^{-0.04(2/3)}(0.30488) = 5.57847.$$

Finally, Tidy's profit is given by

$$
\begin{aligned}
\text{Holding profit} &= \text{Payoff} - \text{FV(Investment)} \\
&= 1{,}000\{[-\Delta_P(0)\boxed{e^{\delta t}}S(1/12) + P(1/12)] \\
&\quad - [-\Delta_P(0)S(0) + P(0)]e^{rt}\} \\
&= 1{,}000\{[0.36698e^{0.04/12}(65) + 5.57847] \\
&\quad - [0.36698(60) + 7.57451]e^{0.1/12}\} \\
&= \boxed{-329}.
\end{aligned}
$$

$\square$

Remark. (i) Note that Tidy bought the put option as well as the stock for delta-hedging, incurring a rather high initial investment.

(ii) Because of the reinvestment of dividends, 36.698 shares bought one month ago grow to $36.698e^{0.04/12}$ shares currently. .

In practice, a marker-maker engaged in delta-hedging needs to rebalance his/her portfolio periodically in response to changes in the value of delta as time elapses and the price of the underlying stock changes. Such a multi-period delta-hedging strategy can be decomposed into a series of one-period delta-hedging strategies, each of which can be analyzed by techniques introduced thus far and evaluated by (7.1.1). Here is an illustrative example.

Example 7.1.4. (CAS Exam 8 Spring 2005 Question 36: Two-period delta-hedging) Assume you have purchased European put options for 100,000 shares of a nondividend-paying stock and you are given the following information.

- Price of stock = $49.16

- Strike price = $50.00

- Continuously compounded risk-free interest rate = 5% per annum

- Volatility = 20% per annum

- There are 20 weeks remaining until maturity.

(a) Determine the initial position you should take in the underlying stock to implement a delta hedging strategy.

(b) You now have the following information.

T (weeks)	Stock Price	d_1
1	$49.33	0.10
2	$49.09	0.05

You decide to readjust the delta hedging strategy on a weekly basis.

Calculate the cumulative cost, including interest, of the hedge at the end of week 2.

Solution. (a) Since

$$d_1 = \frac{\ln(49.16/50) + (0.05 + 0.2^2/2)(20/52)}{0.2\sqrt{20/52}} = 0.08046,$$

$$N(-d_1) = 0.46794,$$

the delta of each put option is $\Delta_P = -N(-d_1) = -0.46794$. In total, we need to $\boxed{\text{buy}}$ $100,000(0.46794) = \boxed{46,794}$ shares of the stock to implement a delta-hedging strategy.

(b)
- Initially, the delta-hedging strategy costs $46,794(49.16) = 2,300,393.04$.
- At the end of week 1, the delta becomes $\Delta_P = -N(-d_1) = -0.46017$, so to remain delta-neutral, we need to hold $100,000(0.46017) = 46,017$ shares in the portfolio. Thus we sell $46,794 - 46,017 = 777$ shares for an income of $777(49.33) = 38,329.41$.
- At the end of week 2, the delta becomes $\Delta_P = -N(-d_1) = -0.48006$, so we need to hold $100,000(0.48006) = 48,006$ shares in the portfolio. Thus we buy $48,006 - 46,017 = 1,989$ shares for a cost of $1989(49.09) = 97,640.01$.

After taking time value of money into account, the cumulative cost of the hedge at the end of week 2 is

$$2,300,393.04e^{0.05(2/52)} - 38,329.41e^{0.05(1/52)} + 97,640.01 = \boxed{2,364,095}.$$

□

An alternative view on the holding profit formula.

The formula of the delta-hedged trader's holding profit as given in (7.1.1) can be recast in an alternative way to cast light on the factors contributing to the holding profit. This can be achieved by rearranging (7.1.1) as (assuming no dividends)

$$
\begin{aligned}
\text{Profit} \ &= \ [\Delta(0)S(t) - V(t)] - e^{rt}[\Delta(0)S(0) - V(0)] \\
&= \ [\Delta(0)S(t) - V(t)] - [\Delta(0)S(0) - V(0)] - (e^{rt} - 1)[\Delta(0)S(0) - V(0)] \\
&= \ \underbrace{\Delta(0)[S(t) - S(0)]}_{(1)} - \underbrace{[V(t) - V(0)]}_{(2)} - \underbrace{(e^{rt} - 1)[\Delta(0)S(0) - V(0)]}_{(3)}. \quad (7.1.2)
\end{aligned}
$$

This formula decomposes the profit into three components:

(1) Gain on the shares (*not* taking time value of money into account)

(2) Gain on the derivative (again, *not* taking time value of money into account; also, recall that the trader is short with respect to the derivative, hence the negative sign before (2))

(3) Interest expense on the initial investment

The difference $(1) - (2)$ is sometimes called the *capital gain*.

Which holding profit formula to use, (7.1.1) or (7.1.2)?

As a matter of fact, (7.1.2) is mainly of theoretical significance, and you will mostly use (7.1.1) for practical computations for two reasons:

- The derivations of (7.1.2) are unnecessarily complicated and rather difficult to follow, whereas the ideas underlying (7.1.1) are not only substantially more transparent, but also consistent with the usual definitions of payoff and profit. When asked to calculate the holding profit in a specific problem, you need not dwell on (7.1.2). Rather, simply apply (7.1.1), which is much more convenient, taking care of dividends if necessary.

- Apart from its relative simplicity, an additional merit of (7.1.1) compared to (7.1.2) is its wider applicability. Not only does the rationale underlying (7.1.2) hold true when the trader employs no hedging (see Example 7.1.5 below), it can also be applied to more complex hedging strategies such as delta-gamma-hedging, while the analogs of (7.1.2) in these cases are not immediately clear.

Example 7.1.5. (SOA Exam MFE Spring 2009 Question 13: Holding profit of a plain long call position) Assume the Black-Scholes framework.

Eight months ago, an investor borrowed money at the risk-free interest rate to purchase a one-year 75-strike European call option on a nondividend-paying stock. At that time, the price of the call option was 8.

Today, the stock price is 85. The investor decides to close out all positions.

You are given:

(i) The continuously compounded risk-free interest rate is 5%.

(ii) The stock's volatility is 26%.

Calculate the eight-month holding profit.

(A) 4.06

(B) 4.20

(C) 4.27

(D) 4.33

(E) 4.47

Solution. Eight months after the purchase of the call, the remaining time to expiration is 4 months. With

$$d_1 = \frac{\ln(85/75) + (0.05 + 0.26^2/2)(4/12)}{0.26\sqrt{4/12}} = 1.01989,$$

$$d_2 = d_1 - 0.26\sqrt{4/12} = 0.86978,$$

$$N(d_1) = 0.84611,$$

$$N(d_2) = 0.80779,$$

the current call price is

$$C = 85(0.84611) - 75e^{-0.05(4/12)}(0.80779) = 12.33647.$$

The 8-month holding profit is today's call price (payoff) less the future value of the old call price (initial investment), or $12.33647 - 8e^{0.05(8/12)} = \boxed{4.0653}$. (**Answer: (A)**) □

Remark. Because we are not doing delta-hedging in this example, there is no need to know the stock price eight months ago.

Example 7.1.6. (CAS Exam 3 Spring 2007 Question 32: Given the change in option price) A market-maker has sold 100 call options, each covering 100 shares of a dividend-paying stock, and has delta-hedged by purchasing the underlying stock.
    You are given the following information about the market-maker's investment:

- The current stock price is $40.

- The continuously compounded risk-free rate is 9%.

- The continuous dividend yield of the stock is 7%.

- The time to expiration of the options is 12 months.

- $N(d_1) = 0.5793$

- $N(d_2) = 0.5000$

The price of the stock quickly jumps to $41 before the market-maker can react. This change causes the price of one call option to increase by $56.08. Calculate the net profit on the market-maker's investment associated with this price move.

(A) Less than −$1,600

(B) At least $-\$1,600$, but less than $-\$800$

(C) At least $-\$800$, but less than $\$0$

(D) At least $\$0$, but less than $\$800$

(E) At least $\$800$

Solution. Because $\Delta(0) = e^{-\delta T} N(d_1) = e^{-0.07(1)}(0.5793) = 0.540136$ for each call, $0.540136(100)(100) = 5,401.36$ shares of the stock are bought at time 0. If we ignore the interest expense because everything happens instantaneously, then the overnight profit, by (7.1.1) (or (7.1.2)), is

$$5,401.36(41 - 40) - 56.08(100) = \boxed{-206.64}. \quad \textbf{(Answer: (C))}$$

□

Profit of an unhedged position vs profit of a delta-hedged position.

We close this section by turning the algebraic results we have developed thus far for delta-hedging into a diagram that visualizes the financial effects of delta-hedging. Specifically, we contrast the profit of an unhedged written option with the profit of a delta-hedged written option for a range of values of stock prices in an attempt to show not only the superiority of delta-hedging, but also its potential deficiencies. For concreteness, we consider a written European call option with the following parameters:

$$S(0) = K = 50, \quad r = 0.08, \quad \delta = 0, \quad \sigma = 0.25, \quad T = 1.$$

The time-0 delta of the call is $\Delta_C(50, 0) = 0.67184$, which means that a seller of the call should buy 0.67184 shares of the stock for the purpose of delta-hedging.

In Figure 7.1.1, we assume that the stock price changes from its initial value of 50 to any value between $\$0$ to $\$100$ in one day, say $S(1/365)$, and plot the overnight profit of the unhedged written call, equal to the future value of initial call price less the new call price evaluated at $S(1/365)$, as well as the overnight profit of the delta-hedged written call (i.e., the written call coupled with 0.67184 shares of the stock) computed according to (7.1.1) with $t = 1/365$ (i.e., 1 day). That the change of the stock price from $S(0) = 50$ to $S(1/365)$ occurs in one day allows us to concentrate mostly on the effect of stock price changes alone on the holding profit of the two positions. It can be seen that as the call becomes more in-the-money, the profit of the (unhedged) written call, which appears as a smoothed version of the short call payoff, decreases inexorably, seriously jeopardizing the financial well-being of the call seller. This corroborates our concern that the call seller is exposed to significant upside stock price risk and justifies the need for delta-hedging. In contrast, the profit of the delta-hedged written call, albeit still exhibiting a decreasing trend as the stock price rises above the initial price of 50, declines at a much more controllable rate. Meanwhile, when the call becomes more out-of-the-money, the profit of the delta-hedged written call decreases quite sharply, due to the shrinking value of the stock holding, and is much below the profit of the unhedged written call. This is another manifestation of the phenomenon we first observed in Subsection 3.1.4 that no position can always outperform another in terms of profit, in the absence of arbitrage opportunities. Parenthetically, although it is not clearly evident in Figure 7.1.1, the profit of the delta-hedged written call is positive when the 1-day stock price is close to 50.

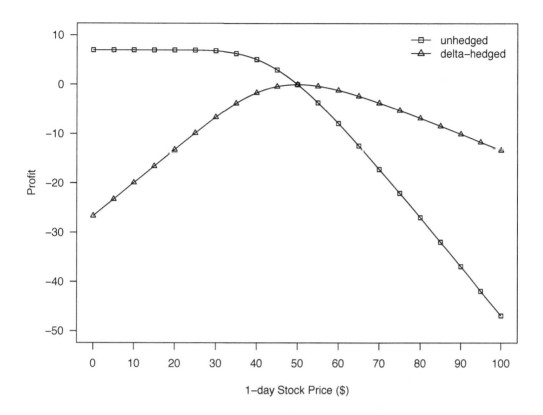

FIGURE 7.1.1

Comparison of the overnight profit of an unhedged written call and the overnight profit of a delta-hedged written call.

You may wonder:

> Shouldn't the delta-hedged written call, with a zero delta, be shielded from the stock price risk? Why does the profit of the delta-hedged written call still vary quite vigorously with the current stock price?

The sad truth is that the delta-hedged written call is only *locally* protected against stock price movements—the tangent to the profit function of the delta-hedged written call at $S(1/365) = 50$ is almost[ii] horizontal. As the stock price changes, however, so does delta itself (recall from Subsection 6.3.1 that the delta of a call increases with the current stock price, or equivalently, the gamma of a call is positive). When $S(1/365) > 50$ (resp. $S(1/365) \leq 50$), having $\Delta_C(50, 0) = 0.67184$ shares at hand is no longer sufficient (resp. more than sufficient) to offset the new delta of the call. In fact, the overall delta of the delta-hedged written is, ignoring the passage of one day,

$$-\Delta_C(S(1/365), 1/365) + \Delta_C(S(0), 0) \begin{cases} < 0, & \text{if } S(1/365) < S(0), \\ = 0, & \text{if } S(1/365) = S(0), \\ > 0, & \text{if } S(1/365) > S(0), \end{cases}$$

[ii]The tangent is not exactly horizontal because one day has passed, i.e., $\Delta_C(50, 1/365) \neq \Delta_C(50, 0)$.

which explains the downward-opening shape of the profit of the delta-hedged written call in Figure 7.1.1.

The above discussions suggest that a delta-hedged written call suffers from huge losses owing to large stock price moves in either direction. These undesirable characteristics of the delta-hedged written call motivate us to bring in additional option Greeks to our portfolio in an effort to hedge against the stock price risk more effectively. This will be the subject of the next section.

7.2   Hedging Multiple Greeks

Why hedge multiple Greeks?

We have illustrated in Figure 7.1.1 that a delta-hedged portfolio does not eliminate risk and can sustain large losses in the case of large stock price moves. In fact, the susceptibility to changes in the stock price is expected from the negativity of the *gamma* of the delta-hedged written call. Its delta is decreasing in $S(1/365)$, being strictly positive when $S(1/365) < 50$, (almost) zero at $S(1/365) = 50$, and strictly negative when $S(1/365) > 50$. This suggests neutralizing not only the delta of a portfolio, but also its gamma, to eliminate any systematic behavior in the portfolio delta and to further control the stock price risk. Because the gamma of the underlying stock is always zero ($\partial^2 S/\partial S^2 = 0$), a *delta-gamma-neutral* portfolio cannot be constructed with the stock alone. Other financial instruments such as options need to be traded to offset the gamma of the prior position.

Furthermore, delta and gamma are measures of the sensitivity of the option price to only one source of uncertainty, namely the stock price. A delta-gamma-hedged position can remain vulnerable to many external factors, including but not limited to stock price volatility, changes in interest rates, the passage of time, etc. To formulate a prudent risk management system, it therefore makes sense to hedge several option Greeks that are of most practical concern simultaneously.

Hedging multiple Greeks: Solving a system of linear equations.

Consider a given position which is to be hedged with respect to m option Greeks. Hedging these m Greeks in general requires m additional financial instruments, the number of units to buy or sell in each of which is to be determined. Since the Greek of a portfolio is simply the sum of the individual Greeks (recall (6.3.7) on page 217), zeroing the m Greeks boils down to solving a system of m simultaneous linear equations, one for each Greek, for the compositions in the m instruments. In a conceivable pen-and-paper problem, you are most likely asked to neutralize $m = 2$ option Greeks via solving two linear equations.

Example 7.2.1. (SOA Exam MFE Spring 2007 Question 10: Delta-gamma-hedging I) For two European call options, Call-I and Call-II, on a stock, you are given:

Greek	Call-I	Call-II
Delta	0.5825	0.7773
Gamma	0.0651	0.0746
Vega	0.0781	0.0596

Suppose you just sold 1000 units of Call-I.

Determine the numbers of units of Call-II and stock you should buy or sell in order to both delta-hedge and gamma-hedge your position in Call-I.

(A) Buy 95.8 units of stock and sell 872.7 units of Call-II

(B) Sell 95.8 units of stock and buy 872.7 units of Call-II

(C) Buy 793.1 units of stock and sell 692.2 units of Call-II

(D) Sell 793.1 units of stock and buy 692.2 units of Call-II

(E) Sell 11.2 units of stock and buy 763.9 units of Call-II

Ambrose's comments:

Be sure to adopt a consistent sign convention. If you buy (resp. sell) an asset, the contribution of that asset to the overall portfolio Greek will be positive (resp. negative) of the constituent option Greek.

Solution. Let x and y be, respectively, the number of units of Call-II and stock to be *bought* (if x or y turns out to be negative, this means that you *sell*). To maintain delta-neutrality and gamma-neutrality, we solve

$$\begin{cases} -1000(0.0651) + 0.0746x = 0 & \text{(gamma-neutrality)} \\ -1000(0.5825) + 0.7773x + y = 0 & \text{(delta-neutrality)} \end{cases}.$$

This gives $x = \boxed{872.6542}$ and $y = \boxed{-95.8141}$. In other words, 872.7 units of Call-II should be bought and 95.8 units of the stock should be sold. (**Answer: (B)**) □

Remark. (i) It is enough to calculate the value of x to conclude that (B) is the correct answer.

(ii) In this example, the values of x and y are determined in turn. We first choose to buy $x = 872.7$ units of Call-II solely to neutralize the gamma of the 1000 units of the short Call-I, then buy $y = -95.8141$ units of the stock to offset the delta of the resulting portfolio.

(iii) The row of "vega" is redundant in this example.

Example 7.2.2. (CAS Exam 3 Fall 2007 Question 24: Delta-gamma-hedging II) An investor has a portfolio consisting of 100 put options on stock A, with a strike price of 40, and 5 shares of stock A. The investor can write put options on stock A with a strike price of 35. The deltas and gammas of the options are listed below:

	Put (Strike = 35)	Put (Strike = 40)
Delta	−0.10	−0.05
Gamma	0.50	0.25

Which one of the following actions would delta and gamma neutralize this portfolio?

(A) Write 100 put options with a strike price of 35.

(B) Write 50 put options with a strike price of 35.

(C) Write 100 put options with a strike price of 35, and buy 5 shares of stock.

(D) Write 100 put options with a strike price of 35, and sell 5 shares of stock.

(E) Write 50 put options with a strike price of 35, and sell 5 shares of stock.

Solution. To gamma-hedge the portfolio, we need to *sell* $100(0.25)/0.50 = 50$ put options. The overall delta is

$$100\Delta_P^{\text{40-strike}} + 5 - 50\Delta_P^{\text{35-strike}} = 100(-0.05) + 5 - 50(-0.10) = 5,$$

so to delta-hedge the resulting portfolio, we also need to sell 5 shares of stock. **(Answer: (E))** □

Remark. The numbers in this example seem pathological. It appears that the 35-strike put is equivalent to two units of the 40-strike put. Thus for delta-gamma-hedging, the 100 units of the 40-strike put, it suffices to sell $100/2 = 50$ units of the 35-strike put without recourse to the stock.

Delta-gamma-hedging vs delta-hedging.

Figure 7.2.1 depicts the overnight profits of the delta-hedged written call and the delta-gamma-hedged written call using the same set of parameters as Figure 7.1.1. The delta-gamma-hedged written call is set up by coupling the written call with 0.94308 units of an otherwise identical 60-strike call and 0.30583 shares of the stock (check!). Whereas stock price risk, be it upside or downside, is not completely eliminated even with delta-gamma-hedging, the fluctuation of the overnight profit with the 1-day stock price is much milder than that of the delta-hedged written call. This suggests that delta-gamma-hedging is a much more prudent risk management strategy than delta-hedging alone. In fact, the delta-gamma-hedged written call presents a decent amount of profit if the stock price rises. (Question: How to compute the overnight profit of the delta-gamma-hedged written call? See Problem 7.4.9)

7.3 Delta-Gamma-Theta Approximation

One of the usefulness of option Greeks is that they indicate the sensitivity of the option price to different risk factors, especially the prevailing stock price and time to maturity, both of which inevitably change over the life of the option. Such sensitivity, once gauged, can be reduced by appropriate hedging strategies, as the previous two sections show. Another application of option Greeks is to approximate option prices computed at new stock prices and at new points of time by virtue of a Taylor series approximation. Interestingly, this simple approximation tool emanating from multi-variable calculus, when applied to delta-hedging, a strategy driven by practical considerations, has surprisingly far-reaching implications for the theory of derivative pricing in general.

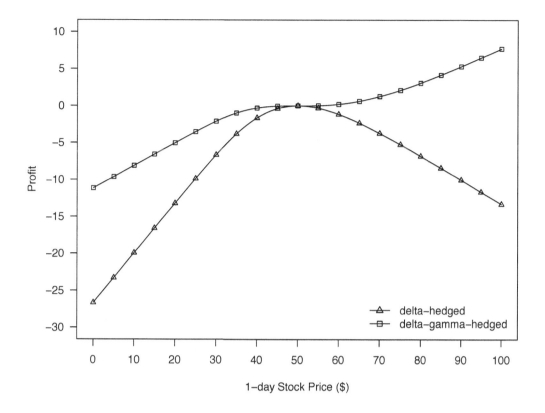

FIGURE 7.2.1
Comparison of the overnight profit of a delta-hedged written call and the overnight profit of a delta-gamma-hedged written call.

Taylor series expansion.

For $t \geq 0$, let $V(S(t), t)$ be the time-t price of a generic option corresponding to the time-t stock price of $S(t)$. Suppose that the time-0 derivative price, $V(S(0), 0)$, and the associated option Greeks are known, and we are interested in the time-h derivative price for some positive h. One solution is to repeat the Black-Scholes valuation procedure at time h and at the new realized stock price $S(h)$. This can be cumbersome (especially in a pen-and-paper exam environment!). A somewhat crude but easier alternative is to approximate the new option price by a Taylor series expansion of the bivariate function $V(s, t)$ at the time-0 arguments $(S(0), 0)$ and capitalizing on the time-0 information about the option price and option Greeks, yielding

$$\underbrace{V(S(h), h)}_{\text{new arguments}} \approx \underbrace{V(S(0), 0)}_{\text{old arguments}} + \underbrace{\Delta(S(0), 0)}_{\text{old arguments}} \epsilon + \frac{1}{2} \underbrace{\Gamma(S(0), 0)}_{\text{old arguments}} \epsilon^2 + \underbrace{\theta(S(0), 0)}_{\text{old arguments}} h, \quad (7.3.1)$$

where $\epsilon = S(h) - S(0)$ is the change in the stock price, and the option price and option Greeks on the right-hand side are all evaluated at the original stock price $S(0)$ and time 0. With the use of the "approximation trinity," namely delta, gamma, and theta of the option, (7.3.1), is naturally termed the *delta-gamma-theta approximation* of the new option price (at the new stock price $S(h)$ and new time h). It is a Taylor series approximation of second order (because of the use of the first and second partial derivatives) in the variable s and first order in the variable t. Note that the units of h and θ should always be consistent: If θ is expressed in days, then so should h be. The two truncated versions,

$$V(S(h), h) \approx V(S(0), 0) + \Delta(S(0), 0)\epsilon, \qquad (7.3.2)$$

and

$$\boxed{V(S(h), h) \approx V(S(0), 0) + \Delta(S(0), 0)\epsilon + \frac{1}{2}\Gamma(S(0), 0)\epsilon^2,} \qquad (7.3.3)$$

are referred to as the *delta approximation* and *delta-gamma approximation*, respectively.

Example 7.3.1. (SOA Exam MFE Spring 2007 Question 19: Delta-gamma approximation) Assume that the Black-Scholes framework holds. The price of a nondividend-paying stock is \$30.00. The price of a put option on this stock is \$4.00.
   You are given:

(i) $\Delta = -0.28$

(ii) $\Gamma = 0.10$

Using the delta-gamma approximation, determine the price of the put option if the stock price changes to \$31.50.

(A) \$3.40

(B) \$3.50

(C) \$3.60

(D) \$3.70

(E) \$3.80

Solution. Using the delta-gamma approximation with $\epsilon = 31.50 - 30 = 1.50$, we estimate the price of the put option if the stock price changes to \$31.50 as

$$
\begin{aligned}
P(31.5) &= P(30) + \Delta(30)\epsilon + \frac{1}{2}\Gamma(30)\epsilon^2 \\
&= 4.00 + (-0.28)(1.50) + \frac{1}{2}(0.10)(1.50)^2 \\
&= \boxed{3.6925}. \qquad \textbf{(Answer: (D))}
\end{aligned}
$$

$\square$

Example 7.3.2. (SOA Exam MFE Spring 2009 Question 20: Given the estimated price, deduce $S(0)$) Assume that the Black-Scholes framework holds. Consider an option on a stock.
   You are given the following information at time 0:

(i) The stock price is $S(0)$, which is greater than 80.

(ii) The option price is 2.34.

(iii) The option delta is -0.181.

(iv) The option gamma is 0.035.

The stock price changes to 86.00. Using the delta-gamma approximation, you find that the option price changes to 2.21.
    Determine $S(0)$.

(A) 84.80

(B) 85.00

(C) 85.20

(D) 85.40

(E) 85.80

Ambrose's comments:

Unlike the previous example, this time you are directly given the delta-gamma approximation, which is quadratic in $\epsilon = S^{\text{new}} - S^{\text{old}}$, and asked to back out ϵ via solving a quadratic equation.

Solution. The delta-gamma approximation says that

$$2.21 = 2.34 + (-0.181)\epsilon + \frac{1}{2}(0.035)\epsilon^2 \text{ or } 0.0175\epsilon^2 - 0.181\epsilon + 0.13 = 0.$$

The solutions to this quadratic equation are

$$\epsilon = \frac{0.181 \pm \sqrt{(-0.181)^2 - 4(0.0175)(0.13)}}{2(0.0175)} = 9.5663 \text{ or } 0.7765.$$

To see which one(s) is/are acceptable values of ϵ, recall that $\epsilon = S^{\text{new}} - S^{\text{old}} = 86 - S^{\text{old}}$, which is less than 6, because $S^{\text{old}} > 80$. Thus we can only take $\epsilon = 0.7765$, which in turn implies that $S^{\text{old}} = 86 - \epsilon = \boxed{85.2235}$. **(Answer: (C))** $\square$

Example 7.3.3. (SOA Exam FETE Fall 2010 Question 16 (c)–(d)) You hold a trading book consisting of many long/short positions in Tempranillo Corp. stock and options. You are analyzing two possible scenarios for Tempranillo Corp. stock and want to make money by adjusting your trading book.

Scenario	Market Conditions
1	Swift downward price movement; rising implied volatility
2	No price movement; falling implied volatility

(c) For each of the above scenarios, determine whether a positive or negative delta, gamma, or vega in your trading book would produce a profit.

Now consider the table below which summarizes your positions at the prior day's close of trading:

Option	Position (# of units)	$delta/$	$gamma/$	$theta/day	$vega/point
Call totals	−600	−736	−19.6	+0.5	−0.8
Put totals	400	−618	+2.2	+0.2	−0.1
Stock	1,200	1,200	0.0	0.0	0.0

where $delta/$ is the change in the $ value of your portfolio per $1 change in the underlying asset or index.

Today Tempranillo Corp.'s stock price dropped $4 and implied volatility on all options rose 2 percentage points.

(d) Estimate the value change in your trading book at closing today.

Solution. (c) The signs of delta, gamma and vega leading to a profit are tabulated below:

Scenario	Delta	Gamma	Vega
1	Negative	Positive	Positive
2	Does not matter	Does not matter	Negative

(d) The portfolio Greeks are given by:

Option	Position	$delta/$	$gamma/$	$theta/day	$vega/point
Call totals	−600	−736	−19.6	+0.5	−0.8
Put totals	400	−618	+2.2	+0.2	−0.1
Stock	1,200	1,200	0.0	0.0	0.0
Total	?	−154	−17.4	+0.7	−0.9

Hence the estimated change in the value of the trading book at closing today is

$$-154(-4) + \frac{1}{2}(-17.4)(-4)^2 + 0.7(1) + (-0.9)(2) = \boxed{475.7}.$$

□

Geometric meaning of delta- and delta-gamma approximations.

Geometrically, the delta approximation is represented by a straight line which is tangent to the option price function at the current stock price, as illustrated in Figure 7.3.1 in the case of a call with the same parameters as page 240. Because the delta of a call increases with the current stock price, the delta approximation as represented by the tangent will understate the increase in the call price when the stock price rises and overstate the decrease in the call price when the stock price declines. In other words, the tangent will be always lying below the call price function. The delta-gamma approximation improves the accuracy of the delta approximation by approximating the call price function by a quadratic function that shares the same first and second derivatives at the current stock price. Incorporating the gamma correction term $\Gamma(S(0), 0)\epsilon^2/2$, which is always positive regardless of the direction of the stock price move, the delta-gamma approximation will always be larger in value than the delta approximation. As shown in Figure 7.3.1, the delta-gamma approximation dramatically enhances the precision of the delta approximation and is much closer to the true call price function.

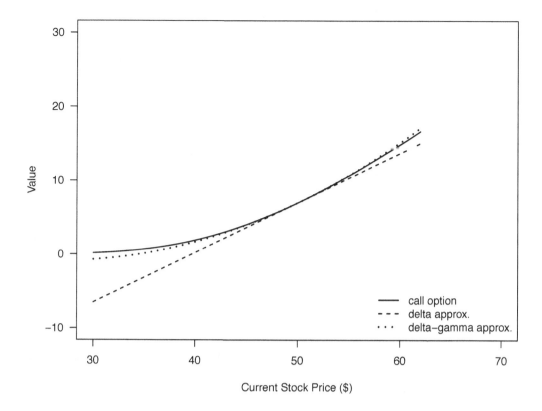

FIGURE 7.3.1

Geometric meaning of the delta- and delta-gamma approximations of the option price.

Application of delta-gamma-theta approximation: The Black-Scholes equation.

> **Warning!**
>
> The material in what follows, which may be skipped on a first reading without losing continuity, is of a more advanced nature and is only given a semi-formal treatment. Actuarial students taking Exam IFM should know that a "one standard deviation move" (see (7.3.5) below) in the stock price causes the overnight profit of a delta-hedged position to be (approximately) zero. Then study Example 7.3.4. The Black-Scholes equation, (7.3.7), despite being a fundamental result in the theory of option pricing, is not required for Exam IFM.

A cursory glance at (7.1.2) shows that three option Greeks play a role in determining the holding profit of a delta-hedged market-maker who sells a European derivative on a nondividend-paying stock. They are delta and gamma (associated with stock price moves), and theta (quantifying the passage of time) of the concerned derivative. Via the use of the delta-gamma-theta approximation to derivative prices, it will be shown that these three seemingly unrelated Greeks are governed by a partial differential equation.

To begin with, we consider a generic time t before the expiration of a T-year derivative and specialize (7.1.2) to the two time points, t and $t + dt$ for some infinitesimally small dt. The holding profit formula then reads

$$\text{Profit} = \Delta(t)[S(t + dt) - S(t)] - [V(t + dt) - V(t)] - (e^{r\,dt} - 1)[\Delta(t)S(t) - V(t)]. \quad (7.3.4)$$

When dt is small enough, the following approximations work reasonably well:

- *Interest:* $e^{r\,dt} - 1 \approx (1 + r\,dt) - 1 = r\,dt$.

- *Delta-gamma-theta approximation:* By (7.3.1),

$$V(t + dt) - V(t) \approx \Delta(t)[S(t + dt) - S(t)] + \frac{1}{2}\Gamma(t)[S(t + dt) - S(t)]^2 + \theta(t)\,dt.$$

- *One-standard-deviation move:*[iii]

$$[S(t + dt) - S(t)]^2 \approx \boxed{[\sigma S(t)]^2\,dt}. \quad (7.3.5)$$

Inserting these approximations into (7.3.4), we have

$$\text{Profit} \approx -\left\{\frac{1}{2}[\sigma S(t)]^2\Gamma(t) + \theta(t) + r[\Delta(t)S(t) - V(t)]\right\}dt. \quad (7.3.6)$$

When we stand at time t, the time-t stock price, $S(t)$, is known. Thus (7.3.6) is non-random. If there are no arbitrage opportunities, then the quantity within the braces in (7.3.6) must be zero—or else we must either always earn (positive) profits or always lose. Therefore,

$$\boxed{rS(t)\Delta(t) + \frac{1}{2}[\sigma S(t)]^2\Gamma(t) + \theta(t) = rV(t), \quad 0 \le t \le T.} \quad (7.3.7)$$

This equation is referred to as the celebrated *Black-Scholes partial differential equation*, or the Black-Scholes *equation* (not to be confused with the "formula!") in short, which must be satisfied by the time-t price of a derivative, $V = V(s,t)$ under the no-arbitrage assumption. Different derivatives are distinguished by different boundary and terminal conditions that the function $V = V(s,t)$ should satisfy as $s \downarrow 0$, $t \downarrow 0$ and $t \uparrow T$. In the case of a K-strike T-year European call option on the stock, the terminal and boundary conditions are

$$V(0,t) = 0 \text{ for all } 0 \le t \le T, \quad V(s,T) = (s - K)_+.$$

Note that α, the continuously compounded expected rate of return on the stock, does not enter (7.3.7), just as it has no role to play in the Black-Scholes pricing formula.

[iii]To understand the term "one standard deviation move," observe that

$$\text{Var}[\ln S(t + dt)|S(t)] = \sigma^2 dt.$$

Applying the delta method in statistics to the exponential function $f(x) = \exp(x)$, we have

$$\text{Var}[S(t + dt)|S(t)] = \text{Var}\{\exp[\ln S(t + dt)]|S(t)\} = \{\exp[\ln S(t)]\}^2 \times \sigma^2\,dt = [\sigma S(t)]^2\,dt.$$

Students with a background in stochastic calculus may also understand the one-standard-deviation move as

$$[S(t + dt) - S(t)]^2 \approx [dS(t)]^2 = [\alpha S(t)\,dt + \sigma S(t)\,dZ(t)]^2 = [\sigma S(t)]^2\,dt$$

using the multiplication rules $dtdt = dtdZ(t) = dZ(t)dt = 0$ and $[dZ(t)]^2 = dt$ for Brownian motion.

Example 7.3.4. (SOA Exam IFM Advanced Derivatives Sample Question 9: Zero-profit stock price) Consider the Black-Scholes framework. A market-maker, who delta-hedges, sells a three-month at-the-money European call option on a nondividend-paying stock.

You are given:

(i) The continuously compounded risk-free interest rate is 10%.

(ii) The current stock price is 50.

(iii) The current call option delta is 0.61791.

(iv) There are 365 days in the year.

If, after one day, the market-maker has zero profit or loss, determine the stock price move over the day.

(A) 0.41

(B) 0.52

(C) 0.63

(D) 0.75

(E) 1.11

Solution. By (7.3.5), the approximate change in the stock price move is given by $\pm \sigma S(0)\sqrt{h}$, where $h = 1/365$ (one day) and $S(0) = 50$. To deduce the value of σ, we use (iii) and get $\Delta = N(d_1) = 0.61791$, or $d_1 = N^{-1}(0.61791) = 0.3$. As

$$d_1 = \frac{\ln(50/50) + (0.1 + \sigma^2/2)(0.25)}{\sigma\sqrt{0.25}} = 0.3,$$

or

$$0.5\sigma^2 - 0.6\sigma + 0.1 = 0.5(\sigma - 1)(\sigma - 0.2) = 0.$$

The two roots are $\sigma = 1$ and $\sigma = 0.2$. We reject $\sigma = 1$ because such a volatility seems too large (and none of the five answers fit). Hence, the answer is $0.2(50)\sqrt{1/365} = \boxed{0.5234}$.
(Answer: (B)) $\qquad\square$

Example 7.3.5. (SOA Exam MFE Sample Question 36: Given the form of $V(s,t)$) Assume the Black-Scholes framework. Consider a derivative security of a stock.

You are given:

(i) The continuously compounded risk-free interest rate is 0.04.

(ii) The volatility of the stock is σ.

(iii) The stock does not pay dividends.

(iv) The derivative security also does not pay dividends.

(v) $S(t)$ denotes the time-t price of the stock.

(vi) The time-t price of the derivative security is $[S(t)]^{-k/\sigma^2}$, where k is a positive constant.

Find k.

(A) 0.04

(B) 0.05

(C) 0.06

(D) 0.07

(E) 0.08

Solution. For notational convenience, let $a = -k/\sigma^2$ and the time-t price of the derivative security can be written as $V(s,t) = s^a$ when the stock price at that time is s. Then $V_t(s,t) = 0$, $V_s(s,t) = as^{a-1}$ and $V_{ss}(s,t) = a(a-1)s^{a-2}$. The derivative security is a tradable asset, so the function $V(s,t)$ must satisfy the Black-Scholes partial differential equation. Plugging the above partial derivatives into the Black-Scholes equation yields

$$rs(as^{a-1}) + \frac{1}{2}\sigma^2 s^2[a(a-1)s^{a-2}] = rs^a,$$

or, upon canceling s^a on both sides,

$$ra + \frac{1}{2}\sigma^2 a(a-1) = r,$$

which is a quadratic equation in a. The two solutions are $a = 1$ and $a = -2r/\sigma^2$. Thus $k = 2r = 2(0.04) = \boxed{0.08}$. (**Answer: (E)**)   □

7.4 Problems

Delta-hedging and calculations of holding profits

Problem 7.4.1. (Calculation of holding profit for long calls) Assume the Black-Scholes framework. Four months ago, Eric *bought* 100 units of a one-year 45-strike European call option on a nondividend-paying stock. He immediately delta-hedged his position with shares of the stock, but has not ever re-balanced his portfolio. He now decides to close out all positions.

You are given:

(i)

	Four Months Ago	Now
Stock price	$40.00	$50.00
Call option price	$4.45539	?
Call option delta	?	0.73507

(ii) The continuously compounded risk-free interest rate is 5%.

(iii) The volatility of the stock is less than 50%.

(a) Calculate the volatility of the stock.

(b) Calculate the four-month holding profit for Eric.

Problem 7.4.2. [HARDER!] (Holding profit for a long straddle) One year ago, Jacky *bought* 10 units of a 2-year at-the-money European straddle on a nondividend-paying stock. He immediately delta-hedged his position with shares of the stock, but has not ever re-balanced his portfolio. He now decides to close out all positions.

You are given:

(i) The risk-free interest rate is a positive constant.

(ii) The current stock price and the stock price one year ago are the same.

(iii) The following information about the European call and put options constituting the straddle:

	One Year Ago	Now
Call option price	$8.29391	$5.59651
Put option price	$4.85116	$3.83641
Call option delta	0.66431	?
Put option delta	?	-0.38209

Calculate Jacky's 1-year holding profit.

(Hint: The given table contains lots of useful information about the market parameters!)

Problem 7.4.3. (Given the delta-hedging strategy, find the price of the option)
Assume the Black-Scholes framework. You are given:

(i) The current price of a nondividend-paying stock is 80.

(ii) An investor has sold 1,000 units of a one-year at-the-money European call option on
the stock. He immediately delta-hedges the commitment with 750 shares of the stock.

(iii) The continuously compounded risk-free interest rate is 7%.

(iv) The volatility of the stock is less than 100%.

Calculate the price of each call option.
    (Hint: Infer from (ii) the volatility of the stock.)

Problem 7.4.4. (Calculation of holding profit for short calls with dividends)
Assume the Black-Scholes framework. You are given:

(i) The current stock price is 50.

(ii) The stock pays dividends continuously at a rate proportional to its price. The dividend
yield is 3%.

(iii) The volatility of the stock is 16%.

(iv) The prices of 1-year at-the-money European call and put options are 4.348 and 1.981,
respectively.

Timothy has just written 200 units of the call in (iv), and he delta-hedged his position
immediately.
    After 3 months, the stock price rises to 55 and the call price increases to 7.316.
    Calculate the three-month holding profit for Timothy.
    (Hint: Remember to take care of dividends.)

**Problem 7.4.5. [HARDER!] (Delta-hedging a call using an otherwise identical
put)** Assume the Black-Scholes framework. You are given:

(i) The current price of a nondividend-paying stock is 20.

(ii) The stock's volatility is 28%.

(iii) The continuously compounded risk-free interest rate is 2%.

(iv) The following information about a 3-month at-the-money European call option on the
stock:

Current Price	Current Delta	Current Gamma	Current Theta
1.16393	0.54210	0.14169	−2.41519

Suppose you have just sold 1,000 units of the call option above. You immediately delta-
hedge your position by trading appropriate units of a European put option having the same
underlying stock, strike price, and time to expiration as the call option.
    Calculate the theta of your overall position.

Problem 7.4.6. (Testing your conceptual understanding of Figure 7.1.1) Assume the Black-Scholes framework. Yesterday, you sold a European call option on a nondividend-paying stock. You immediately delta-hedged the commitment with shares of the stock. Today, you decide to close out all positions.

Which of the following statements about your delta-hedged portfolio today is *incorrect*?

(A) Your delta when today's stock price is $50 is (approximately) zero.

(B) Stock price risk is not completely eliminated.

(C) You lose from large stock price moves in either direction.

(D) The larger the stock price today, the smaller your delta.

(E) Your gamma is a negative constant.

Hedging multiple Greeks

Problem 7.4.7. (Implementing delta-gamma-hedging given raw information) Assume the Black-Scholes framework. You are given:

(i) The current price of a stock is 60.

(ii) The stock pays no dividends.

(iii) The stock's volatility is 30%.

(iv) The continuously compounded risk-free interest rate is 5%.

Suppose you have just *bought* 200 1-year 60-strike European call options.

Determine the numbers of units of a 1-year 65-strike European *put* option and the stock you should buy or sell in order to both delta-hedge and gamma-hedge your position in the 60-strike European calls.

Problem 7.4.8. (Based on Example 7.2.1: Delta-vega-hedging given summarized information) For two European call options, Call-I and Call-II, on a stock, you are given:

Greek	Call-I	Call-II
Delta	0.5825	0.7773
Gamma	0.0651	0.0746
Vega	0.0781	0.0596

Suppose you just sold 1,000 units of Call-I. You buy or sell appropriate units of the stock and Call-II in order to both delta-hedge and vega-hedge your position in Call-I.

Calculate the gamma of your hedged portfolio.

Problem 7.4.9. (Holding profit of a delta-gamma-hedged portfolio) Assume the Black-Scholes framework. You are given:

(i) The current price of a stock is $50.

(ii) The stock pays no dividends.

(iii) The stock's volatility is 25%.

(iv) The continuously compounded risk-free interest rate is 5%.

Suppose you have just sold 1,000 1-year 50-strike European call options.

(a) Determine the numbers of units of a 1-year 60-strike European call option and the stock you should buy or sell in order to both delta-hedge and gamma-hedge your position in the 50-strike European calls.

(b) You are further given:

(v) The original (i.e., time-0) prices of the 1-year 50-strike call and 1-year 60-strike call are 6.1680 and 2.5127, respectively.

(vi) If the one-month stock price remains unchanged at $50, then the one-month prices of the 50-strike call and 60-strike call (both of which will expire in 11 months) are 5.8611 and 2.2591, respectively.

Calculate your profit after one month if the delta-gamma-hedging strategy in part (a) is implemented and the one-month stock price remains unchanged at $50.

(Hint: We don't have a formula of the holding profit for a delta-gamma-hedging strategy. Reason flexibly and adapt the rationale behind (7.1.1).)

Problem 7.4.10. (Delta-, gamma-, and theta-hedge given option Greeks table) Assume the Black-Scholes framework. You are given:

(i) The current price of a nondividend-paying stock is 50.

(ii) The stock's volatility is 30%.

(iii) The continuously compounded risk-free interest rate is 8%.

(iv) The following information about two European call options on the stock:

	Call A	Call B
Price	10.0618	6.0214
Delta	0.6951	0.5056
Gamma	0.0191	0.0217
Theta (Per Year)	−4.1201	−3.9835

In each of the following cases, calculate the amount of the net investment you make today (including the sale of the 1,000 options in the first place):

(a) You have just sold 1,000 units of Call A. You immediately delta-hedge your position with shares of the stock.

(b) You have just sold 1,000 units of Call A. You immediately delta-hedge and gamma-hedge your position with shares of the stock and Call B.

(c) You are now further given that Call A is 50-strike and Call B is 60-strike, and both of them are 18-month call options. You have just sold 1,000 units of a 50-strike put otherwise identical to Call A. You immediately delta-hedge and theta-hedge your position with Call A and Call B.

Delta-gamma-theta approximation

Problem 7.4.11. (Direct application of delta-gamma approximation) Assume the Black-Scholes framework. The current prices of a stock and a call option on the stock are $10 and $2, respectively.

You are given:

(i) $\Delta = 0.6$

(ii) $\Gamma = 0.2$

Use the delta-gamma approximation to estimate the option value if the stock price jumps to $10.50.

Problem 7.4.12. (Given the delta approximation, find the delta-gamma approximation) Assume the Black-Scholes framework. For a 3-month 80-strike European put option on a nondividend-paying stock, you are given:

(i) The current price of the stock is 75.

(ii) The current price of the put option is 6.168.

(iii) The continuously compounded risk-free interest rate is 5%.

The price of the stock suddenly increases to 78. Using the delta approximation, you find that the put price decreases to 4.253.
   Using the delta-gamma approximation, calculate the price of the put.

Problem 7.4.13. [HARDER!] (Given the delta and delta-gamma approximations, find the exact option price) Assume the Black-Scholes framework. Consider a 9-month at-the-money European put option on a futures contract.
   The continuously compounded risk-free interest rate is 8%.
   The futures price instantaneously decreases by 10. You are given:

(i) Using the delta approximation, you find that the option price increases by 4.148.

(ii) Using the delta-gamma approximation, you find that the option price increases by 4.231.

Calculate the *exact* price of the put option at the new futures price, i.e., after the initial futures price drops by 10.

Problem 7.4.14. (Parameter-dependent candidate price) Assume the Black-Scholes framework. Determine all value(s) of a, in terms of r, σ, γ, such that $V(S(t), t) := AS(t)^a e^{\gamma t}$ represents the time-t price of a derivative security on a nondividend-paying stock.

Problem 7.4.15. (Delta-theta-hedging) Assume the Black-Scholes framework. You are given:

(i) The current price of a nondividend-paying stock is 70.

(ii) The stock's volatility is 25%.

(iii) The continuously compounded risk-free interest rate is 5%.

(iv) The following information about two European put options on the stock:

Put	Current Price	Current Delta	Current Gamma	Current Theta
A	6.9389	-0.2867	0.0112	?
B	9.0062	-0.3433	0.0121	-0.2060

Suppose you have just sold 1,000 units of Put A. You immediately delta-hedge and theta-hedge your position by trading appropriate units of Put B and the stock.

Calculate the amount of net investment you make today (including the sale of the 1,000 Put A).

Problem 7.4.16. (Calculation of holding profit given various partial derivatives) Assume the Black-Scholes framework. For $t \geq 0$, let $S(t)$ be the time-t price of a nondividend-paying stock and $V(s,t)$ be the time-t price of a European derivative when the price of the underlying stock at that time is s. You are given:

(i) $S(0) = 10$ and $S(2) = 12$.

(ii) The continuously compounded risk-free interest rate is 5%.

(iii) The stock's volatility is 40%.

(iv) The following partial derivatives of $V(s,t)$ for various s and t:

s	t	$V_t(s,t)$	$V_s(s,t)$	$V_{ss}(s,t)$	$V(s,t)$
10	0	-0.5861	?	0.0491	3.2738
12	0	-0.6310	?	0.0341	4.7877
10	2	-0.9804	0.6274	0.0946	?
12	2	-1.0142	0.7825	0.0613	?

At time 0, Jason *bought* 100 units of the derivative. He immediately delta-hedged his position with shares of the stock, but has not ever re-balanced his portfolio. Two years later, he decided to close out all positions.

Calculate the two-year holding profit for Jason.

Problem 7.4.17. (Holding profit calculations given a table of option Greeks – I) Assume the Black-Scholes framework. One year ago, Kelvin bought 1,000 units of a European call option on a nondividend-paying stock. He immediately delta-hedged his position with appropriate number of shares of the stock, but has not ever re-balanced his portfolio. He now decides to close out all positions.

You are given:

(i) The stock's volatility is 20%.

(ii)

	One Year Ago	Now
Stock price	40	?
Call price	5.6295	5.7653
Call delta	?	0.7296
Call gamma	0.0331	0.0385
Call theta (per year)	-1.6569	-2.1911
Call elasticity	?	5.4417

Calculate Kelvin's 1-year holding profit.

Problem 7.4.18. [HARDER!] (Holding profit calculations given a table of option prices and Greeks – II) Assume the Black-Scholes framework. Three months ago, Tyler bought 100 units of an at-the-money European call option on a nondividend-paying stock. He immediately delta-hedged his position with appropriate units of an otherwise identical European put option, but has not ever re-balanced his portfolio. He now decides to close out all positions.

You are given:

(i) The stock's volatility is 36%.

(ii)

	Three Months Ago	Now
Stock price	45	42
Call option price	?	3.22341
Call delta	0.58082	0.4584
Call gamma	0.02785	0.03711
Call theta (per year)	−4.05975	−4.56265

Calculate Tyler's three-month holding profit.

(Hint: Make good use of the information presented in the table. First deduce the continuously compounded risk-free interest rate from the "Now" column. Then infer the time to maturity of the call from the original "Call delta." Finally, use put-call parity to find the prices and delta of the put. Be patient—you can do this problem!)

Problem 7.4.19. (Holding profit calculations given a table of option Greeks – III) Assume the Black-Scholes framework. Three months ago, you sold 1,000 units of a 1-year European put option on a nondividend-paying stock. You immediately delta-hedged your position with appropriate number of shares of the stock, but have not ever re-balanced his portfolio. You now decide to close out all positions.

You are given:

(i) The stock's volatility is 30%.

(ii) The put option is at-the-money *currently*.

(iii) Your careless secretary has provided you with the following values. However, she is not sure about whether these values are for the put option you sold or for a call option with the same strike, time to maturity, and underlying stock:

	Three Months Ago	Now
Stock price	50	55
Price	5.2121	4.5111
Delta	0.5129	−0.3809
Gamma	0.0266	0.0267
Theta (per year)	−4.2164	−2.1069

(Note: Values for each column are for the same option. The two columns, however, may or may not correspond to the same option.)

Calculate your three-month holding profit.

8

Exotic Options

Chapter overview: This chapter provides a coherent introduction to a wide variety of *exotic options* (or exotics, in short), which are options whose payoff structures or exercise features differ from those of standard, also known as *plain vanilla* calls and puts (which are options we have studied in this book), in practically important ways that appeal to investors of considerable diversity. Being highly customized in nature, these exotics cater to different hedging needs and are actively traded over the counters. In this chapter, you will be exposed to a number of the most typical exotic options.

Exotic options can be broadly categorized into two groups. The first group consists of *path-independent* exotics, whose payoffs at expiration (for European ones) or at the time of exercise (for American ones) depend only on the price of the underlying asset at that point of time, but not on the prior asset prices. Essentially, they are independent of the path by which the asset arrived at the final price. These exotics include all-or-nothing options, gap options, exchange options, and compound options, which are explored in Sections 8.1 to 8.3 in turn. The second group comprises *path-dependent* exotics, whose payoff at expiration or the time of exercise depends on at least some of the intermediate asset prices. The dependence on the asset price path leading to the terminal asset price can be as simple as whether the asset price ever crosses a certain level, or be captured more globally by means of more complex quantities such as the average asset price and maximum asset price over the life of the exotics. Sections 8.4 to 8.7 investigate Asian, lookback, shout, and barrier options, which are quintessential path-dependent exotic options. We end in Section 8.8 with two examples of path-dependent exotics that can be regarded as path-independent with the right perspective.

For each exotic option, we discuss its typical use, cash flow characteristics (with focus on how the exotic departs from and is motivated from a plain vanilla option), and pricing in the Black-Scholes framework whenever a closed-form pricing formula is available or simple enough, or in the binomial framework where all derivatives can be easily priced. We also devote special attention to how and why exotic options share properties not possessed by plain vanilla options. Unexpected anomalies include the possibility of negative option prices and negative gamma.

8.1 Gap Options

8.1.1 Introduction

Cash flow characteristics.

If you contemplate carefully, the strike price of a plain vanilla option plays two delicate roles. It determines not only (i) whether a positive payoff arises, but also (ii) the magnitude of the payoff. To see this, one can write the payoff of a general K-strike T-year plain vanilla

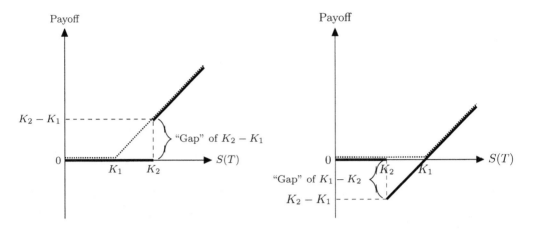

FIGURE 8.1.1
The payoff diagram of a K_1-strike K_2-trigger European gap call option (in bold) when (i) $K_1 < K_2$ (left), and (ii) $K_1 \geq K_2$ (right).

call as

$$\text{Call payoff} = (S(T) - K)_+ = \underbrace{[S(T) - K]}_{(ii)} \times \underbrace{1_{\{S(T) > K\}}}_{(i)}, \tag{8.1.1}$$

where the two K's are playing distinct roles. In a gap option, these two functions are separated. The option must be exercised when and only when the stock price exceeds (for a call) or is less than (for a put) the *payment trigger* (it is the price that "triggers" the payoff of the option), denoted by K_2. In random variable form, the payoff of a generic European gap call option is given by

$$\text{Gap call payoff} = [S(T) - K_1]1_{\{S(T) > K_2\}} = \begin{cases} S(T) - K_1, & \text{if } S(T) > K_2, \\ 0, & \text{if } S(T) \leq K_2, \end{cases}$$

and that of a gap put option is

$$\text{Gap put payoff} = [K_1 - S(T)]1_{\{S(T) < K_2\}} = \begin{cases} K_1 - S(T), & \text{if } S(T) < K_2, \\ 0, & \text{if } S(T) \geq K_2, \end{cases}$$

where K_1, still called the strike price, determines the size of the payoff—it is the price that is paid to acquire the underlying stock in the case of a call and the price that is received in exchange for the stock in the case of a put. Generally, $K_1 \neq K_2$, which creates a conspicuous discontinuity in the payoff function. The payoff diagram of a K_1-strike K_2-trigger T-year gap call option is sketched in Figure 8.1.1, where the "gap" in the payoff function is evident.

Note that exercising a gap option is compulsory when $S(T) > K_2$ for a call and $K_2 > S(T)$ for a put, even if this may result in *negative* payoffs (by extension, a gap option may have a *negative* price). As such, the term "option" is a bit of a misnomer.

The following example compares a gap option with its plain vanilla counterpart and sheds light on a common reason why gap options are traded in practice.

Example 8.1.1. (Optimal payment trigger) For a fixed strike price K_1, determine the value of the payment trigger that maximizes the payoff of a European gap option.

Solution. Without loss of generality, we consider the case of a gap call option.

Case 1. If $K_1 < K_2$, then the payoff of a K_1-strike K_2-trigger gap call can be written as

$$
\begin{aligned}
[S(T) - K_1]1_{\{S(T) > K_2\}} &= [S(T) - K_1]1_{\{S(T) > K_1\}} \\
&\quad - \underbrace{[S(T) - K_1]1_{\{K_1 \leq S(T) < K_2\}}}_{\geq 0} \\
&\leq [S(T) - K_1]1_{\{S(T) > K_1\}} \\
&= (S(T) - K_1)_+.
\end{aligned}
$$

Case 2. If $K_2 \leq K_1$, then we have

$$
\begin{aligned}
[S(T) - K_1]1_{\{S(T) > K_2\}} &= \underbrace{[S(T) - K_1]1_{\{K_2 < S(T) \leq K_1\}}}_{\leq 0} \\
&\quad + [S(T) - K_1]1_{\{S(T) > K_1\}} \\
&\leq [S(T) - K_1]1_{\{S(T) > K_1\}} \\
&= (S(T) - K_1)_+.
\end{aligned}
$$

Combining both cases, we conclude that the payoff of a gap call option is maximized when $\boxed{K_2 = K_1}$, in which case the gap call reduces to a plain vanilla call.

The same conclusion can also be drawn geometrically. In Figure 8.1.1, the payoff function of the plain vanilla K_1-strike call is sketched using a dotted line in both panels. It can be seen that the payoff of the plain vanilla call always dominates that of the K_1-strike K_2-trigger gap call. The domination vanishes when and only when $\boxed{K_2 = K_1}$. That the payoff of a gap options is always no larger than that of the corresponding plain vanilla option with the same strike price means that investors may pursue a gap option as a cheaper alternative to a plain vanilla option, if they are convinced that it is unlikely for the stock price to fall between K_1 and K_2 at maturity. $\square$

Put-call parity for gap options.

We know that the prices of otherwise identical plain vanilla European calls and puts are related by put-call parity. It is possible to develop an analogous put-call parity for otherwise identical European gap options.

Example 8.1.2. (Put-call parity for gap options) Let $C^{\text{gap}}(K_1, K_2)$ and $P^{\text{gap}}(K_1, K_2)$ be the current prices of a T-year European gap call option and a T-year European gap put option, respectively, both with a strike price of K_1 and a payment trigger of K_2. Prove that

$$
C^{\text{gap}}(K_1, K_2) - P^{\text{gap}}(K_1, K_2) = F_{0,T}^P(S) - K^* e^{-rT}.
$$

Identify K^*.

Solution. The difference between the payoffs of the two gap options is

$$[S(T) - K_1]1_{\{S(T)>K_2\}} - [K_1 - S(T)]1_{\{S(T)<K_2\}}$$
$$= [S(T) - K_1]1_{\{S(T)>K_2\}} + [S(T) - K_1]1_{\{S(T)<K_2\}} = S(T) - K_1.$$

It follows from this time-T payoff identity that we have the time-0 price identity

$$C^{\text{gap}}(K_1, K_2) - P^{\text{gap}}(K_1, K_2) = F^P_{0,T}(S) - K_1 e^{-rT}, \tag{8.1.2}$$

with $\boxed{K^* = K_1}$.   □

8.1.2 All-or-Nothing Options[i]

All-or-nothing options: Motivation and taxonomy.

Before delving into the valuation of gap options, we digress briefly to discuss somewhat toy examples of exotic options which are useful vehicles for constructing and pricing gap options and are sometimes objects of independent interest. To see what all-or-nothing options are and how they arise, we observe from (8.1.1) that the payoff formula of a plain vanilla European call option can be decomposed, with the aid of indicator functions, as

$$\begin{aligned}
\text{Payoff} &= [S(T) - K] \times 1_{\{S(T)>K\}} \\
&= \underbrace{S(T) \times 1_{\{S(T)>K\}}}_{(A)} - \underbrace{K \times 1_{\{S(T)>K\}}}_{(B)}.
\end{aligned} \tag{8.1.3}$$

Exotic options whose payoffs are given by (A) or (B) are examples of *all-or-nothing options* (also known variously as *binary* and *digital* options) because they can be thought of as 0 or 1—either you receive the share or cash ("all"), or you do not ("nothing"). In general, all-or-nothing options can be categorized according to the following criteria:

- *Cash-or-nothing:* The payoff at expiration is a fixed amount ("cash").

- *Asset-or-nothing:* The payoff at expiration is based on the value of an asset, usually a stock ("asset").

- *Call or put:* If the payoff is made if $S(T) > K$ (resp. $S(T) < K$), which is the same triggering condition for a plain vanilla call (resp. put) to be exercised, then the option is called an all-or-nothing *call* (resp. *put*) option.

Let's look at the characteristics and pricing of all-or-nothing options.

Cash-or-nothing options.

The payoff of a T-year K-strike European *cash-or-nothing call* option that pays \$1 (or another fixed amount) at expiration is given by the indicator function $1_{\{S(T)>K\}}$. They are straight bets on the market—the holder of such a cash-or-nothing call option believes that the price of the underlying asset will rise above the level of K. Analogously, the payoff of

---

[i]Actuarial students studying for Exam IFM are reminded that this subsection is beyond the exam syllabus. However, the material in this subsection is useful for understanding gap options.

an otherwise identical European cash-or-nothing put option is $1_{\{S(T) \leq K\}}$[ii] and its holder is convinced of the price of the underlying asset falling below the level of K.

Recall that the risk-neutral probability that $S(T) > K$ is given by $N(d_2)$, where

$$d_2 = \frac{\ln[F_{0,T}^P(S)/K] + (r - \sigma^2/2)T}{\sigma\sqrt{T}}.$$

It follows by risk-neutral pricing that the Black-Scholes price of a T-year K-strike European cash-or-nothing call option that pays \$1 at expiration, denoted by $C^{\mathrm{C/N}}$ ("C/N" stands for "cash-or-nothing"), is given by

$$\boxed{C^{\mathrm{C/N}} = \mathrm{e}^{-rT} N(d_2),}$$

which constitutes the second term in the Black-Scholes pricing formula for a plain vanilla European call option. Analogously, the price of an otherwise identical cash-or-nothing put option, written as $P^{\mathrm{C/N}}$, is

$$\boxed{P^{\mathrm{C/N}} = \mathrm{e}^{-rT} N(-d_2).}$$

Of course, the sum of the two cash-or-nothing option prices above equals

$$C^{\mathrm{C/N}} + P^{\mathrm{C/N}} = \mathrm{e}^{-rT} = \mathrm{PV}_{0,T}(1),$$

which comes as no surprise to us because $1_{\{S(T)>K\}} + 1_{\{S(T)\leq K\}} = 1$, or, equivalently, \$1 is paid no matter what happens at time T.

Example 8.1.3. (Cash-or-nothing spread) Assume the Black-Scholes framework. A European cash-or-nothing spread on a stock pays \$1 at time T if and only if the stock price at time T lies in the interval $[a, b]$, where a and b are positive numbers with $a < b$.

Determine an expression for the price of such a spread using common symbols.

Solution. The payoff of the cash-or-nothing spread can be expressed as the difference between that of an a-strike cash-or-nothing call of \$1 and that of a b-strike cash-or-nothing call of \$1. Its price is the difference between the prices of the two cash-or-nothing calls and equals

$$V = \mathrm{e}^{-rT}[N(d_2^{@a}) - N(d_2^{@b})],$$

where

$$d_2^{@a} = \frac{\ln[F_{0,T}^P(S)/a] + (r - \sigma^2/2)T}{\sigma\sqrt{T}} \quad \text{and} \quad d_2^{@b} = \frac{\ln[F_{0,T}^P(S)/b] + (r - \sigma^2/2)T}{\sigma\sqrt{T}}.$$

$\square$

Problem 8.1.4. (Valuing a path-dependent cash-or-nothing option) Assume the Black-Scholes framework. For $t \geq 0$, let $S(t)$ be the time-t price of a nondividend-paying stock. You are given:

[ii]It does not matter whether we use $1_{\{S(T)<K\}}$ or $1_{\{S(T)\leq K\}}$, because stock prices are continuous random variables in the Black-Scholes framework.

(i) $S(0) = 180$.

(ii) The stock's volatility is 20%.

(iii) The continuously compounded expected rate of return on the stock is 8%.

(iv) The continuously compounded risk-free interest rate is 5%.

Consider a 1-year European partial cash-or-nothing option on the stock. The option's payoff is

$$\text{Payoff} = \begin{cases} 1000, & \text{if } S(0.5) > 200 \text{ and } S(1) > 1.5S(0.5), \\ 500, & \text{if } S(0.5) < 200 \text{ and } S(1) > 1.5S(0.5), \\ 0, & \text{otherwise.} \end{cases}$$

Calculate the time-0 price of this option.

Ambrose's comments:

This question illustrates that the deceptively simple cash-or-nothing option can form a not-so-simple option. In fact, this special cash-or-nothing option has the features of a *Bermudan* option and a path-dependent option. If you can fully conquer this example, then no cash-or-nothing questions can probably challenge you!

Solution. By risk-neutral pricing, the 6-month price of the special 1-year derivative is given by

$$
\begin{aligned}
V(0.5) &= \overbrace{1{,}000\mathrm{e}^{-0.05(0.5)}N(d_2)}^{\substack{\text{price of a 6-month}\\\text{C/N call of \$1{,}000}}} 1_{\{S(0.5)>200\}} + \overbrace{500\mathrm{e}^{-0.05(0.5)}N(d_2)}^{\substack{\text{price of a 6-month}\\\text{C/N call of \$500}}} 1_{\{S(0.5)<200\}} \\
&= 1{,}000\mathrm{e}^{-0.025}N(d_2)1_{\{S(0.5)>200\}} + 500\mathrm{e}^{-0.025}N(d_2) \underbrace{1_{\{S(0.5)<200\}}}_{1-1_{\{S(0.5)>200\}}}, \quad (8.1.4) \\
&= 500\mathrm{e}^{-0.025}N(d_2)(1 + 1_{\{S(0.5)>200\}}),
\end{aligned}
$$

where

$$
d_2 = \frac{\ln[\overbrace{S(0.5)/1.5S(0.5)}^{S(0.5)\text{ gets canceled}}] + (0.05 - 0.2^2/2)(0.5)}{0.2\sqrt{0.5}} = -2.76101,
$$

which is a constant free of $S(0.5)$. Thus, the special derivative is equivalent to $\$500\mathrm{e}^{-0.025}N(d_2)$ payable in 6 months (without uncertainty) and a 6-month 200-strike cash-or-nothing call of $\$500\mathrm{e}^{-0.025}N(d_2)$. With

$$
d_2' = \frac{\ln(180/200) + (0.05 - 0.2^2/2)(0.5)}{0.2\sqrt{0.5}} = -0.63895,
$$

the time-0 price of the derivative, by risk-neutral pricing again, is

$$
\begin{aligned}
V(0) &= 500\mathrm{e}^{-0.025}N(d_2) \times \mathrm{e}^{-0.025}[1 + N(d_2')] \\
&= 500\mathrm{e}^{-0.05} \underbrace{N(-2.76101)}_{0.00288}[1 + \underbrace{N(-0.63895)}_{0.26143}] \\
&= \boxed{1.7279}.
\end{aligned}
$$

$\square$

Remark. From (8.1.4), you can also view the special derivative as a 6-month 200-strike cash-or-nothing call of $\$1,000e^{-0.025}N(d_2)$, plus a 6-month 200-strike cash-or-nothing put of $\$500e^{-0.025}N(d_2)$.

Asset-or-nothing options.

An *asset-or-nothing option* pays one share of the underlying asset if the terminal price of the asset is higher than (an asset-or-nothing *call*) or lower than (an asset-or-nothing *put*) the strike price K. By risk-neutral pricing, the Black-Scholes price of a K-strike European asset-or-nothing call option, $C^{A/N}$, is given by the first term of the plain vanilla Black-Scholes call formula, namely,

$$\boxed{C^{A/N} = F^P_{0,T}(S)N(d_1),} \qquad (8.1.5)$$

where

$$d_1 = \frac{\ln[F^P_{0,T}(S)/K] + (r + \sigma^2/2)T}{\sigma\sqrt{T}}.$$

The corresponding Black-Scholes price of an asset-or-nothing put option is

$$\boxed{P^{A/N} = F^P_{0,T}(S)N(-d_1).} \qquad (8.1.6)$$

Of course, adding (8.1.5) and (8.1.6) yields $F^P_{0,T}(S)$, as expected from the time-T payoff identity

$$S(T) \times 1_{\{S(T)>K\}} + S(T) \times 1_{\{S(T)\leq K\}} = S(T).$$

Example 8.1.5. (SOA Exam MFE Spring 2009 Question 4: Price of an asset-or-nothing put) Your company has just written one million units of a one-year European asset-or-nothing put option on an equity index fund.

The equity index fund is currently trading at 1000. It pays dividends continuously at a rate proportional to its price; the dividend yield is 2%. It has a volatility of 20%.

The option's payoff will be made only if the equity index fund is down by more than 40% at the end of one year.

The continuously compounded risk-free interest rate is 2.5%.

Using the Black-Scholes model, determine the price of the asset-or-nothing put options.

(A) 0.2 Million

(B) 0.9 Million

(C) 2.7 Million

(D) 3.6 Million

(E) 4.2 Million

Solution. The asset-or-nothing put's payoff will be made only if the equity index fund decreases to $1000(1 - 40\%) = 600$. As

$$d_1 = \frac{\ln(1,000/600) + (0.025 - 0.02 + 0.2^2/2)(1)}{0.2\sqrt{1}} = 2.67913,$$

the price of each 600-strike asset-or-nothing put option is given by

$$S(0)e^{-\delta T}N(-d_1) = 1,000e^{-0.02}\underbrace{N(-2.67913)}_{0.00369} = 3.616933105.$$

The total price is

$$1 \text{ million} \times 3.616933105 = \boxed{3.6 \text{ million}}. \quad (\textbf{Answer: (D)})$$

$\square$

8.1.3   Pricing and Hedging Gap Options

Black-Scholes pricing formula: First form.

With the notion of all-or-nothing options at our disposal, pricing gap options becomes an almost immediate task. In the language of all-or-nothing options, a K-strike plain vanilla call can be decomposed into a long K-strike asset-or-nothing call coupled with a short K-strike cash-or-nothing call of $\$K$. More generally, a K_1-strike K_2-trigger gap call is composed of a long K_2-strike asset-or-nothing call coupled with a short K_2-strike cash-or-nothing call of $\$K_1$:

$$
\begin{aligned}
\text{Gap call payoff} &= [S(T) - K_1]1_{\{S(T)>K_2\}} \\
&= \underbrace{S(T) \times 1_{\{S(T)>K_2\}}}_{K_2\text{-strike A/N call}} - \underbrace{K_1 \times 1_{\{S(T)>K_2\}}}_{K_2\text{-strike C/N call of }\$K_1}. \quad (8.1.7)
\end{aligned}
$$

It follows from this time-T payoff identity and the results in Subsection 8.1.2 that the price of such a gap call, denoted by $C^{\text{gap}}(K_1, K_2)$, is

$$C^{\text{gap}}(K_1, K_2) = C^{\text{A/N}} - K_1 \times C^{\text{C/N}} = F_{0,T}^P(S)N(d_1) - F_{0,T}^P(\boxed{K_1})N(d_2), \quad (8.1.8)$$

where

$$d_1 = \frac{\ln[F_{0,T}^P(S)/F_{0,T}^P(\boxed{K_2})] + (\sigma^2/2)T}{\sigma\sqrt{T}} \quad \text{and} \quad d_2 = d_1 - \sigma\sqrt{T}$$

both of which depend on the payment trigger K_2, but not the strike price K_1. Similarly, the price of the corresponding gap put option is

$$P^{\text{gap}}(K_1, K_2) = F_{0,T}^P(\boxed{K_1})N(-d_2) - F_{0,T}^P(S)N(-d_1). \quad (8.1.9)$$

Of course, subtracting (8.1.9) from (8.1.8) yields

$$C^{\text{gap}}(K_1, K_2) - P^{\text{gap}}(K_1, K_2) = F_{0,T}^P(S) - F_{0,T}^P(K_1),$$

which is in agreement with (8.1.2) (recall that put-call parity holds irrespective of the pricing model).

Which one to use? K_1 or K_2?

You should never confuse the positions of K_1 and K_2. Notice carefully that K_1, which is used to determine the *magnitude* of the payoff, enters the pricing formula outside the $N(\cdot)$ function, whereas K_2, which serves to determine the *probability* of making a payment, appears inside d_1 and d_2. If you keep (8.1.7) and the pricing formulas of all-or-nothing options in mind, you will not have any difficulty writing the pricing formula of a gap option.

Example 8.1.6. (Direct computation of gap option price) Assume the Black-Scholes framework. Let $S(t)$ denote the price at time t of a nondividend-paying stock.

Consider a European gap option which matures in one year. If the one-year stock price is greater than \$100, the payoff is

$$S(1) - 90;$$

otherwise, the payoff is zero.

You are given:

(i) $S(0) = \$80$.

(ii) The stock's volatility is 30%.

(iii) The continuously compounded risk-free interest rate is 8%.

Calculate the price of the gap option.

Solution. The gap option is a 90-strike 100-trigger 1-year European gap call. With $K_1 = 90$, $K_2 = 100$, we have

$$d_1 = \frac{\ln(80/\boxed{100}) + (0.08 + 0.3^2/2)(1)}{0.3\sqrt{1}} = -0.32715,$$
$$d_2 = d_1 - 0.3\sqrt{1} = -0.62715,$$

the price of the gap call is

$$
\begin{aligned}
C^{\text{gap}}(90, 100) &= 80N(-0.32715) - \boxed{90}\,e^{-0.08}N(-0.62715) \\
&= 80(0.37178) - 90e^{-0.08}(0.26528) \\
&= \boxed{7.7028}.
\end{aligned}
$$

$\square$

Black-Scholes pricing formula: second form.

An alternative but instructive way to price a gap option is to decompose it into a combination of a plain vanilla option and a cash-or-nothing option, both of which can be easily priced by what we learned before. In the case of a European gap call, this can be achieved by writing its payoff as

$$[S(T) - K_1]1_{\{S(T)>K_2\}} = [S(T) - K_2]1_{\{S(T)>K_2\}} + (K_2 - K_1)1_{\{S(T)>K_2\}}$$
$$= \underbrace{(S(T) - K_2)_+}_{K_2\text{-strike plain vanilla call}} + \underbrace{(K_2 - K_1)1_{\{S(T)>K_2\}}}_{\substack{K_2\text{-strike cash-or-nothing} \\ \text{call of } \$(K_2-K_1)}},$$

from which it easily follows that

$$\boxed{C^{\text{gap}}(K_1, K_2) = \text{Price of a } K_2\text{-strike call} + (K_2 - K_1)e^{-rT}N(d_2),} \qquad (8.1.10)$$

where the d_2 inside $N(\cdot)$ is computed using K_2. An analogous decomposition for a European gap put is

$$\boxed{P^{\text{gap}}(K_1, K_2) = \text{Price of a } K_2\text{-strike put} + (K_1 - K_2)e^{-rT}N(-d_2).} \qquad (8.1.11)$$

As a check, we subtract (8.1.11) from (8.1.10), leading to

$$C^{\text{gap}}(K_1, K_2) - P^{\text{gap}}(K_1, K_2)$$
$$= [C(K_2) - P(K_2)] + [(K_2 - K_1)e^{-rT}N(d_2) - (K_1 - K_2)e^{-rT}N(-d_2)]$$
$$= \underbrace{[F_{0,T}^P(S) - K_2e^{-rT}]}_{\text{plain vanilla put-call parity}} + (K_2 - K_1)e^{-rT}[N(d_2) + N(-d_2)]$$
$$= [F_{0,T}^P(S) - K_2e^{-rT}] + (K_2 - K_1)e^{-rT}$$
$$= F_{0,T}^P(S) - K_1e^{-rT},$$

which is what put-call parity for gap options, (8.1.2), says.

An ingenious problem can force you to use these alternative pricing formulas (in preference to (8.1.8) or (8.1.9)) by providing you only with information about the plain vanilla option and cash-or-nothing option, but not the values of market ingredients such as $S(0), r, \delta, \sigma$, etc. You cannot just blindly calculate!

Example 8.1.7. (SOA Exam MFE Spring 2007 Question 17: Calculation of gap call price) Let $S(t)$ denote the price at time t of a stock that pays dividends continuously at a rate proportional to its price. Consider a European gap option with expiration date $T, T > 0$.

If the stock price at time T is greater than \$100, the payoff is

$$S(T) - 90;$$

otherwise, the payoff is zero.

You are given:

(i) $S(0) = \$80$.

(ii) The price of a European call option with expiration date T and strike price \$100 is \$4.

(iii) The delta of the call option in (ii) is 0.2.

Calculate the price of the gap option.

(A) $3.60

(B) $5.20

(C) $6.40

(D) $10.80

(E) There is not enough information to solve the problem.

Solution. Here $K_1 = 90$ and $K_2 = 100$. The gap call option is equivalent to a 100-strike European call option and a 100-strike cash-or-nothing call option of $10. Considering the price and delta of the 100-strike European call, we have

$$Se^{-\delta T} N(d_1) - Ke^{-rT} N(d_2) = 80(0.2) - 100e^{-rT} N(d_2) = 4,$$

which gives $e^{-rT} N(d_2) = 0.12$. By (8.1.10), the price of the gap option is $4 + 10(0.12) = \boxed{5.2}$. (**Answer: (B)**)

$\square$

Delta-hedging.

To compute the delta of a gap option for the purpose of delta-hedging, one may differentiate $C^{\mathrm{gap}}(K_1, K_2)$ with respect to the current stock price S directly. However, we do *not* end up with the simple formula

$$\boxed{\text{WRONG!}} \quad \Delta_C^{\mathrm{gap}} = e^{-\delta T} N(d_1) \quad \boxed{\text{WRONG!}}$$

because the strike price K_1 and the payment trigger K_2 are generally not identical, which makes the cancellation identity (see (6.3.2) on page 204)

$$F_{0,T}^P(S) N'(d_1) = F_{0,T}^P(K) N'(d_2)$$

not valid. To derive the delta of a gap call, it is easier to exploit the alternative form of the pricing formula in (8.1.10) and (8.1.11), where the plain vanilla option and cash-or-nothing option have a mathematically tractable delta, yielding

$$
\begin{aligned}
\Delta_C^{\mathrm{gap}} &= e^{-\delta T} N(d_1) + (K_2 - K_1) e^{-rT} N'(d_2) \frac{\partial d_2}{\partial S} \\
&= e^{-\delta T} N(d_1) + (K_2 - K_1) \times \frac{e^{-rT}}{S\sigma\sqrt{T}} N'(d_2),
\end{aligned}
\tag{8.1.12}
$$

where d_1 and d_2 are again computed using K_2. Similarly, the delta of a gap put is

$$\Delta_P^{\mathrm{gap}} = -e^{-\delta T} N(-d_1) - (K_1 - K_2) \times \frac{e^{-rT}}{S\sigma\sqrt{T}} N'(-d_2).$$

Example 8.1.8. (SOA Exam IFM Advanced Derivatives Sample Question 18: Delta-hedging for gap option) A market-maker sells 1,000 1-year European gap call options, and delta-hedges the position with shares.

You are given:

(i) Each gap call option is written on 1 share of a nondividend-paying stock.

(ii) The current price of the stock is 100.

(iii) The stock's volatility is 100%.

(iv) Each gap call option has a strike price of 130.

(v) Each gap call option has a payment trigger of 100.

(vi) The risk-free interest rate is 0%.

Under the Black-Scholes framework, determine the initial number of shares in the delta-hedge.

(A) 586

(B) 594

(C) 684

(D) 692

(E) 797

Solution. Because the market-maker's delta is $-1,000\Delta_C^{\text{gap}}$, he/she requires $1,000\Delta_C^{\text{gap}}$ shares to delta-hedge. As in (8.1.12), we write the gap call price as $C(100) - 30N(d_2)$. With

$$d_1 = \frac{\ln(100/100) + (0 - 0 + 1^2/2)(1)}{1\sqrt{1}} = 0.5$$
$$d_2 = d_1 - 1\sqrt{1} = -0.5$$
$$N(d_1) = 0.69146$$
$$N(d_2) = 0.30854,$$

the delta of the gap call option is

$$\Delta_C^{\text{gap}} = N(d_1) - 30 \times \frac{1}{\sqrt{2\pi}} e^{-d_2^2/2} \times \frac{1}{100(1)(\sqrt{1})} = 0.58584,$$

so the required number of shares in the delta-hedge is $1,000(0.58584) = \boxed{586}$.
(**Answer: (A)**)   □

Example 8.1.9. (**SOA Exam MFE Advanced Derivatives Sample Question 53: Gamma of a gap option**) Assume the Black-Scholes framework. For a European put option and a European gap call option on a stock, you are given:

(i) The expiry date for both options is T.

(ii) The put option has a strike price of 40.

(iii) The gap call option has strike price 45 and payment trigger 40.

(iv) The time-0 gamma of the put option is 0.07.

(v) The time-0 gamma of the gap call option is 0.08.

Consider a European cash-or-nothing call option that pays 1000 at time T if the stock price at that time is higher than 40.

Find the time-0 gamma of the cash-or-nothing call option.

(A) -5

(B) -2

(C) $\ \ 2$

(D) $\ \ 5$

(E) $\ \ 8$

Solution. In terms of payoff,

$$\text{Gap call} = \text{Plain vanilla call} - 5 \times \text{Cash-or-nothing call.}$$

Differentiating twice with respect to the stock price yields

$$
\begin{aligned}
\text{Gap gamma} &= \text{Plain-vanilla call gamma} - 5 \times \text{Cash-or-nothing call gamma} \\
\text{Gap gamma} &= \text{Plain-vanilla put gamma} - 5 \times \text{Cash-or-nothing call gamma} \\
0.08 &= 0.07 - 5 \times \underbrace{\text{Cash-or-nothing call gamma}}_{-0.002}
\end{aligned}
$$

The gamma of the required cash-or-nothing option is $1000(-0.002) = \boxed{-2}$.
(**Answer: (B)**)

8.2   Exchange Options

8.2.1   Introduction

A unifying view on options – Exchange option.

The options discussed thus far mostly involve the receipt or delivery of an asset (usually a stock) in exchange for cash. More precisely, a plain vanilla call gives you the right, but not the obligation, to exchange cash for the underlying stock, while a plain vanilla put entitles you to exchange the stock for cash. An *exchange option* unifies and generalizes plain vanilla options by giving its holder the opportunity to "exchange" a risky asset for another risky asset (e.g., stock, commodity, currency, futures, etc.). It is a surprisingly fruitful concept which furnishes us with useful insights into the genuine meaning of an option. Notation-wise, we let, for $i = 1, 2$,

$S_i(t)$ be the time-t price of Asset i,

$F_{t,T}^P(S_i)$ be the time-t price of a prepaid forward on Asset i maturing at time T.

Consider a T-year European exchange option which gives you the right to exchange Asset 2 for Asset 1 at time T. More precisely, with the exchange option you have the right to give up one unit of Asset 2 in return for one unit of Asset 1 at time T. Its time-T payoff is the difference between the time-T price of Asset 1 and the time-T price of Asset 2, if this difference is positive:

$$\text{Exchange option payoff} = (S_1(T) - S_2(T))_+ \,.$$

In order that the exchange option finishes in-the-money, it is necessary and sufficient that $S_1(T) > S_2(T)$, or Asset 1 outperforms Asset 2. For this reason, an exchange option is also called an *outperformance option*.

Call or put?

The concept of an exchange option blurs the traditional distinction between a call and a put. To see this, consider a hypothetical executive compensation option with a payoff of

$$\text{Payoff} = (S_{\text{Apple}} - S_{\text{Samsung}})_+,$$

where S_{Apple} and S_{Samsung} are, respectively, the share prices of Apple and Samsung at the same reference point of time. Note that the option pays off if and only if Apple outperforms Samsung (i.e., when the share price of Apple is higher than that of Samsung). Endowed with this compensation option, the executive of Apple will have a higher incentive to work truly in Apple's interest.

Note that the payoff function of the compensation option can be viewed from two completely different angles:

1. If you think of Apple as the underlying asset and Samsung as the (random) strike, the compensation option is a call on Apple. This is how the executives of Apple view the compensation option.

2. If you regard Samsung as the underlying asset and Apple as the (random) strike, the compensation option becomes a put on Samsung.

The interesting observation is that whether the compensation option is a call or put depends critically on the specification of the underlying asset and strike asset. To avoid confusion, we will describe an exchange option by explicitly specifying which asset will be given up in exchange for which asset, without the use of the ambiguous terms "exchange call" and "exchange put."

8.2.2 Pricing Exchange Options

Black-Scholes pricing formula and its assumption.

Despite the co-existence of two risky assets, an exchange option enjoys a Black-Scholes-type pricing formula which is a simple variant of the plain vanilla Black-Scholes formula we have seen in Chapter 6. To price exchange options in the Black-Scholes framework, the most general assumption with respect to the future behavior of the two risky assets is:

Pricing Assumption

The ratios of the two prepaid forward prices, $\dfrac{F_{t,T}^P(S_1)}{F_{t,T}^P(S_2)}$ for $t \in [0,T]$, are lognormally distributed with volatility σ. That is,

$$\text{Var}\left[\ln\left(\frac{F_{t,T}^P(S_1)}{F_{t,T}^P(S_2)}\right)\right] = \sigma^2 t, \quad \text{for } 0 \le t \le T. \tag{8.2.1}$$

Under the above assumption, the time-0 price of the T-year European exchange option with payoff $(S_1(T) - S_2(T))_+$ takes the following familiar Black-Scholes form:[iii]

$$\text{BS}\left(F_{0,T}^P(S_1), 0; F_{0,T}^P(S_2), 0; \sigma, T\right) = F_{0,T}^P(S_1)N(d_1) - F_{0,T}^P(S_2)N(d_2), \tag{8.2.2}$$

where

$$d_1 = \frac{\ln[F_{0,T}^P(S_1)/F_{0,T}^P(S_2)] + (\sigma^2/2)T}{\sigma\sqrt{T}},$$

$$d_2 = d_1 - \sigma\sqrt{T} = \frac{\ln[F_{0,T}^P(S_1)/F_{0,T}^P(S_2)] - (\sigma^2/2)T}{\sigma\sqrt{T}}.$$

It should be stressed that the only distributional assumption that is required for the validity of (8.2.2) is the *ratio* of the two prepaid forward prices being lognormally distributed. The individual prepaid prices $\{F_{t,T}^P(S_1)\}_{t\in[0,T]}$ and $\{F_{t,T}^P(S_2)\}_{t\in[0,T]}$ themselves need not be lognormal random variables. If $\{F_{t,T}^P(S_1)\}_{t\in[0,T]}$ and $\{F_{t,T}^P(S_2)\}_{t\in[0,T]}$ indeed follow lognormal distributions, as is often the case in practice, with respective volatilities σ_1 and σ_2, and ρ is the constant correlation coefficient between the natural logarithms of the returns on the two prepaid forward prices, then we have the following explicit computational formula for σ:

By definition,[iv]

$$\begin{aligned}
\sigma^2 t &= \text{Var}\left(\ln[F_{t,T}^P(S_1)/F_{t,T}^P(S_2)]\right) \\
&= \text{Var}\left(\ln[F_{t,T}^P(S_1)] - \ln[F_{t,T}^P(S_2)]\right) \\
&= \text{Var}\left(\ln[F_{t,T}^P(S_1)]\right) + \text{Var}\left(\ln[F_{t,T}^P(S_2)]\right) - 2\text{Cov}\left(\ln[F_{t,T}^P(S_1)], \ln[F_{t,T}^P(S_2)]\right) \\
&= \sigma_1^2 t + \sigma_2^2 t - 2\rho\sigma_1\sigma_2 t
\end{aligned}$$

for $0 \le t \le T$, or

$$\sigma = \sqrt{\sigma_1^2 + \sigma_2^2 - 2\rho\sigma_1\sigma_2}, \tag{8.2.3}$$

[iii]Because we can always view an exchange option as being written on the (nondividend-paying) prepaid forwards on the two assets maturing at the same time as the option and the volatility parameter enters the pricing formula only through the term $\sigma\sqrt{T}$, which is the volatility applicable to the T-year horizon, we can suppress the second, fourth, and sixth arguments in (6.1.3) and produce the following minimalist version of the Black-Scholes pricing function:

$$\text{BS}(s_1; s_2; \sigma) := s_1 N(d_1) - s_2 N(d_2),$$

where

$$d_1 = \frac{\ln(s_1/s_2) + \sigma^2/2}{\sigma} \quad \text{and} \quad d_2 = d_1 - \sigma.$$

[iv]Recall that $\text{Var}(X - Y) = \text{Var}(X) + \text{Var}(Y) - 2\text{Cov}(X,Y)$ for two random variables X and Y.

which is sometimes known as the "blended volatility."

(Warning: In (8.2.3), we have $-2\rho\sigma_1\sigma_2$, not $+2\rho\sigma_1\sigma_2$. The negative sign stems from taking the logarithm of the *ratio* of the two prepaid forward prices, or the *difference* between the natural logarithms of the two prepaid forward prices.)

A "change-of-unit" proof of (8.2.2).

There are different ways to establish the pricing formula of exchange options (see, for example, pages 483 and 484 of Panjer (1998) for a direct calculus-based proof). A swift and instructive proof of (8.2.2) is due to the pioneering paper by Margrabe (1978) (in recognition of Margrabe (1978)'s contributions, exchange options are also known as Margrabe's options) which challenges our conventional notion of what a risk-free asset is. So far, we have been quantifying the value of assets in terms of a common benchmark asset, which is usually a currency, say dollars. Defying convention, here we liberate the notion of a risk-free, benchmark asset from cash to a risky asset, which we take as the prepaid forward on Asset 2 in this context.

We begin by noticing that the payoff of the exchange option, in terms of dollars, is $\$(S_1(T) - S_2(T))_+$. In terms of the time-$T$ value of the prepaid forward on Asset 2, this becomes

$$\frac{\$(S_1(T) - S_2(T))_+}{\$S_2(T)} = \left(\frac{F_{T,T}^P(S_1)}{F_{T,T}^P(S_2)} - 1\right)_+ \quad \text{units of prepaid forward on Asset 2,}$$

which is the payoff of a 1-strike (in units of the prepaid forward on Asset 2) *plain vanilla* call option with the underlying asset being one unit of the prepaid forward on Asset 1 measured in units of the prepaid forward on Asset 2. By the plain vanilla Black-Scholes formula with $S(0) = F_{0,T}^P(S_1)/F_{0,T}^P(S_2)$ (the current value of the underlying asset with respect to the benchmark), $K = 1$, $\delta = 0$, and $r = 0$ (the dividend yield of a prepaid forward is zero), the time-0 price of this call option, in units of the prepaid forward on Asset 2 again, is

$$\text{BS}(F_{0,T}^P(S_1)/F_{0,T}^P(S_2), 0; 1, 0; \sigma, T) = \left[\frac{F_{0,T}^P(S_1)}{F_{0,T}^P(S_2)}\right] N(d_1) - N(d_2),$$

where

$$d_1 = \frac{\ln[F_{0,T}^P(S_1)/F_{0,T}^P(S_2)] + (\sigma^2/2)T}{\sigma\sqrt{T}},$$

$$d_2 = d_1 - \sigma\sqrt{T},$$

with σ being the volatility of the prepaid forward on Asset 1 in units of the prepaid forward on Asset 2. Finally, multiplying the time-0 price of the call in units of the prepaid forward on Asset 2 by $F_{0,T}^P(S_2)$ gives the traditional dollar price

$$F_{0,T}^P(S_1)N(d_1) - F_{0,T}^P(S_2)N(d_2),$$

which is nothing but (8.2.2).

Special cases.

It should be noted that (8.2.2) is the most general Black-Scholes-type pricing formula in this entire book. It closely resembles the ordinary Black-Scholes formula presented in Section 6.1 and is the "master formula" which incorporates several important special cases:

1. A plain vanilla European call on a stock can be regarded as an exchange option giving its holder the right to exchange Asset 2, which is cash always worth the strike price K at time T (i.e., $S_2(T) = K$), for Asset 1, which is the underlying risky stock worth $S(T)$ at time T. Under the assumption that prepaid forward prices on the stock are lognormally distributed with volatility σ_1, we can specialize (8.2.2) with $F_{0,T}^P(S_1) = F_{0,T}^P(S)$, $F_{0,T}^P(S_2) = F_{0,T}^P(K)$, and $\sigma_2 = 0$ (because asset 2 is risk-free), to retrieve the plain vanilla call price formula:

$$C = \text{BS}\left(F_{0,T}^P(S), 0; F_{0,T}^P(K), 0; \sigma, T\right) = F_{0,T}^P(S)N(d_1) - F_{0,T}^P(K)N(d_2),$$

where the blended volatility is

$$\sigma = \sqrt{\sigma_1^2 + \sigma_2^2 - 2\rho\sigma_1\sigma_2} = \sqrt{\sigma_1^2 + 0^2 - 2\rho\sigma_1(0)} = \sigma_1,$$

which is the volatility of the prepaid forward on the stock, and

$$
\begin{aligned}
d_1 &= \frac{\ln[F_{0,T}^P(S)/F_{0,T}^P(K)] + (\sigma^2/2)T}{\sigma\sqrt{T}}, \\
d_2 &= d_1 - \sigma\sqrt{T}.
\end{aligned}
$$

2. A plain vanilla European put can also be viewed as an exchange option giving its holder the right to give up the underlying risky stock (Asset 2) in return for K units of cash (Asset 1). Applying (8.2.2) with $F_{0,T}^P(S_1) = F_{0,T}^P(K)$, $F_{0,T}^P(S_2) = F_{0,T}^P(S)$, $\sigma_1 = 0$ (because asset 1 is risk-free) and σ being the volatility of the prepaid forward on the stock recovers the plain vanilla put price formula:

$$P = \text{BS}\left(F_{0,T}^P(K), 0; F_{0,T}^P(S), 0; \sigma, T\right) = F_{0,T}^P(K)N(-d_2) - F_{0,T}^P(S)N(-d_1),$$

where d_1 and d_2 are defined above.

Example 8.2.1. (How does ρ affect the price of an exchange option?) Assume the Black-Scholes framework. Consider two stocks, each of which pays dividends continuously at a rate proportional to its price. For $j = 1, 2$ and $t \geq 0$, let $S_j(t)$ be the time-t price of one share of stock j.

You are given:

(i) $S_1(0) = S_2(0) = 200$.

(ii) Stock 1 and Stock 2 share the same dividend yield of 5%.

(iii) Stock 1's volatility is 30%.

(iv) Stock 2's volatility is 20%.

Consider a 6-month European exchange option to exchange Stock 2 for Stock 1.

(a) Calculate the price of the exchange option when $\rho = 0$.

(b) Calculate the price of the exchange option when $\rho = 0.7$.

(c) In the light of the results in parts (a) and (b), describe how the price of an exchange option behaves as a function of the correlation ρ.

Solution. (a) The blended volatility is

$$\sigma = \sqrt{0.3^2 + 0.2^2 - 2(0.3)(0.2)(0)} = 0.36056.$$

As

$$d_1 = \frac{\ln(200/200) + (0.05 - 0.05 + 0.36056^2/2)(0.5)}{0.36056\sqrt{0.5}} = 0.12748,$$

$$d_2 = d_1 - 0.36056\sqrt{0.5} = -0.12748,$$

$$N(d_1) = 0.55072,$$

$$N(d_2) = 0.44928,$$

the price of the exchange option is

$$200e^{-0.05(0.5)}(0.55072) - 200e^{-0.05(0.5)}(0.44928) = \boxed{19.79}.$$

(b) Now the blended volatility becomes

$$\sigma = \sqrt{0.3^2 + 0.2^2 - 2(0.3)(0.2)(0.7)} = 0.21448.$$

As

$$d_1 = \frac{\ln(200/200) + (0.05 - 0.05 + 0.21448^2/2)(0.5)}{0.21448\sqrt{0.5}} = 0.07583,$$

$$d_2 = d_1 - 0.21448\sqrt{0.5} = -0.07583,$$

$$N(d_1) = 0.53022,$$

$$N(d_2) = 0.46978,$$

the price of the exchange option is

$$200e^{-0.05(0.5)}(0.53022) - 200e^{-0.05(0.5)}(0.46978) = \boxed{11.79}.$$

(c) The results in parts (a) and (b) suggest that the price of an exchange option decreases as the value of ρ increases. This comes as no surprise to us because the blended volatility $\sigma = \sqrt{\sigma_1^2 + \sigma_2^2 - 2\rho\sigma_1\sigma_2}$ is decreasing in ρ, and the Black-Scholes (call or put) price function is increasing in σ (i.e., vega is positive). $\qquad\square$

Exchanging unequal volumes of risky assets.

Thus far, we have focused on the simple case when one unit of risky asset can be exchanged for one unit of another risky asset at the maturity of the exchange option. We now generalize our discussion to the case when the exchange ratio is not necessarily one-to-one. More precisely, suppose that a European exchange option allows its holder to obtain c_1 units of Asset 1 by giving up c_2 units of Asset 2 at time T, with $c_1 \neq c_2$. In practice, an exchange option can be structured this way (e.g., $c_1 = 1$ and $c_2 = S_1(0)/S_2(0)$) to make it at-the-money at inception.

In this more general setting, one may be tempted to compute the blended volatility σ

by scaling up σ_1^2 and σ_2^2 in (8.2.3) to $c_1^2\sigma_1^2$ and $c_2^2\sigma_2^2$, respectively. This turns out to be completely incorrect! The truth is that σ should remain unchanged at

$$\sigma = \sqrt{\sigma_1^2 + \sigma_2^2 - 2\rho\sigma_1\sigma_2}. \quad \text{(why?)}$$

However, the components of the pricing formula do need to be scaled up by c_1 and c_2 appropriately:

$$\text{BS}\left(c_1 F_{0,T}^P(S_1), 0; c_2 F_{0,T}^P(S_2), 0; \sigma, T\right) = \boxed{c_1} \times F_{0,T}^P(S_1)N(d_1) - \boxed{c_2} \times F_{0,T}^P(S_2)N(d_2),$$

where

$$d_1 = \frac{\ln\left[\dfrac{c_1 F_{0,T}^P(S_1)}{c_2 F_{0,T}^P(S_2)}\right] + (\sigma^2/2)T}{\sigma\sqrt{T}} \quad \text{and} \quad d_2 = d_1 - \sigma\sqrt{T}.$$

Example 8.2.2. (An exchange option exchanging unequal volumes of assets)
Consider two nondividend-paying stocks. For $j = 1, 2$, and $t \geq 0$, let $S_j(t)$ denote the price of one share of stock j at time t (in years). Under the Black-Scholes framework, you price a four-month European exchange option that provides the right to obtain $S_1(0)/S_2(0)$ shares of stock 2 in exchange for one share of stock 1. You are given:

(i) The current price of stock 1 is \$100. The current price of stock 2 is \$300.

(ii) The volatility of stock 1 is 10%. The volatility of stock 2 is 20%.

(iii) The correlation between the continuously compounded returns on the two stocks is 0.5.

(iv) The continuously compounded risk-free interest rate is 15%.

Determine the current price of the exchange option.

Solution. Let's first calculate the blended volatility σ:

$$\sigma = \sqrt{\sigma_1^2 + \sigma_2^2 - 2\rho\sigma_1\sigma_2} = \sqrt{0.1^2 + 0.2^2 - 2(0.5)(0.1)(0.2)} = \sqrt{0.03}.$$

Note that the dollar value of stock 2 that can be exchanged for stock 1 is $[S_1(0)/S_2(0)]S_2(0) = S_1(0)$, meaning that the exchange option is at-the-money. With

$$
\begin{aligned}
d_1 &= \frac{\ln(1) + (\sigma^2/2)T}{\sigma\sqrt{T}} = \frac{\sigma\sqrt{T}}{2} = \frac{\sqrt{0.03(1/3)}}{2} = 0.05, \\
d_2 &= d_1 - \sigma\sqrt{T} = 0.05 - \sqrt{0.03/3} = -0.05, \\
N(d_1) &= 0.51994, \\
N(d_2) &= 0.48006,
\end{aligned}
$$

the price of the exchange option is $100(0.51994) - 100(0.48006) = \boxed{3.988}$. □

8.2.3   Pricing Maximum and Minimum Contingent Claims

In this subsection, we present two exotic derivatives whose payoff is the maximum or minimum of the prices of two assets over a fixed horizon. These derivatives are intimately connected to exchange options and can be priced using the techniques in Subsection 8.2.2.

Maximum claims.

Consider a T-year *maximum contingent claim*[v] whose time-T payoff is

$$\max\left(S_1(T), S_2(T)\right).$$

This contingent claim can be of value to an investor who is undecided about which stock to hold at maturity. In terms of the payoff of an exchange option, the payoff of the maximum claim can be rewritten either as

$$\max\left(S_1(T), S_2(T)\right) = S_1(T) + \underbrace{\left(S_2(T) - S_1(T)\right)_+}_{\text{exchange Asset 1 for Asset 2}}$$

or as

$$\max\left(S_1(T), S_2(T)\right) = S_2(T) + \underbrace{\left(S_1(T) - S_2(T)\right)_+}_{\text{exchange Asset 2 for Asset 1}}.$$

These two representations imply that the maximum claim is the same as holding one of the two assets alongside a T-year European option to exchange this asset for the other asset. It follows that the Black-Scholes price of the maximum claim, denoted by $V^{\max}$, is

$$
\begin{aligned}
V^{\max} &= F_{0,T}^P(S_1) + \mathrm{BS}\left(F_{0,T}^P(S_2), 0; F_{0,T}^P(S_1), 0; \sigma, T\right) \\
&= F_{0,T}^P(S_2) + \mathrm{BS}\left(F_{0,T}^P(S_1), 0; F_{0,T}^P(S_2), 0; \sigma, T\right),
\end{aligned}
\tag{8.2.4}
$$

depending on which way of valuation is easier (see Problem 8.9.25 for the explicit pricing formula).

Example 8.2.3. (SOA Exam MFE Spring 2007 Question 6: Pricing a maximum claim) Consider a model with two stocks. Each stock pays dividends continuously at a rate proportional to its price. $S_j(t)$ denotes the price of one share of stock j at time t.

Consider a claim maturing at time 3. The payoff of the claim is

$$\text{Maximum}\left(S_1(3), S_2(3)\right).$$

You are given:

(i) $S_1(0) = \$100$.

(ii) $S_2(0) = \$200$.

(iii) Stock 1 pays dividends of amount $0.05 S_1(t)\,\mathrm{d}t$ between time t and time $t + \mathrm{d}t$.

(iv) Stock 2 pays dividends of amount $0.1 S_2(t)\,\mathrm{d}t$ between time t and time $t + \mathrm{d}t$.

---

[v]We prefer not to use the term "maximum option," which commonly refers to an option on the maximum of two asset prices (e.g., $(\max(S_1(T), S_2(T)) - K)_+$).

(v) The price of a European option to exchange Stock 2 for Stock 1 at time 3 is $10.

Calculate the price of the claim.

(A) $96

(B) $145

(C) $158

(D) $200

(E) $234

Solution. We begin by decomposing the payoff of the maximum claim in terms of the payoff of the exchange option described in (v):

$$\max \left(S_1(3), S_2(3)\right) = S_2(3) + \underbrace{\left(S_1(3) - S_2(3)\right)_+}_{\text{exchange option in (v)}}$$

Thus the price of the maximum claim is

$$V^{\max} = F^P_{0,3}(S_2) + \underbrace{10}_{\text{from (v)}} = 200e^{-0.1(3)} + 10 = \boxed{158.16}. \quad (\textbf{Answer: (C)})$$

$\square$

Remark. (i) This question does not require the Black-Scholes assumptions.

(ii) The current price and dividend yield of Stock 1 are not needed for this question.

Minimum claims.

In a similar vein, for a T-year *minimum contingent claim* with time-T payoff given by

$$\min \left(S_1(T), S_2(T)\right),$$

one can rewrite its payoff in two different ways:

1. In terms of the payoff of the maximum claim, we have

$$\min \left(S_1(T), S_2(T)\right) = S_1(T) + S_2(T) - \max \left(S_1(T), S_2(T)\right)$$

because of the identity $\max(x, y) + \min(x, y) = x + y$, which may be familiar to students who have learned multiple-life theory in life contingencies. It follows that the current price of the minimum claim, written as $V^{\min}$, is

$$V^{\min} = F^P_{0,T}(S_1) + F^P_{0,T}(S_2) - V^{\max},$$

with $V^{\max}$ given in (8.2.4). A downside to this form of the pricing formula is that the price of the maximum claim is a prerequisite to using this formula.

2. Pulling either $S_1(T)$ or $S_2(T)$ outside the minimum operator, we have

$$\begin{aligned}
\min \left(S_1(T), S_2(T)\right) &= S_1(T) + \min \left(S_2(T) - S_1(T), 0\right) \\
&= S_1(T) - \max \left(S_1(T) - S_2(T), 0\right) \\
&= S_1(T) - \left(S_1(T) - S_2(T)\right)_+
\end{aligned}$$

or

$$\begin{aligned}
\min\left(S_1(T), S_2(T)\right) &= S_2(T) + \min\left(S_1(T) - S_2(T), 0\right) \\
&= S_2(T) - \left(S_2(T) - S_1(T)\right)_+,
\end{aligned}$$

where the identity $\min(x, y) = -\max(-x, -y)$ is used. Hence we can state the price of the minimum claim directly in terms of the prices of appropriate exchange options:

$$\begin{aligned}
V^{\min} &= F_{0,T}^P(S_1) - \text{BS}\left(F_{0,T}^P(S_1), 0; F_{0,T}^P(S_2), 0; \sigma, T\right) \\
&= F_{0,T}^P(S_2) - \text{BS}\left(F_{0,T}^P(S_2), 0; F_{0,T}^P(S_1), 0; \sigma, T\right).
\end{aligned}$$

Example 8.2.4. (SOA Exam IFM Advanced Derivatives Sample Question 54: An option on the maximum of two stock prices) Assume the Black-Scholes framework. Consider two nondividend-paying stocks whose time-t prices are denoted by $S_1(t)$ and $S_2(t)$, respectively.

You are given:

(i) $S_1(0) = 10$ and $S_2(0) = 20$.

(ii) Stock 1's volatility is 0.18.

(iii) Stock 2's volatility is 0.25.

(iv) The correlation between the continuously compounded returns of the two stocks is -0.40.

(v) The continuously compounded risk-free interest rate is 5%.

(vi) A one-year European option with payoff $\max\{\min[2S_1(1), S_2(1)] - 17, 0\}$ has a current (time-0) price of 1.632.

Consider a European option that gives its holder the right to sell either two shares of Stock 1 or one share of Stock 2 at a price of 17 one year from now.

Calculate the current (time-0) price of this option.

(A) 0.67

(B) 1.12

(C) 1.49

(D) 5.18

(E) 7.86

Ambrose's comments:

This is a hard problem. It requires careful analysis of complex payoff functions, correct identification of the underlying asset, and an ingenious application of put-call parity.

Solution. The payoff of the European option in question giving its holder the right to sell either two shares of Stock 1 or one share of Stock 2, whichever is worth less, at a price of 17 one year from now is

$$(17 - \underbrace{\min[2S_1(1), S_2(1)]}_{\substack{\text{you want to give up the} \\ \text{one with the lower value}}})_+,$$

which is the payoff of a 1-year 17-strike *put* on the 1-year minimum claim whose payoff is $\min[2S_1(1), S_2(1)]$. The price of the option in (vi) is the *call* counterpart, with 1-year payoff given by

$$(\min[2S_1(1), S_2(1)] - 17)_+.$$

Applying put-call parity with the 1-year minimum claim serving as the underlying asset, we have

$$1.632 - P(17, 1) = F_{0,1}^P[\min(2S_1, S_2)] - 17e^{-0.05},$$

so it remains to evaluate the time-0 price of the minimum claim, $F_{0,1}^P[\min(2S_1, S_2)]$.

To proceed, we write

$$
\begin{aligned}
\min[2S_1(1), S_2(1)] &= 2S_1(1) + \min[S_2(1) - 2S_1(1), 0] \\
&= 2S_1(1) - \max[2S_1(1) - S_2(1), 0]
\end{aligned}
$$

and notice that $\max[2S_1(1) - S_2(1), 0] = (2S_1(1) - S_2(1))_+$ is the payoff of a 1-year at-the-money European exchange option giving you the right to give up one unit of Stock 2 in return for two units of Stock 1. Here are the routine Black-Scholes calculations for valuing this exchange option:

$$
\begin{aligned}
\sigma &= \sqrt{\sigma_1^2 + \sigma_2^2 - 2\rho\sigma_1\sigma_2} \\
&= \sqrt{0.18^2 + 0.25^2 - 2(-0.40)(0.18)(0.25)} \\
&= 0.361801, \\
d_1 &= \frac{\ln[2S_1(0)/S_2(0)] + (\sigma^2/2)T}{\sigma\sqrt{T}} = \frac{\sigma\sqrt{T}}{2} = 0.18090, \\
d_2 &= d_1 - \sigma\sqrt{T} = -0.18090, \\
N(d_1) &= 0.57178, \\
N(d_2) &= 0.42822.
\end{aligned}
$$

Therefore, the price of the exchange option is $20(0.57178) - 20(0.42822) = 2.8712$, and the price of the minimum claim is

$$F_{0,1}^P[\min(2S_1, S_2)] = 2F_{0,1}^P(S_1) - 2.8712 = 2(10) - 2.8712 = 17.1288.$$

Finally, the price of the required put on the minimum claim is

$$P(17, 1) = 1.632 - 17.1288 + 17e^{-0.05} = \boxed{0.6741}. \quad (\textbf{Answer: (A)})$$

$\square$

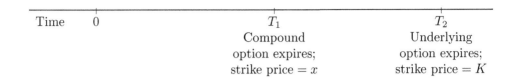

FIGURE 8.3.1
A timeline diagram showing how a compound option works.

8.3   Compound Options

Definition.

A *compound option* (or a compound in short) is an option written on another option, i.e., the underlying asset of the compound is an option in itself. A compound can be more intricate than a plain vanilla option because of the existence of two strikes and two expiration dates, one each for the compound option and the underlying option (see the timeline diagram in Figure 8.3.1). The strike price of the compound option is the price at which the holder may choose to buy (in the case of a compound call) or sell (in the case of a compound put) the underlying option, which matures later. As each of the compound option and the underlying option can be a call or a put, there are four kinds of compound options: a call on a call (CallOnCall), a call on a put (CallOnPut), a put on a call (PutOnCall), and a put on a put (PutOnPut).

Just as a plain vanilla option allows its holder to lock in the price for trading the underlying asset, a compound also enables its holder to lock in the price for the insurance offered by the underlying option while delaying the decision on whether or not to acquire that insurance to a later time. You will see a clearer picture of the mechanics of and motivation behind a compound option in Example 8.3.1 below.

Pricing using a binomial tree model.

Valuing a compound option in the binomial option pricing model is only cosmetically different from pricing a plain vanilla option. All of the essential ideas remain. The key is to transform the binomial stock price tree into a binomial tree for the price of the option that underlies the compound. As soon as this is done, all the usual pricing techniques for binomial trees in Chapter 4 carry over easily. Here is an example.

> **Example 8.3.1. (CAS Exam 3 Fall 2007 Question 28: Pricing a compound option by binomial trees)** You are given the following information on a compound CallOnPut option:
>
> - The continuously compounded risk-free rate is 5%.
>
> - The strike price of the underlying option is 43.
>
> - The strike price of the compound option is 3.

- The compound option expires in 6 months.

- The underlying option expires 6 months after the compound option.

- The underlying option is American.

$$
\begin{array}{ccc}
\underline{\text{Today}} & \underline{\text{6 months}} & \underline{\text{12 months}} \\
 & & S_{HH} = 64.52 \\
 & S_H = 50.80 & \\
S_0 = 40.00 & & S_{HL} = 42.16 \\
 & S_L = 33.20 & \\
 & & S_{LL} = 27.56
\end{array}
$$

Based on the above binomial stock price tree, calculate the value of the compound option.

(A) Less than 3.00

(B) At least 3.00, but less than 3.50

(C) At least 3.50, but less than 4.00

(D) At least 4.00, but less than 4.50

(E) At least 4.50

Solution. We first derive the evolution of the price of the underlying American put option on which the compound call is written. With $u = 1.27$ and $d = 0.83$, the risk-neutral probability of an up move is (assuming no dividends)

$$
p^* = \frac{e^{0.05(0.5)} - 0.83}{1.27 - 0.83} = 0.443898.
$$

- *H node:* If the stock price rises to 50.80 in 6 months, then by risk-neutral pricing, the 6-month price of the American put is

$$
P_H = \max\{e^{-0.05(0.5)}(1 - 0.443898)(43 - 42.16), 43 - 50.80\} = 0.455592.
$$

Because the strike price of the CallOnPut is 3, which is higher than P_H, its payoff is zero: $V_H = (P_H - 3)_+ = 0$.

- *L node:* If the stock price drops to 33.20 in 6 months, then the 6-month price of the American put becomes

$$
P_L = \max\{e^{-0.05(0.5)}(1 - 0.443898)(43 - 27.56), 43 - 33.2\} = 9.8,
$$

with early exercise being optimal. The payoff of the CallOnPut is $V_L = (P_L - 3)_+ = 6.8$.

Here is the one-period binomial tree for the prices of the American put: (the values of the CallOnPut are shown in parentheses)

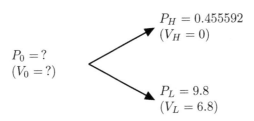

$$P_H = 0.455592$$
$$(V_H = 0)$$

$$P_0 = ?$$
$$(V_0 = ?)$$

$$P_L = 9.8$$
$$(V_L = 6.8)$$

Finally, by risk-neutral pricing again, the current price of the CallOnPut (assumed to be European) is

$$
\begin{aligned}
V_0 &= e^{-0.05(0.5)}[p^* V_H + (1 - p^*) V_L] \\
&= e^{-0.05(0.5)}[0.443898(0) + (1 - 0.443898)(6.8)] \\
&= \boxed{3.6881}. \quad \textbf{(Answer: (C))}
\end{aligned}
$$

$\square$

Why buy a compound option? Why so complex?

If the result of a compound option is to allow its holder to buy or sell the underlying option, why not trade the underlying option directly? The appeal of a compound option is that it provides its holder with not only the ability to lock in a price for insurance, but also the *flexibility* to postpone the decision on whether or not to obtain that insurance. With the compound option, its holder can wait and buy or sell the underlying option at a later time if doing so is to his/her advantage.

In Example 8.3.1 above, you can buy the 1-year American put at the outset for $5.5125 (check this). If you do not want to pay this much now but still desire downside protection in 1 year, you may opt for the 6-month compound call on the American put for a smaller initial investment of $3.6881. Should the stock price drop to $33.2 after 6 months, you can exercise the compound call after 6 months, pay an additional payment of $3 to own the put, and be protected on the downside.

Black-Scholes pricing formulas for compound options.

Pricing a compound option in the Black-Scholes framework is no easy task. The technical difficulty, in brief, stems from the fact that even if future prices of the primitive are lognormally distributed, the prices of the option on which the compound is written do not follow a lognormal distribution. In fact, the generally non-zero probability that the underlying option expires unexercised invalidates the lognormality assumption on its terminal payoff. We refer interested readers to Geske (1979) for the closed-form pricing formulas of compound options. You may be surprised to know that the pricing formula involves the bivariate normal distribution function.

Compound option parity.

In the absence of an option pricing model, statements about the prices of European compound options can still be made if we specialize put-call parity to the case when the underlying asset is an option rather than a stock. Doing so makes the put-call parity equation take the following special forms:

For compound options having the same strike price x and same time to expiration T_1, we have

$$\text{Price of CallOnCall} - \text{Price of PutOnCall} = \text{Price of call} - xe^{-rT_1},$$
$$\text{Price of CallOnPut} - \text{Price of PutOnPut} = \text{Price of put} - xe^{-rT_1}.$$

You should notice that when computing the present value of the strike price, it is the common strike price of the compound options (not that of the underling option) that gets discounted, and that the duration for discount is T_1, which is the "lifespan" of the compound options (but not that of the underlying option, which matures later, say at time T_2 with $T_1 \le T_2$).

Example 8.3.2. (Multiple applications of parity) You are given:

(i) The current price of a stock is 42.

(ii) The stock pays dividends continuously at a rate proportional to its price. The dividend yield is 3%.

(iii) The continuously compounded risk-free interest rate is 5%.

(iv) The following table shows the prices of 3-month 4-strike European compound options, whose underlying options are 45-strike and expire 1 year from now:

	Call	Put
Call on	2.20	3.14
Put on	1.46	?

Calculate the price of the PutOnPut option.

Ambrose's comments:

This problem involves altogether *three* applications of parity – compound option parity twice, and ordinary put-call parity once!

Solution. Note the two different times to expiration, $T_1 = 0.25$ and $T_2 = 1$. Applying compound option parity to the CallOnCall and PutOnCall,

$$\text{CallOnCall price} - \text{PutOnCall price} = \text{Call price} - xe^{-rT_1}$$
$$2.20 - 1.46 = \text{Call price} - 4e^{-0.05(0.25)}$$
$$\text{Call price} = 4.690311.$$

Using put-call parity for the T_2-year plain vanilla options, we have

$$\text{Put price} = \text{Call price} - S(0)e^{-\delta T_2} + Ke^{-rT_2}$$
$$= 4.690311 - 42e^{-0.03(1)} + 45e^{-0.05(1)}$$
$$= 6.736923.$$

Applying compound option parity once more, this time to the CallOnPut and PutOnPut, we have

$$\text{PutOnPut price} = \text{CallOnPut price} - \text{Put price} + xe^{-rT_1}$$
$$= 3.14 - 6.736923 + 4e^{-0.05(0.25)}$$
$$= \boxed{0.3534}.$$

□

8.4   Asian Options

8.4.1   Introduction

Asian options[vi] are exotic options whose payoffs depend on the *average* price of the under-lying stock over the life of the option, where the manner in which the average is determined is specified in the contract. Due to this averaging feature, Asian options are *path-dependent* options, meaning that the payoff of the option at expiration depends not only on the ter-minal price of the stock, but also on the *path* by which the stock arrives at that terminal price. In general, path dependence renders the pricing and hedging of exotic options com-plex because of the need for treating stock price paths separately even if they lead to the same terminal price.

Taxonomy.

There are eight($= 2^3$) basic types of Asian options, described by the following three dimen-sions (see Figure 8.4.1):

1. Is the average defined as an arithmetic average or a geometric average? For given stock prices $S(h), S(2h), \ldots, S(nh)$ observed every h years over a horizontal of length T for a total of $n = T/h$ observations (values of h and n are specified in the contract), these two averages are defined, respectively, as

$$A(T) = \frac{1}{n} \sum_{i=1}^{n} S(ih) \quad \text{and} \quad G(T) = \left[\prod_{i=1}^{n} S(ih) \right]^{1/n}.$$

   Whereas arithmetic averages are used more often in practice, they are not as mathemat-ically tractable as their geometric counterparts when it comes to pricing and hedging.

2. Does the average defined in the preceding dimension substitute the price of the under-lying stock or the strike price in the plain vanilla payoff formula? When the average replaces the stock price, the option is called an *average price* option. When the average is applied to the strike price, the option is known as an *average strike* option.

3. Is the option a call or a put in nature? The plain vanilla call or put payoff formula, $(S(T) - K)_+$ or $(K - S(T))_+$, is then modified in the above two dimensions.

   For instance, the payoffs of the four European T-year Asian options based on geometric

---

[vi] Asian options are called "Asian" not because their inventors were Asian, but because they (Mark Standish and David Spaughton) were in Asia (more precisely, Tokyo) when they developed these average-dependent options.

"Arithmetic average'		"price"		"Call"
or	×	(if average replaces stock "price")	×	or
		or		
"Geometric average"		"strike"		"Put"
		(if average replaces "strike")		

FIGURE 8.4.1
The three-dimensional descriptions of Asian options.

averages are:

$$\text{Geometric average price } K\text{-strike call} \; = \; (\; \overbrace{G(T)}^{\text{replaces } S(T)} \; - K)_+,$$

$$\text{Geometric average price } K\text{-strike put} \; = \; (K - \overbrace{G(T)}^{\text{replaces } S(T)})_+,$$

$$\text{Geometric average strike call} \; = \; (S(T) - \underbrace{G(T)}_{\text{replaces } K})_+,$$

$$\text{Geometric average strike put} \; = \; (\underbrace{G(T)}_{\text{replaces } K} - S(T))_+.$$

The four analogous expressions for arithmetic Asian options are obtained by replacing $G(T)$ above by $A(T)$.

Example 8.4.1. (CAS Exam 3 Fall 2007 Question 27: Payoff of a geometric average price call) At the beginning of the year, a speculator purchases a six-month geometric average price call option on a company's stock. The strike price is 3.5. The payoff is based on an evaluation of the stock price at each month's end.

Date	Stock Price
January 31	1.27
February 28	4.11
March 31	5.10
April 30	5.50
May 31	5.13
June 30	4.70

Based on the above stock prices, calculate the payoff of the option.

(A) Less than .3

(B) At least .3, but less than .75

(C) At least .75, but less than 1.00

(D) At least 1.00, but less than 1.75

(E) At least 1.75

Solution. The geometric average is

$$G = (1.27 \times 4.11 \times 5.10 \times 5.50 \times 5.13 \times 4.70)^{1/6} = 3.902105.$$

The payoff of the 3.5-strike six-month geometric average price call option is

$$(3.902105 - 3.5)_+ = \boxed{0.4021}. \quad (\textbf{Answer: (B)})$$

$\square$

Why trade Asian options?

Arguably the most common use of Asian options is to address a very practically common hedging need, namely to hedge the exposure to "average" or the "total." This bears particular relevance to investors engaged in *repeated* transactions in the same underlying asset over time. For the purpose of illustration, consider Company A back in Example 6.2.4, which is a US-based international company with an operation in Japan (before selling it to Company B). It receives revenue in Japanese yen on a regular basis, say monthly, as a result of its Japanese operation and the Japanese yen received will be converted back to US dollars. Company A is therefore susceptible to the depreciation of Japanese yen against US dollar not at a particular point of time, but *at the end of every month*.

To hedge such exchange rate risk, Company A may consider several alternatives. For concreteness, we fix a horizon of interest, say one year. An obvious hedging strategy, in the spirit of Section 3.1.1, is to purchase 12 plain vanilla dollar-denominated puts on Japanese yen with the common strike price K, one expiring at the end of each month, to create 12 floors. In effect, this ensures that the revenue Company A will receive in US dollars will be at least K at the end of each month, or $12K$ on a yearly basis (ignoring time value of money for simplicity). A more economical alternative is to purchase 12 arithmetic average price 1-year K-strike dollar-denominated Asian puts on Japanese yen. If we let $X(i/12)$ be the yen/dollar exchange rate at the end of month i, then the year-end payoff of Company A as a result of the use of the Asian puts is (again, ignoring time value of money)

$$\sum_{i=1}^{12} X(i/12) + 12\left(K - \frac{1}{12}\sum_{i=1}^{12} X(i/12)\right)_+ = \max\left(12K, \sum_{i=1}^{12} X(i/12)\right),$$

which is bounded from below by $12K$. Whereas these Asian puts also place a floor on the annual revenue Company A will receive in US dollars, they cost less than the 12 plain vanilla puts. Loosely speaking, the volatility of the average is less than the volatility of a single stock price, rendering the Asian puts cheaper. Furthermore, there are possible instances in which some of the individual plain vanilla puts expire in-the-money, but the average yen/dollar exchange is so high that the Asian puts do not and thus are less valuable.

8.4.2 Pricing Asian Options

Binomial tree models.

Pricing an Asian option using a binomial tree is an instructive exercise that not only sheds light on the cash flow characteristics of an Asian option, but also reinforces our understanding of the mechanics of binomial option pricing we learned back in Chapter 4. We calculate the parameters u and d and build a binomial stock price tree as usual but keep in mind that

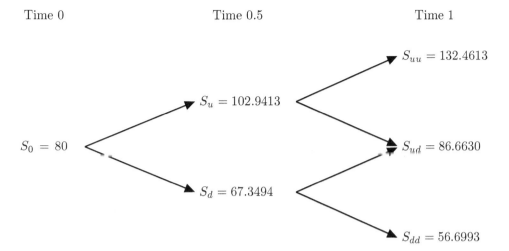

Time 0   Time 0.5   Time 1

$S_{uu} = 132.4613$

$S_u = 102.9413$

$S_0 = 80$

$S_{ud} = 86.6630$

$S_d = 67.3494$

$S_{dd} = 56.6993$

FIGURE 8.4.2
The two-period binomial stock price tree for Example 8.4.2.

different stock price paths with the same terminal stock price may entail different interim stock prices and therefore possibly different payoffs for an Asian option (e.g., in a two-period tree, the ud path and du path should be carefully distinguished). Therefore, for an n-period binomial tree, a total number of 2^n stock price paths need to be dealt with, in contrast to merely $n + 1$ terminal stock prices for path-independent options. As soon as the payoffs have been determined *path by path*, the same valuation principle in Chapter 4 applies:

> Discount the expected value of the payoff under the risk-neutral probability at the risk-free interest rate, *on a path-by-path basis*.

We illustrate these ideas in the following example.

Example 8.4.2. (Valuing an Asian option by a binomial tree) You use the following information to construct a binomial forward tree for modeling the price movements of a nondividend-paying stock:

(i) The length of each period is 6 months.

(ii) The current stock price is 80.

(iii) The stock's volatility is 30%.

(iv) The continuously compounded risk-free interest rate is 8%.

Calculate the price of a 1-year at-the-money geometric average price Asian European call option on the stock, with the average calculated based on ending 6-month stock prices (not including the initial stock price).

Solution. In a forward tree, the up and down parameters are, respectively,

$$u = \exp[(r - \delta)h + \sigma\sqrt{h}] = \exp[0.08(0.5) + 0.3\sqrt{0.5}] = 1.286766,$$
$$d = \exp[(r - \delta)h - \sigma\sqrt{h}] = \exp[0.08(0.5) - 0.3\sqrt{0.5}] = 0.841868,$$

and the risk-neutral probability of an up move is

$$p^* = \frac{1}{1 + e^{\sigma\sqrt{h}}} = \frac{1}{1 + e^{0.3\sqrt{0.5}}} = 0.447165.$$

The resulting stock prices are depicted in Figure 8.4.2, from which we can calculate the geometric average and the payoff $(= (G(1) - 80)_+)$ of the geometric average price Asian call corresponding to different stock price paths:

Path	Geometric Average $G(1)$	Payoff	RN Probability
uu	$\sqrt{(102.9413)(132.4613)} = 116.7722$	36.7722	$(p^*)^2$
ud	$\sqrt{(102.9413)(86.6630)} = 94.4521$	14.4521	$(p^*)(1 - p^*)$
du	$\sqrt{(67.3494)(86.6630)} = 76.3983$	0	$(p^*)(1 - p^*)$
dd	$\sqrt{(67.3494)(56.6993)} = 61.7954$	0	$(1 - p^*)^2$

In the table above, we have no choice but to list the $2^2 = 4$ possible stock price paths, each of which gives rise to different payoffs for the Asian call. Note in particular that the ud and du paths involve different 6-month stock prices although the 1-year stock price is the same (in contrast, for plain vanilla options, the ud and du paths are essentially indistinguishable). By discounting the (risk-neutral) weighted average of the four payoff values back to time 0 at the risk-free interest rate, we can calculate the current price of the geometric average price Asian call option as

$$
\begin{aligned}
V_0 &= \sum_{\text{4 paths}} \text{Discount factor} \times \text{RN probability} \times \text{Payoff} \\
&= e^{-0.08(1)}[0.447165^2(36.7722) + 0.447165(1 - 0.447165)(14.4521)] \\
&= \boxed{10.0855}.
\end{aligned}
$$

$\square$

Black-Scholes framework.

Pricing Asian options in the Black-Scholes framework is, as you can expect, a technically difficult task. The difficulty stems from the complex distributions followed by the stock price averages. It can be shown that while geometric averages of lognormal stock prices remain lognormally distributed, arithmetic stock price averages no longer follow a tractable distribution. It follows that there exist closed-form pricing formulas for geometric average Asian options (see, e.g., Appendix 14.A of McDonald (2013)), but no such analytic formulas have been obtained to date for arithmetic average Asian options.

8.5 Lookback Options

Definition.

First studied by Goldman, Sosin and Gatto (1979), *lookback options* are exotic options that allow their holders to "look back" over time and select the most favorable stock price during the life of the option for the purpose of determining the option payoffs. They share a certain degree of similarity with Asian options in the sense that both are path-dependent options—which stock price is the most favorable depends on the entire stock price path. They differ in terms of how the intermediate stock prices enter the payoff calculations, with Asian options using (arithmetic or geometric) stock price averages and lookback options employing the maximum or minimum levels of the stock price ever realized.

Standard and extrema lookback options.

There are two kinds of lookback options, depending on how the optimal stock price plays its role in the payoff formula. The more popular kind is a *standard*[vii] lookback option, which is an option whose strike price is set to the stock price that maximizes the option payoff—the minimum stock price over the life of the option, $m = m(T) = \min_{t \in [0,T]} S(t)$, in the case of a standard lookback call, and the corresponding maximum stock price, $M = M(T) = \max_{t \in [0,T]} S(t)$, in the case of a standard lookback put. Essentially, a standard lookback call (resp. put) allows its holder to buy (resp. sell) the underlying asset at its lowest price (resp. highest price). The payoffs of standard lookback call and put options are[viii]

$$\text{Payoff of a standard lookback call} \;=\; S(T) - m,$$
$$\text{Payoff of a standard lookback put} \;=\; M - S(T).$$

Because the strike price of standard lookback options varies with time and is not known until their expiration, they are also known as *floating-strike* lookback options.

In the same spirit, an *extrema* (also known as fixed-strike) lookback option is an option for which the terminal stock price is replaced by the most favorable stock price over the life of the option—the maximum stock price over the life of the option in the case of an extrema lookback call and the minimum stock price in the case of an extrema lookback put. The payoffs of K-strike extrema lookback call and put options are

$$\text{Payoff of an extrema lookback call} \;=\; (M - K)_+,$$
$$\text{Payoff of an extrema lookback put} \;=\; (K - m)_+.$$

Example 8.5.1. (SOA Exam IFM Finance and Investment Sample Question 40: Payoffs of different lookback options) Several lookback options are written on the same underlying index. They all expire in 3 years.

Let $S(t)$ denote the value at time t of the index on which the option is written.

The initial index price, $S(0)$, is 150.

The index price when the option expires, $S(3)$, is 200.

The maximum index price over the 3-year period is 210.

The minimum index price over the 3-year period is 120.

Calculate the sum of the payoffs for the following three lookback options:

[vii]The terminology here follows Bean (2018).
[viii]There is no need to express the payoff as $(S(T) - m)_+$ because $S(T) \geq m$ always.

- Standard lookback call

- Extrema lookback call with a strike price of 100

- Extrema lookback put with a strike price of 100

(A) 180

(B) 190

(C) 200

(D) 210

(E) 220

Solution. We tabulate the payoff of each lookback option as follows:

Lookback Option	Payoff
Standard lookback call	$S(3) - m = 200 - 120 = 80$
Extrema 100-strike lookback call	$(M - K)_+ = (210 - 100)_+ = 110$
Extrema 100-strike lookback put	$(K - m)_+ = (100 - 120)_+ = 0$

The sum of the payoffs is $80 + 110 + 0 = \boxed{190}$. (**Answer: (B)**)   □

Typical use.

Lookback options provide their holders with considerable protection. They are, to a certain extent, an improvement on plain vanilla American options. If you hold an American option, you can take advantage of favorable stock prices by exercising the option prior to maturity. Because future stock prices cannot be fully predicted, any act of early exercise will undesirably expose yourself to the possibility of ex-post regret; an even more favorable stock price may occur later. A lookback option eliminates such room for regret by making sure that the optimal stock price over the option's lifetime is used in the computation of its payoff. In fact, standard lookback options are in effect not genuine options because they will always be exercised. As a trade-off, lookback options tend to be much more expensive than their plain vanilla counterparts. This may explain why despite their attractiveness, the trading of lookback options is not overwhelming.

Pricing.

As for most path-dependent options, it is not a straightforward task to derive the explicit pricing formulas for lookback options under the Black-Scholes framework. This requires the determination of the (risk-neutral) joint distribution of the maximum or minimum stock price and the terminal stock price, which is beyond the scope of this book. Interested readers may refer to, for instance, Section 8.7 of Back (2005) and Section 6.7 of Musiela and Rutkowski (2005). Here we content ourselves with the valuation of lookback options by virtue of the binomial option pricing model. In the course of doing so, we gain additional insights into the payoff characteristics of a lookback option.

Example 8.5.2. (Pricing a standard lookback call by a two-period binomial tree) You are given the following information about a binomial stock price model:

(i) The length of each period is 1 month.

(ii) The current stock price is 800.

(iii) The stock pays no dividends.

(iv) $u = 1.0594$, where u is one plus the percentage change in the stock price per period if the price goes up.

(v) $d = 0.9439$, where d is one plus the percentage change in the stock price per period if the price goes down.

(vi) The continuously compounded risk-free interest rate is 6%.

Calculate the price of a standard 2-month European lookback call option.

Solution. The binomial stock price tree is plotted in Figure 8.5.1. The risk-neutral probability of an up move is

$$p^* = \frac{e^{0.06/12} - 0.9439}{1.0594 - 0.9439} = 0.5291.$$

The payoff of the standard 2-month European lookback call is $S(2/12) - m$. We start with the end of the binomial tree and determine the possible terminal payoffs of the lookback call *path by path*:

Path	$S(2/12)$	m	$S(2/12) - m$
uu	897.86	800	97.86
ud	799.97	799.97	0
du	799.97	755.12	44.85
dd	712.76	712.76	0

Observe again that the ud and du paths give rise to different payoffs to the lookback call, a clear manifestation of the path-dependent nature of lookback options. By risk-neutral pricing, the price of the lookback call is

$$
\begin{aligned}
V_0 &= \sum_{\text{4 paths}} \text{Discount factor} \times \text{RN probability} \times \text{Payoff} \\
&= e^{-0.06(2/12)}[97.86(0.5291)^2 + 44.85(1 - 0.5291)(0.5291)] \\
&= \boxed{38.19}.
\end{aligned}
$$

□

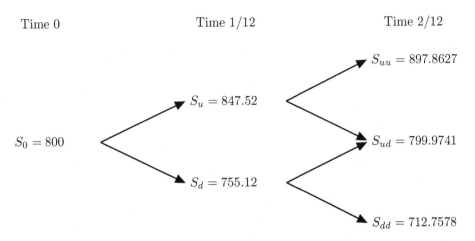

FIGURE 8.5.1
The two-period binomial stock price tree used in Example 8.5.2.

8.6 Shout Options

Definition.

A *shout option* is a European option where the holder is equipped with the right to "shout" once over the life of the option to lock in a minimum payoff. When the shout option expires, the holder receives the maximum of the exercise value of the option at the shout time or that at maturity. The decision to shout is made irrevocably. Consider, for instance, a K-strike T-year shout call option, whose payoff at maturity is

$$\text{Payoff} = \begin{cases} \max\{S(T) - K, S(\tau) - K, 0\}, & \text{if the holder has shouted,} \\ \max\{S(T) - K, 0\} = (S(T) - K)_+, & \text{if the holder has not shouted,} \end{cases}$$

where $\tau \in [0, T]$ is the (random) shout time, which is at the option holder's choosing, and $S(\tau)$ is the then stock price. Analogously, the payoff at maturity of a K-strike T-year shout put option is

$$\text{Payoff} = \begin{cases} \max\{K - S(T), K - S(\tau), 0\}, & \text{if the holder has shouted,} \\ \max\{K - S(T), 0\} = (K - S(T))_+, & \text{if the holder has not shouted.} \end{cases}$$

Because the payoff of a shout option depends on whether shouting has ever taken place, which in turn is induced by the intermediate stock prices, a shout option is intrinsically a path-dependent exotic option.

Pricing shout options in a binomial tree framework.

Because shouting at a time when the shout option in question is in-the-money allows its holder to lock in a positive payoff, it is always beneficial to shout whenever the option is in-the-money. However, as shouting can occur only once during the life of the option, the critical question is *when* to shout to the best advantage of the holder. The answer to this question is closely related to the price of the shout option.

Again, we address the pricing of shout options in the binomial tree framework, paying our prime attention to the optimal time of shouting. The first step of the investigation is to identify all in-the-money nodes prior to expiration[ix]. Then we start from the second to last period of the binomial tree and decide whether shouting at any in-the-money node (implicitly assuming that the right to shout has not been exercised to date) is judicious by comparing the current value of the shout option under two mutually exclusive actions.

Case 1.   If we shout *now* to lock in the current positive payoff, then the shout option becomes a European path-independent claim whose terminal payoff is the maximum of the locked-in payoff and the terminal payoff of the otherwise identical plain vanilla option. In this case, the current value of the shout option can be easily evaluated by taking the discounted risk-neutral expectation of its terminal payoff.

Case 2.   If we do not shout now, but leave the right to shout for potential use in the future, then the current value of the shout option needs to be found via backward induction, the use of which is necessary because the shout option in this case remains path-dependent—its terminal payoffs depend on whether the stock price reaches any further in-the-money node that induces shouting.

Such "shout-or-not-to-shout" comparisons then extend sequentially to the beginning of the binomial tree, whence we obtain the initial price of the shout option. We travel backward in time because whether we shout at a future node can affect the current value of the shout option. The spirit for pricing shout options therefore parallels in some sense the backward inductive procedure for pricing American options, with one noticeable difference: early exercise immediately extinguishes an American option, with the holder walking away with the exercise value, whereas with shouting the holder of the shout option has to wait until maturity to receive the payoff, with the potential to earn more than the immediate exercise value if the option ends up more in-the-money.

As a matter of fact, it is a no-brainer to shout whenever the option is in-the-money in the second to last period. This is because shouting locks in a minimum payoff while not shouting brings you no benefits, and the option will expire in the following period. However, at prior in-the-money nodes it may be advantageous to wait and retain the opportunity to shout for future use, especially when it is relatively likely (in the risk-neutral sense) that the option will become even more in-the-money, as shown by the following illustrative example.

Example 8.6.1. (Pricing a shout option in a 2-period binomial tree) You use the following information to construct a binomial forward tree for modeling the price movements of a nondividend-paying stock:

(i) The length of each period is 6 months.

(ii) The current stock price is 100.

(iii) The stock's volatility is 20%.

(iv) The continuously compounded risk-free interest rate is 6%.

Calculate the price of a 1-year 105-strike shout put option on the stock.

[ix]Technically, shouting is also allowed when the option matures, which has the same effect of locking in the exercise value to be paid right away. It is somewhat too late to shout!

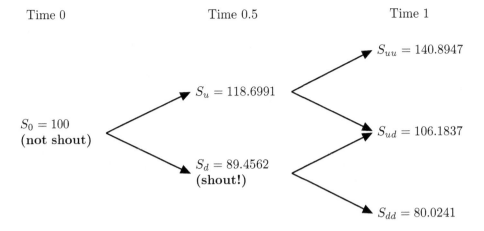

Time 0   Time 0.5   Time 1

$S_0 = 100$
(not shout)

$S_u = 118.6991$

$S_d = 89.4562$
(shout!)

$S_{uu} = 140.8947$

$S_{ud} = 106.1837$

$S_{dd} = 80.0241$

FIGURE 8.6.1
The two-period binomial stock price tree for Example 8.6.1.

Solution. The forward tree parameters are

$$u = \exp[(r - \delta)h + \sigma\sqrt{h}] = \exp[(0.06)(0.5) + 0.2\sqrt{0.5}] = 1.186991,$$
$$d = \exp[(r - \delta)h - \sigma\sqrt{h}] = \exp[(0.06)(0.5) - 0.2\sqrt{0.5}] = 0.894562.$$

The risk-neutral probability of an up move is

$$p^* = \frac{e^{(r-\delta)h} - d}{u - d} = \frac{e^{(0.06)(0.5)} - 0.894562}{1.186991 - 0.894562} = 0.464703,$$

or

$$p^* = \frac{1}{1 + e^{\sigma\sqrt{h}}} = \frac{1}{1 + e^{0.2\sqrt{0.5}}} = 0.464703.$$

The stock prices are depicted in Figure 8.6.1.

Note that if we never shout during the life of the shout put, then its possible terminal payoffs are

$$P_{uu} = P_{ud} = 0, \quad \text{and} \quad P_{dd} = 24.9759.$$

We now investigate when to shout over the 1-year life of the put optimally. Prior to maturity, the put option is in-the-money only at the initial node and the d node, so it suffices to focus on these two nodes.

- *d node:* If we shout at the d node (implicitly, this means that we did not shout at the initial node), then we can lock in a minimum payoff of $105 - 89.4562 = 15.5438$. With shouting, the two possible terminal payoffs stemming from the d node are

$$V_{ud} = \max(0, 15.5438) = 15.5438 \quad \text{and} \quad V_{dd} = \max(24.9759, 15.5438) = 24.9759.$$

By risk-neutral pricing, the price of the shout put at the d node is

$$V_d = e^{-0.06(0.5)}[0.464703(15.5438) + (1 - 0.464703)(24.9759)] = 19.9842.$$

- *Initial node:* Now we further roll back one period to the initial node, where we have to decide whether to shout at the outset or to wait until we hit the d node.

Case 1. If we shout at the initial node, then we guarantee a minimum payoff of $105 - 100 = 5$ and the three possible terminal payoffs become

$$
\begin{aligned}
V_{uu} &= \max(0, 5) = 5, \\
V_{ud} &= \max(0, 5) = 5, \\
V_{dd} &= \max(24.9759, 5) = 24.9759.
\end{aligned}
$$

By risk-neutral pricing, the current price of the shout put, regarded as a 1-year European derivative with the three possible payoffs above, is

$$
\begin{aligned}
V_0 &= e^{-0.06(1)}[0.464703^2(5) + 2(0.464703)(1 - 0.464703)(5) \\
&\qquad + (1 - 0.464703)^2(24.9759)] \\
&= 10.0994.
\end{aligned}
$$

Case 2. If we do not shout at the initial node, then the holding value of the shout put is $V_u = 0$ at the u node (because $P_{uu} = P_{ud} = 0$) and $V_d = 19.9842$ at the d node (in the presence of shouting). By backward induction, the price of the shout put at the initial node is

$$
\begin{aligned}
V_0 &= e^{-0.06(0.5)}[0.464703(0) + (1 - 0.464703)(19.9842)] \\
&= 10.3813.
\end{aligned}
$$

Since Case 2 brings a higher value to the shout put at the initial node, it makes more sense to wait and shout at the d node, if we happen to enter that node. To conclude, the initial price of the shout put is $\boxed{10.3813}$.

$\square$

Remark. (i) Loosely speaking, in this example it is more valuable to defer shouting to the d node because shouting at the d node locks in a much higher payoff than the initial node (15.5438 versus 5) and that it is relatively more likely to enter the d node than the u node. In some cases, however, it is more beneficial to shout at the outset.

(ii) By comparison, the price of an otherwise identical plain vanilla put is 6.7399. The value of the right to shout is worth $10.3813 - 6.7399 = 3.6414$. As a check, observe that the value of the option at the d node, V_d, increases from $e^{-0.06(0.5)}(1 - 0.464703)(24.9759) = 12.9744$ for the plain vanilla put to 19.9842 for the shout put. The risk-neutral expected present value of such an increase realized at the d node is $e^{-0.06(0.5)}(1 - 0.464703)(19.9842 - 12.9744) = 3.6414$.

8.7 Barrier Options

A *barrier option*, as its name suggests, has a payoff which depends on whether the price of the underlying asset reaches a specified level, called the *barrier*, over the option's life.

"Up" (if $B > S(0)$)		"in" (alive if barrier is hit)		"Call"
or	–and–	or	×	or
"Down" (if $B < S(0)$)		"out" (dead if barrier is hit)		"Put"

FIGURE 8.7.1
The three-dimensional descriptions of barrier options.

The option can either come into life or perish whenever the barrier is hit. It is yet another example of a path-dependent exotic option.

Taxonomy.

Just like Asian options, barrier options are described in three dimensions (see Figure 8.7.1):

1. They are called an "up" or "down" option, depending on whether the barrier B is higher than or lower than the initial stock price $S(0)$, respectively.

2. They are further categorized into "in" and "out" options, according to whether they come into existence and become a plain vanilla option, or cease to exist and become worthless, respectively, when the stock price reaches the barrier.

3. Is the option a call or a put in nature? The plain vanilla call or put payoff formula, $(S(T) - K)_+$ or $(K - S(T))_+$, is then multiplied by an indicator function carrying the event that the maximum stock price $M = M(T) = \max_{t \in [0,T]} S(t)$ (in the case of "up" options) or the minimum stock price $m = m(T) = \min_{t \in [0,T]} S(t)$ (in the case of "down" options) over the life of the option is greater than or smaller than[x] the barrier B. Such an event, which is independent of whether the option is a call or a put, reflects whether the barrier is reached, from above or from below, and whether the payoff is triggered (for "in" options) or extinguished (for "out" options) at the point the barrier is breached.

For example, the payoff of a T-year K-strike up-and-in European call is given by

$$\underbrace{(S(T) - K)_+}_{\text{plain vanilla call}} \times \underbrace{\mathbf{1}_{\{M(T) > B\}}}_{\text{represents whether the call is "in"}} ,$$

with $B > S(0)$. The event $\{M(T) > B\}$ inside the indicator function suggests that the stock price has to rise above B some time over the coming T years in order to trigger the payoff of the call at maturity. Likewise, the payoff of a T-year K-strike down-and-out European put is

$$\underbrace{(K - S(T))_+}_{\text{plain vanilla put}} \times \underbrace{\mathbf{1}_{\{m(T) > B\}}}_{\text{represents whether the put is "in"}} ,$$

with $B < S(0)$. Whenever the stock price drops below B, in which case the event $\{m(T) > B\}$ is false and the indicator function becomes zero, the put will be knocked out. That the stock price stays above B throughout the life of the barrier put is a necessary condition for the put to pay off.

---

[x]It is only a matter of convention whether to use a strict inequality ("<" or ">") or a weak inequality ("≤" or "≥") for the event inside the indicator function.

Example 8.7.1. (SOA Exam MFE Spring 2009 Question 2: Payoffs of some Asian and barrier options) You have observed the following monthly closing prices for stock XYZ:

Date	Stock Price
January 31, 2008	105
February 29, 2008	120
March 31, 2008	115
April 30, 2008	110
May 31, 2008	115
June 30, 2008	110
July 31, 2008	100
August 31, 2008	90
September 30, 2008	105
October 31, 2008	125
November 30, 2008	110
December 31, 2008	115

The following are one-year European options on stock XYZ. The options were issued on December 31, 2007.

(i) An arithmetic average Asian call option (the average is calculated based on monthly closing stock prices) with a strike of 100.

(ii) An up-and-out call option with a barrier of 125 and a strike of 120.

(iii) An up-and-in call option with a barrier of 120 and a strike of 110.

Calculate the difference in payoffs between the option with the largest payoff and the option with the smallest payoff.

(A) 5

(B) 10

(C) 15

(D) 20

(E) 25

Solution. We need to calculate the payoff of each option.

(i) The arithmetic average is $A = (105 + 120 + \cdots + 115)/12 = 110$, so the payoff of the 100-strike arithmetic average price Asian call is $(110 - 100)_+ = 10$.

(ii) Because the call is knocked-out on Oct 31, 2008, when the stock price is 125, its payoff is 0.

(iii) The call is knocked-in on Feb 29, 2008, when the stock price is 120. The payoff of the barrier call is $(115 - 110)_+ = 5$.

The difference in payoffs between the option with the largest payoff and the option with the smallest payoff is $10 - 0 = \boxed{10}$. (**Answer: (B)**) □

Example 8.7.2. (Pricing barrier options without pricing!) The current stock price is 60. The price of a 6-month 65-strike European call has a price of 5.

Consider the following 6-month European barrier options:

(1) Down-and-in call option with a barrier of 58 and a strike of 65

(2) Down-and-out call option with a barrier of 58 and a strike of 65

(3) Up-and-in call option with a barrier of 62 and a strike of 65

(4) Up-and-out call option with a barrier of 62 and a strike of 65

Which of the following pairs contains an option with a price of 5 and an option with zero price?

(A) (1) and (2)

(B) (1) and (3)

(C) (2) and (3)

(D) (2) and (4)

(E) (3) and (4)

Ambrose's comment:

If the barrier and the strike price are ordered in a "nice" way, a barrier option may have the same price as the corresponding plain vanilla option.

Solution. Let's focus on options (3) and (4) (because they are the answers!).

(3)   Intuitively, in order that the 65-strike 62-barrier up-and-in call pays off, the 6-month stock price must be higher than 65. To reach this high, the barrier of 62 will necessarily be crossed. In other words, the existence of the barrier plays no role in the option payoff. This makes the up-and-in call identical to a plain vanilla call, with both sharing the same price of 5.

Algebraically, the payoff of the 65-strike 62-barrier up-and-in call is

$$\text{Payoff} = (S(1/2)-65)_+ \times 1_{\{M(1/2)>62\}} = (S(1/2)-65)\times 1_{\{S(1/2)>65, M(1/2)>62\}}$$

Because $S(1/2) > 65$ implies that $M(1/2) > 62$, the payoff can be further written as

$$\text{Payoff} = (S(1/2) - 65) \times 1_{\{S(1/2)>65\}} = (S(1/2) - 65)_+,$$

which is the same as the payoff of the 65-strike plain vanilla call. Both calls therefore have the same price of 5.

(4)   Since it is impossible for the up-and-out call to pay off (i.e., the 6-month stock price exceeds 65) without passing through the barrier of 62—the barrier, when hit, will make the dead—the call is always worthless at expiration. It follows that its price must be zero.

That the up-and-out call is worth nothing can also be seen algebraically. Its payoff is

$$\text{Payoff} = (S(1/2)-65)_+ \times 1_{\{M(1/2)<62\}} = (S(1/2)-65)\times 1_{\{S(1/2)>65,M(1/2)<62\}}.$$

The two events $\{S(1/2) > 65\}$ and $\{M(1/2) < 62\}$ are mutually exclusive, so $1_{\{S(1/2)>65,M(1/2)<62\}} = 0$ and the payoff is always zero.

In conclusion, option (3) has a price of 5 while option (4) has a zero price. **(Answer: (E))** □

Remark. (i) The only conclusion we can make concerning options (1) and (2) is that their prices are between 0 and 5. Since the stock price can go below 58 before ending up being higher than 65 in 6 months, the barrier of 58 does affect the payoffs of the two barrier calls.

(ii) In general, if $S(0) < B < K$, then an up-and-in call always costs the same as an otherwise identical plain vanilla call, while an up-and-out call is always worthless; if $K < B < S(0)$, then a down-and-in put always has the same price as the corresponding plain vanilla put, whereas a down-and-out call always has no value.

Parity relation for barrier options.

There is a simple, model-independent relationship between otherwise identical (i.e., same barrier B, strike K, expiration time T) knock-in and knock-out European barrier options:

$$
\boxed{\begin{array}{ccc} \text{Price of} \\ \text{knock-in option} \end{array} + \begin{array}{c} \text{Price of} \\ \text{knock-out option} \end{array} = \begin{array}{c} \text{Price of} \\ \text{plain vanilla option} \end{array}} \quad (8.7.1)
$$

To prove this, without loss of generality we consider otherwise identical up-and-in and up-and-out European call options. Then the parity relation follows directly by translating the time-T payoff identity

$$\underbrace{(S(T) - K)_+ 1_{\{M(T)>B\}}}_{\text{up-and-in call payoff}} + \underbrace{(S(T) - K)_+ 1_{\{M(T)\le B\}}}_{\text{up-and-out call payoff}} = (S(T) - K)_+.$$

into the time-0 *price* identity (8.7.1).

Given the price of either the knock-in option or the knock-out option, we can price the other barrier option by valuing the plain vanilla option using a given stock price model (e.g., the Black-Scholes formula[xi]) and applying the above parity relation.

Example 8.7.3. (CAS Exam 3 Spring 2007 Question 34: Given the prices of several barrier options) Barrier call option prices are shown in the table below. Each option has the same underlying asset and the same strike price.

Type of Option	Price	Barrier
down-and-out	$25	30,000
up-and-out	$15	50,000
down-and-in	$30	30,000
up-and-in	$X	50,000
down rebate	$25	30,000
up rebate	$20	50,000

[xi]In the binomial option pricing model, you can obtain the price of any barrier option directly.

Calculate X, the price of the up-and-in option.

(A) $20

(B) $25

(C) $30

(D) $35

(E) $40

Solution. This question involves two applications of the barrier option parity (8.7.1).

- Summing the prices of the 30,000-barrier down-and-out call (Option 1) and the 30,000-barrier down-and-in call (Option 3), we find that the price of a plain vanilla call is $55.

- Focusing on the 50,000-barrier up-and-out call (Option 2) and the 50,000-barrier up-and-in call (Option 4), we have

$$X = 55 - 15 = \boxed{40}. \quad \textbf{(Answer: (E))}$$

□

Remark. The given information about rebate options, which pay a constant amount when the barrier is hit, with the payment made either at the time of reaching the barrier or at the maturity time of the option, is not needed.

Example 8.7.4. (SOA Exam IFM Advanced Derivatives Sample Question 42: Partial barrier option) Prices for 6-month 60-strike European up-and-out call options on a stock S are available. Below is a table of option prices with respect to various B, the level of the barrier. Here, $S(0) = 50$.

B	Price of up-and-out call
60	0
70	0.1294
80	0.7583
90	1.6616
∞	4.0861

Consider a special 6-month 60-strike European "knock-in, partial knock-out" call option that knocks in at $B_1 = 70$, and "partially" knocks out at $B_2 = 80$. The strike price of the option is 60. The following table summarizes the payoff at the exercise date:

B_1 Not Hit	B_1 Hit	
	B_2 Not Hit	B_2 Hit
0	$2 \times \max[S(0.5) - 60, 0]$	$\max[S(0.5) - 60, 0]$

Calculate the price of the option.

(A) 0.6289

(B) 1.3872

(C) 2.1455

(D) 4.5856

(E) It cannot be determined from the information given above.

Solution. The key observation is that the given "knock-in, partial knock-out" call is equivalent to

  Two long 60-strike up-and-*in* calls with a barrier of $B_1 = 70$, plus

  one short 60-strike up-and-*in* call with a barrier of $B_2 = 80$

This can be seen either by inspection or algebraically by writing the payoff formula of the "knock-in, partial knock-out" call as

$$
\begin{aligned}
\text{Payoff} \; = \; & 2(S(0.5) - 60)_+ \times 1_{\{70 \le M(0.5) < 80\}} + (S(0.5) - 60)_+ \times 1_{\{M(0.5) \ge 80\}} \\
= \; & 2(S(0.5) - 60)_+ \times \left(1_{\{M(0.5) \ge 70\}} - 1_{\{M(0.5) \ge 80\}}\right) \\
& + (S(0.5) - 60)_+ \times 1_{\{80 \le M(0.5)\}} \\
= \; & 2 \underbrace{(S(0.5) - 60)_+ \times 1_{\{M(0.5) \ge 70\}}}_{\text{60-strike up-and-in call with barrier } B_1 = 70} \; - \; \underbrace{(S(0.5) - 60)_+ \times 1_{\{M(0.5) \ge 80\}}}_{\text{60-strike up-and-in call with barrier } B_2 = 80} \, .
\end{aligned}
$$

Notice that the up-and-out call with an infinite barrier is identical to a plain vanilla call—the barrier will never be reached! By the barrier option parity relation (8.7.1), the price of the 60-strike up-and-in call with a barrier of 70 is $4.0861 - 0.1294 = 3.9567$, and that of the 60-strike up-and-in call with a barrier of 80 is $4.0861 - 0.7583 = 3.3278$. Therefore the price of the "knock-in, partial knock-out" call equals

$$2(3.9567) - 3.3278 = \boxed{4.5856}. \qquad \textbf{(Answer: (D))}$$

Alternatively, we may express the payoff formula of the "knock-in, partial knock-out" call directly in terms of the payoffs of appropriate up-and-out calls:

$$
\begin{aligned}
\text{Payoff} \; = \; & 2(S(0.5) - 60)_+ \times 1_{\{70 \le M(0.5) < 80\}} + (S(0.5) - 60)_+ \times 1_{\{M(0.5) \ge 80\}} \\
= \; & 2(S(0.5) - 60)_+ \times \left(1_{\{M(0.5) < 80\}} - 1_{\{M(0.5) < 70\}}\right) \\
& + (S(0.5) - 60)_+ \times \left(1 - 1_{\{M(0.5) < 80\}}\right) \\
= \; & (S(0.5) - 60)_+ + (S(0.5) - 60)_+ \times 1_{\{M(0.5) < 80\}} \\
& - 2(S(0.5) - 60)_+ \times 1_{\{M(0.5) < 70\}} \\
= \; & 4.0861 + 0.7583 - 2(0.1294) \\
= \; & \boxed{4.5856}. \qquad \textbf{(Answer: (D))}
\end{aligned}
$$

□

Why trade barrier options?

A direct consequence of the barrier options parity relation (8.7.1) is that a barrier option must be cheaper than an otherwise identical plain vanilla option. In fact, one reason why barrier options are appealing is that they provide a more economical alternative for investors to engage in hedging and speculation than plain vanilla options. Consider again Company A back in Example 6.2.4, which is a US-based international company with an operation in Japan and vulnerable to the depreciation of Japanese yen against US dollar. The most natural solution to hedge against the US/yen exchange rate risk is to purchase plain vanilla puts on yen. To lower the hedging cost, Company A may elect to buy barrier puts on yen. What kinds of barrier options suit Company A? Down-and-out puts are worthless when they are most needed, while up-and-in puts pay off only when the dollar/yen exchange first rises sufficiently to breach the barrier, then falls below the put strike. Down-and-in puts and up-and-out puts, however, can satisfy Company A's needs and are worth considering.

Example 8.7.5. (CAS Exam 3 Fall 2007 Question 26: True-or-false statements about exotic options) Which one of the following statements is true about exotic options?

(A) Asian options are worth more than European options.

(B) Barrier options have a lower premium than standard options.

(C) Gap options cannot be priced with the Black-Scholes formula.

(D) Compound options can be priced with the Black-Scholes formula.

(E) Asian options are path-independent options.

Solution. **(Answer: (B) or (D))**

(B)   As discussed above, barrier options never pay more than a plain vanilla option. This cost consideration also partly explains why barrier options are traded in practice.

(D)   Please refer to Geske (1979) for details.   □

Pricing a barrier option by binomial methods.

Pricing barrier options in the Black-Scholes framework is not easy and, again, beyond the scope of this book; that requires the study of the (risk-neutral) joint distribution of the terminal stock price $S(T)$ and the maximum or minimum stock price, $M(T)$ or $m(T)$. Pricing in the binomial model is instructive and not difficult, however. As for other path-dependent options we have seen in earlier sections, when valuing a barrier option make sure that you consider each stock price path separately, keep the level of the barrier in mind, and inspect whether the barrier is crossed or not in each path. Then the same discounted risk-neutral expectation methodology applies.

Example 8.7.6. (Pricing a barrier option in a binomial tree model) For a binomial model for the price of a nondividend-paying stock, you are given:

(i) The length of each period is one month.

(ii) The current stock price is 50.

(iii) $u = 1.122401$, where u is one plus the percentage change in the stock price per period if the price goes up.

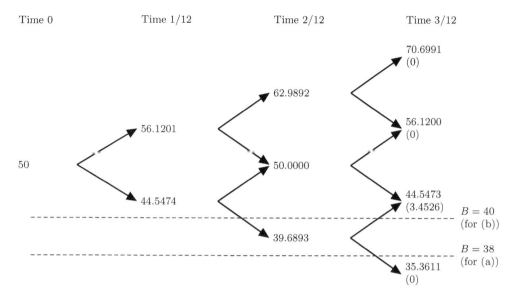

FIGURE 8.7.2
The three-period binomial stock price tree for Example 8.7.6.

(iv) $d = 0.890947$, where d is one plus the percentage change in the stock price per period if the price goes down.

(v) The continuously compounded risk-free interest rate is 10%.

Calculate the price of a 3-month 48-strike down-and-out European put option:

(a) With a barrier of 38

(b) With a barrier of 40

Solution. The stock price evolution is depicted in Figure 8.7.2 with the payoffs of the down-and-out put shown in parenthesis. The risk-neutral probability of an up move is

$$p^* = \frac{e^{0.1/12} - 0.890947}{1.122401 - 0.890947} = 0.507319.$$

(a) Observe that the barrier of 38 is crossed and the down-and-out put gets knocked out only when the stock price goes from the dd node to the ddd node. The only node in which the barrier put pays off is the udd node with a payoff of $V_{udd} = 48 - 44.5474 = 3.4526$, regardless of from where the node is entered. By risk-neutral pricing, the current price of the barrier put option is

$$V_0 = e^{-0.1(0.25)}[\ \underbrace{3}_{\text{3 possible paths}}\ p^*(1-p^*)^2 V_{udd}] = \boxed{1.2440}.$$

(b) When the barrier is changed from 38 to 40, the ddu path will cross the barrier and knock out the down-and-out put. With the same payoff of $V_{udd} = 3.4526$ at the udd node, attainable by only the udd and dud paths, the current price of the barrier put option becomes

$$V_0 = e^{-0.1(0.25)}[\ \underbrace{2}_{\text{2 possible paths}}\ p^*(1-p^*)^2 V_{udd}] = \boxed{0.8293}.$$

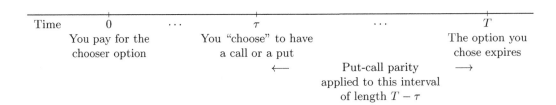

FIGURE 8.8.1
A timeline diagram showing how a chooser option works.

□

8.8   Other Exotic Options

We conclude this chapter with two special exotic options, both of which involve an intermediate point of time where the design of the option changes dramatically and epitomize how
a seemingly path-dependent option can be viewed in the correct light as a path-independent
option and priced easily even in the Black-Scholes framework.

8.8.1   Chooser Options

What is a chooser option and why trade it?

A *chooser option*, also known as an as-you-like-it option or U-Choose option, allows its
holder to "choose," at a fixed[xii] intermediate time τ, whether the option is a European
call option or a European put option expiring at a later time $T\,(\geq \tau)$, both with the same
underlying asset and typically[xiii] the same strike price K. The decision made at time τ is
irrevocable. See Figure 8.8.1, which chronicles the important events surrounding a chooser
option.

Since the chooser can turn out to be either a call or a put, it bears a resemblance to
a straddle in the sense that it can benefit from the upside due to the call as well as from
the downside due to the put, depending on which option you choose. However, there is a
chance that you make the wrong choice at the decision date. You may, for example, have
picked a call at time τ, but the time-T stock price drops so much that the call ends up being
out-of-the-money and it is the put that pays off. Due to the mismatch between what you
choose at time τ and what you actually desire at time T, a chooser is a cheaper alternative
to a τ-year straddle. We will see very shortly how the decision date impacts on the price of
the chooser relative to the corresponding straddle.

[xii] Unlike Section 8.6, where the holder of the shout option can freely choose when to shout, in this section
τ is a fixed point of time.
[xiii] Refer to Section 8.4 of Back (2005) for the general case when the call and put have non-identical strike
prices and maturity dates.

What exactly is the payoff of a chooser option and how to deal with it?

In the remainder of this subsection, we will write $C[S(t), K, t, T]$ (resp. $P[S(t), K, t, T]$) for the time-t price of a K-strike European call (resp. put) maturing at time T when the time-t stock price is $S(t)$. This seemingly cumbersome notation will be handy when it comes to a precise and concise description of our chooser option pricing formulas.

At time τ, the holder of the chooser option is free to "choose" whether it becomes a call or a put, depending on which one is worth more. Therefore, he/she will take the more valuable one, so that the time-τ value of the chooser is

$$\text{Payoff of chooser} = \max\{C[S(\tau), K, \tau, T], P[S(\tau), K, \tau, T]\}, \qquad (8.8.1)$$

which involves the time-τ price of a call and that of a put, both having the same strike price and time to expiration. To further simplify the payoff function, put-call parity naturally comes to our rescue. Assume for simplicity that the stock pays dividends continuously at a rate proportional to its price, with a dividend yield of δ (try Problem 8.9.47 for discrete dividends). There are two ways to manipulate the payoff of the chooser option.

- *Taking the put price outside first:* If we first take the put price out of the maximum function in (8.8.1), then

$$\text{Payoff of chooser} = P[S(\tau), K, \tau, T] + \max\{C[S(\tau), K, \tau, T] - P[S(\tau), K, \tau, T], 0\}.$$

By put-call parity, the time-τ price of the call and that of the otherwise identical put are related by

$$C[S(\tau), K, \tau, T] - P[S(\tau), K, \tau, T] = S(\tau)e^{-\delta(T-\tau)} - Ke^{-r(T-\tau)},$$

so

$$\text{Payoff of chooser} = P[S(\tau), K, \tau, T] + \left(S(\tau)e^{-\delta(T-\tau)} - Ke^{-r(T-\tau)} \right)_+.$$

The positive part function reminds us of the payoff of a call, but the stock price is multiplied by a factor of $e^{-\delta(T-\tau)}$. To make the call payoff on a per unit basis, we take this factor out of the positive part function, yielding

$$\text{Payoff of chooser} = P[S(\tau), K, \tau, T] + e^{-\delta(T-\tau)} \left(S(\tau) - Ke^{-(r-\delta)(T-\tau)} \right)_+,$$

which shows that the chooser option is essentially composed of two plain vanilla (path-independent) options:

(1) A K-strike put which expires at time T.
(2) A $Ke^{-(r-\delta)(T-\tau)}$-strike call which expires at time τ.

It follows directly from this time-τ payoff identity that the time-0 price of the chooser option is

$$\text{Price of chooser} = P[S(0), K, 0, \boxed{T}] + e^{-\delta(T-\tau)}C[S(0), Ke^{-(r-\delta)(T-\tau)}, 0, \boxed{\tau}]. \quad (8.8.2)$$

Notice that the two option prices have generally different strike prices and different times to maturity, and can be calculated separately in a given pricing model (e.g., binomial, Black-Scholes).

- *Taking the call price outside first:* Similarly, if we first pull the call price out of the maximum function, then it can be shown that the price of the chooser option can be expressed as (exercise!)

$$\text{Price of chooser} = C[S(0), K, 0, \boxed{T}] + e^{-\delta(T-\tau)}P[S(0), Ke^{-(r-\delta)(T-\tau)}, 0, \boxed{\tau}]. \quad (8.8.3)$$

The equality of (8.8.2) and (8.8.3) is an easy consequence of put-call parity. Which formula should we use, (8.8.2) or (8.8.3)? It all depends on what information is provided in a given setting!

Example 8.8.1. (SOA Exam IFM Advanced Derivatives Sample Question 25: Given the price of the chooser) Consider a chooser option (also known as an as-you-like-it option) on a nondividend-paying stock. At time 1, its holder will choose whether it becomes a European call option or a European put option, each of which will expire at time 3 with a strike price of $100.

The chooser option price is $20 at time $t = 0$.

The stock price is $95 at time $t = 0$. Let $C(T)$ denote the price of a European call option at time $t = 0$ on the stock expiring at time T, $T > 0$, with a strike price of $100. You are given:

(i) The risk-free interest rate is 0.

(ii) $C(1) = \$4$.

Determine $C(3)$.

(A) $ $9

(B) $11

(C) $13

(D) $15

(E) $17

Solution. We are given in (ii) the price of a τ-year call with $\tau = 1$, so we use (8.8.2). With $r = \delta = 0$, the price of the chooser option is

$$P(95, 100, 0, 3) + C(95, 100, 0, 1) = P(95, 100, 0, 3) + 4 = 20,$$

which implies that $P(95, 100, 0, 3) = 16$. To calculate $C(3) = C(95, 100, 0, 3)$, we use put-call parity and get

$$C(95, 100, 0, 3) = P(95, 100, 0, 3) + 95 - 100 = \boxed{11}. \quad \textbf{(Answer: (B))}$$

□

Remark. This question does not require any pricing model.

Effect of the decision date.

Apart from providing a model-free basis for pricing and hedging a chooser, (8.8.2) and (8.8.3) also shed some light on the structure of a chooser relative to a straddle; the latter can be regarded as a chooser with $\tau = T$ (i.e., the call or put expires immediately as soon as the holder of the chooser option makes his/her decision). For simplicity, we consider the

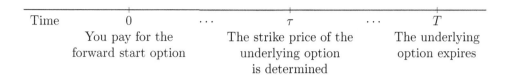

FIGURE 8.8.2
A timeline diagram showing how a forward start option works.

nondividend-paying case (i.e., $\delta = 0$), so that

$$\text{Price of chooser} \overset{(8.8.2)}{=} P[S(0), K, 0, T] + C[S(0), Ke^{-r(T-\tau)}, 0, \tau]$$

$$\overset{(8.8.3)}{=} C[S(0), K, 0, T] + P[S(0), Ke^{-r(T-\tau)}, 0, \tau].$$

Then (8.8.3) shows that the difference between the price of the chooser and that of a T-year straddle is due to the difference between the prices of the concerned puts, with the chooser comprised of a put with a lower strike and a short time to maturity. It will be shown in Section 9.3 that put prices are non-decreasing with respect to the strike price and, when the underlying stock pays no dividends, non-decreasing with respect to the time to maturity. Therefore, the price of the chooser is non-decreasing in the decision date τ. This makes intuitive sense because the closer τ is to the maturity date T, the less uncertainty the holder of the chooser has to face after making his/her choice and the lower the probability of a wrong decision. In the limiting case as $\tau \uparrow T$, we have $P[S(0), Ke^{-r(T-\tau)}, 0, \tau] \uparrow P[S(0), K, 0, T]$, and the chooser coincides with the straddle.

8.8.2 Forward Start Options

Definition.

A *forward start option* is another example of a path-dependent exotic option which can be transformed into a path-independent derivative and priced easily and analytically. Like a chooser option, it involves two fixed time points, say τ and T with $\tau \le T$. At time 0, the holder of the forward start option pays a fixed amount (this is the price of the forward start to be determined), which entitles him/her to receive, at time τ, a plain vanilla European option. The underlying option itself matures at time T, and has a strike price of K, which is not determined at the outset, but is a positive multiple of the time-τ stock price, i.e., $K = c \times S(\tau)$ for some positive number c (if $c = 1$, then the underlying option is at-the-money when issued at time τ). Figure 8.8.2 depicts the mechanics of a forward start option. Essentially, a forward start is a T-year option that comes to life at time τ. Its path dependence stems from the fact that its time-T payoff depends on not only the time-T stock price, but also the time-τ stock price.

Pricing in the Black-Scholes framework.

Suppose that the underlying option is a European call (a put is handled analogously) and that the stock pays dividends continuously at a rate proportional to its price, with a dividend yield of δ.[xiv] In symbolic form, the time-T payoff of the forward start option is given by

$$V(T) = (S(T) - cS(\tau))_+ .$$

The key to pricing such a forward start option is not to directly evaluate the time-0 discounted risk-neutral expectation

$$V(0) = \mathbb{E}^*[e^{-rT}(S(T) - cS(\tau))_+],$$

where $S(\tau)$ and $S(T)$ are both random (from the point of view of time 0), but to rewrite it using iterated expectation (also known as double expectation) into

$$V(0) = \mathbb{E}^*\{e^{-r\tau}\mathbb{E}^*[e^{-r(T-\tau)}(S(T) - cS(\tau))_+|S(\tau)]\} = \mathbb{E}^*[e^{-r\tau}V(\tau)],$$

where the inner conditional expectation is the time-τ price of the forward start option:

$$V(\tau) = \mathbb{E}^*[e^{-r(T-\tau)}(S(T) - cS(\tau))_+|S(\tau)].$$

At time τ, the prevailing stock price $S(\tau)$ is observed, so the forward start option is equivalent to a plain vanilla $cS(\tau)$-strike call option which has $T - \tau$ more years to live. By the plain vanilla Black-Scholes call price formula, we have

$$V(\tau) = S(\tau)[e^{-\delta(T-\tau)}N(d_1) - ce^{-r(T-\tau)}N(d_2)], \qquad (8.8.4)$$

where

$$
\begin{aligned}
d_1 &= \frac{\ln[S(\tau)/(c \times S(\tau))] + (r - \delta + \sigma^2/2)(T - \tau)}{\sigma\sqrt{T - \tau}} \\
&= \frac{-\ln c + (r - \delta + \sigma^2/2)(T - \tau)}{\sigma\sqrt{T - \tau}},
\end{aligned}
$$

and

$$d_2 = d_1 - \sigma\sqrt{T - \tau}.$$

Quite intriguingly, both d_1 and d_2 do not depend on $S(\tau)$, the stock price at the valuation date. It follows that the T-year forward start option is equivalent to a τ-year *path-independent* derivative (in fact, a prepaid forward) which pays an amount given in (8.8.4), which is a constant multiple of $S(\tau)$, at time τ (and nothing at time T). With this point of view, we deduce that the time-0 price of the forward start option should be

$$V(0) = F^P_{0,\tau}(S)[e^{-\delta(T-\tau)}N(d_1) - ce^{-r(T-\tau)}N(d_2)], \qquad (8.8.5)$$

which can be put into better perspective as[xv]

$$\boxed{V(0) = e^{-\delta\tau} \times C(S(0), cS(0), T - \tau),} \qquad (8.8.6)$$

with $C(S(0), cS(0), T - \tau)$ denoting the current price of a $(T - \tau)$-year $cS(0)$-strike call when the current stock price is $S(0)$. In particular, when $\delta = 0$, the forward start option has the same price as this call with the translated expiration time $T - \tau$.

Instead of memorizing (8.8.5) and (8.8.6) by heart, you will be better off mastering the "sequential" pricing technique (i.e., going from time T to time τ, then from time τ to time 0), which is intuitive and applicable to even more complex situations (see the last few end-of-chapter problems).

[xiv]The pricing formula does not work when the stock pays discrete, non-random dividends. Do you know why?

[xv]In fact, (8.8.6) is true even outside the Black-Scholes framework. For details, see pages 440 and 441 of Sundaram and Das (2016).

Example 8.8.2. (SOA Exam IFM Advanced Derivatives Sample Question 19: Valuing a forward start option) Consider a *forward start option* which, 1 year from today, will give its owner a 1-year European call option with a strike price equal to the stock price at that time.

You are given:

(i) The European call option is on a stock that pays no dividends.

(ii) The stock's volatility is 30%.

(iii) The forward price for delivery of 1 share of the stock 1 year from today is 100.

(iv) The continuously compounded risk-free interest rate is 8%.

Under the Black-Scholes framework, determine the price today of the forward start option.

(A) 11.90

(B) 13.10

(C) 14.50

(D) 15.70

(E) 16.80

Solution. With $\tau = 1$, $T = 1 + 1 = 2$,

$$d_1 = \frac{\ln[S(1)/S(1)] + (0.08 + 0.3^2/2)(1)}{0.3\sqrt{1}} = 0.41667,$$

$$d_2 = d_1 - 0.3\sqrt{1} = 0.11667,$$

the time-1 value of the underlying European call is

$$
\begin{aligned}
S(1)N(d_1) - S(1)e^{-0.08}N(d_2) &= S(1)[0.66154 - e^{-0.08}(0.54644)] \\
&= 0.157112S(1).
\end{aligned}
$$

The time-0 value of the forward start option is then

$$0.157112F_{0,1}^P(S) = 0.157112e^{-0.08}F_{0,1}(S) = \boxed{14.5033}. \quad \textbf{(Answer: (C))}$$

□

Example 8.8.3. (SOA Exam IFM Advanced Derivatives Sample Question 33: Rolling insurance) You own one share of a nondividend-paying stock. Because you worry that its price may drop over the next year, you decide to employ a *rolling insurance strategy*, which entails obtaining one 3-month European put option on the stock every three months, with the first one being bought immediately.

You are given:

(i) The continuously compounded risk-free interest rate is 8%.

(ii) The stock's volatility is 30%.

(iii) The current stock price is 45.

(iv) The strike price for each option is 90% of the then-current stock price.

Your broker will sell you the four options but will charge you for their total cost now. Under the Black-Scholes framework, how much do you now pay your broker?

(A) 1.59

(B) 2.24

(C) 2.86

(D) 3.48

(E) 3.61

Solution. Let us first calculate the current price of a 3-month European put with strike price being 90% of the current stock price S. With $K = 0.9S$, $r = 0.08$, $\sigma = 0.3$, and $T = 1/4$, we have

$$d_1 = \frac{\ln(S/0.9S) + (0.08 + 0.3^2/2)(0.25)}{0.3\sqrt{0.25}} = 0.91074,$$

$$d_2 = d_1 - 0.3\sqrt{0.25} = 0.76074,$$

$$N(-d_1) = 0.18122,$$

$$N(-d_2) = 0.22341,$$

and the put price is

$$0.9Se^{-0.08(0.25)}N(-d_2) - S \times N(-d_1) = 0.01587S.$$

The rolling insurance strategy is then equivalent to $0.01587S(0)$ paid at time 0, $0.01587S(0.25)$ at time 0.25, $0.01587S(0.5)$ at time 0.5, and $0.01587S(0.75)$ at time 0.75. Their total price at time 0 is the sum of their prepaid forward prices, or

$$0.01587[S(0) + F^P_{0,0.25}(S) + F^P_{0,0.5}(S) + F^P_{0,0.75}(S)] = 0.01587(4)(45) = \boxed{2.8566}.$$

(Answer: (C))   □

8.9   Problems

Gap options

Problem 8.9.1. (Asset-or-nothing call) Assume the Black-Scholes framework. For $t \geq 0$, let $S(t)$ denote the time-t price of a stock that pays dividends continuously at a rate proportional to its price. The dividend yield δ is 2%.

Let π be the time-0 price of the following 4-year European asset or nothing option on the stock. The option pays $S(4)$ four years from now if $S(4)$ is greater than 110% of $S(0)$, and pays nothing otherwise.

You are given:

(i) $\text{Var}[\ln S(t)] = 0.09t$, for $t > 0$.

(ii) The continuously compounded risk-free interest rate is 6%.

Calculate the ratio $\pi/S(0)$.

Problem 8.9.2. (Deriving the cash-or-nothing option pricing formula from the Black-Scholes formula) This problem shows how the pricing formula for a T-year K-strike European cash-or-nothing call option of $\$1$ can be retrieved from that of a plain vanilla T-year K-strike European call option.

(a) Using the fact that $(x - K)_+ = \int_K^\infty 1_{\{x > c\}}\,dc$ for any real x, express the Black-Scholes price of the plain vanilla European call option in terms of the price(s) of appropriate cash-or-nothing call option(s).

(b) Obtain the pricing formula for the cash-or-nothing call option by differentiating the Black-Scholes call price with respect to K.

Problem 8.9.3. (Valuing a path-dependent partial asset-or-nothing option) Assume the Black-Scholes framework. For $t \geq 0$, let $S(t)$ be the time-t price of a stock. You are given:

(i) $S(0) = 100$.

(ii) The stock's volatility is 25%.

(iii) The stock pays dividends continuously at a rate proportional to its price. The dividend yield is 2%.

(iv) The continuously compounded expected rate of return on the stock is 6%.

(v) The continuously compounded risk-free interest rate is 5%.

Consider a special 3-year European asset-or-nothing option on the stock. The option's 3-year payoff is

$$\text{Payoff} = \begin{cases} 2S(3), & \text{if } S(1) > 120 \text{ and } S(3) > 1.5S(1), \\ S(3), & \text{if } S(1) < 120 \text{ and } S(3) > 1.5S(1), \\ 0, & \text{otherwise.} \end{cases}$$

Calculate the time-0 price of this option.

(Hint: First work out the 1-year price according to whether $S(1) > 120$ or $S(1) < 120$. You will see that the 3-year partial asset-or-nothing option is equivalent to the sum of two appropriate 1-year asset-or-nothing options. From this observation, go back to time 0.)

Problem 8.9.4. (Holding profit for a delta-hedged cash-or-nothing call) Assume the Black-Scholes framework. Three months ago, Embryo (Ambrose's twin brother) *bought* a 1-year 50-strike European cash-or-nothing call option of $1,000 on a nondividend-paying stock. He immediately delta-hedged his position with shares of the stock, but has not ever re-balanced his portfolio. He now decides to close out all positions.

You are given:

(i) The stock's volatility is 30%.

(ii) The continuously compounded risk-free interest rate is 2%.

(iii)

	Three months ago	Now
Stock price	50	55
Cash-or-nothing call option price	457.5501	606.7117

Calculate Embryo's three-month holding profit.

Problem 8.9.5. (Standard computations of gap option price – I: Continuous dividend) Assume the Black-Scholes framework. You are given:

(i) The current stock price is 95.

(ii) The stock's volatility is 10%.

(iii) The stock pays dividends continuously at a rate proportional to its price. The dividend yield is 10%.

(iv) The continuously compounded risk-free interest rate is 5%.

Calculate the price of a 6-month 100-strike European gap call option with a payment trigger of 110.

Problem 8.9.6. (Standard computations of gap option price – II: Discrete dividend) Assume the Black-Scholes framework. Let $S(t)$ denote the price at time t of a stock, which will pay a dividend of $1 after 3 months.

Consider a European gap option which matures in 9 months. If the 9-month stock price is less than $28, the payoff is

$$30 - S(9/12);$$

otherwise, the payoff is zero.

You are given:

(i) $S(0) = \$30$.

(ii) $\mathrm{Var}[\ln F^P_{t,9/12}(S)] = 0.1225t$, for $0 \le t \le 9/12$.

(iii) The continuously compounded risk-free interest rate is 8%.

Calculate the price of the gap option.

Problem 8.9.7. (Gap version of "Rolls Royce car" problem) Assume the Black-Scholes framework.

Michael has ordered a Rolls Royce car for 200,000 British pounds, which he will *pay* when the car is delivered to him in three months. Because Michael has got an A+ in ACTS:4830 from the devilish instructor, Apple Lo, as a reward his insanely rich mum has promised to give him US $320,000 three months from now.

Worried about the US dollar losing value relative to the British pound, Michael now decides to buy appropriate 3-month at-the-money European currency *gap* options to exactly cover the shortfall in case it occurs. To lower his cost, Michael is willing to take a smaller payoff (than the payoff before buying any gap options) if the 3-month dollar/pound exchange rate is between 1.50 and 1.60.

You are given:

(i) The current dollar/pound exchange rate is 1.60.

(ii) The continuously compounded risk-free interest rate in the United States is 1%.

(iii) The continuously compounded risk-free interest rate in the United Kingdom is 2%.

(iv) The volatility of the exchange rate of dollar per pound is 20%.

Describe the gap options that Michael should buy. Hence calculate the total cost of the currency gap options in US dollars.

Problem 8.9.8. (Price of a gap-like contingent claim) Assume the Black-Scholes framework. For $t \geq 0$, let $S(t)$ denote the time-t price of a stock. Consider a 1-year European contingent claim. If the 1-year stock price is *less* than $60, the payoff of the contingent claim is $S(1) - 40$; otherwise, the payoff is zero.

You are given:

(i) $S(0) = 50$.

(ii) The only dividend is 2 to be paid in eight months.

(iii) $\mathrm{Var}[\ln F^P_{t,1}(S)] = 0.04t$, for $0 \leq t \leq 1$.

(iv) The continuously compounded risk-free interest rate is 4%.

Calculate the current price of the contingent claim.

Problem 8.9.9. (Dissecting a "sad" derivative into appropriate gap options) Consider a "sad" 1-year European contingent claim on a stock. You are given:

(i) The time-0 stock price is 70.

(ii) The stock pays dividends continuously at a rate proportional to its price. The dividend yield is 3%.

(iii) The continuously compounded risk-free interest rate is 6%.

(iv) The time-1 payoff diagram (not drawn to scale) of the contingent claim is described by the following "sad" face:

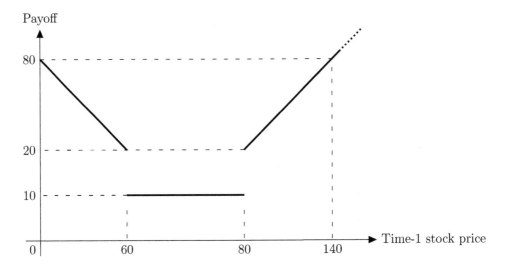

Let $C(K_1, K_2)$ be the time-0 price of a 1-year European gap call option on the above stock with a strike price of K_1 and a payment trigger of K_2. Let $P(K_1, K_2)$ be the time-0 price of an otherwise identical gap put option.

Determine whether each of the following is a correct expression for the time-0 price of the contingent claim.

(A) $P(70, 60) + C(70, 80) + 10e^{-0.06}$

(B) $C(70, 60) + C(70, 80) - 70e^{-0.03} + 80e^{-0.06}$

(C) $C(70, 60) + C(70, 80) - 70e^{-0.03} + 70e^{-0.06}$

(D) $P(70, 60) + P(70, 80) + 70e^{-0.03} - 60e^{-0.06}$

(E) $P(70, 60) + P(70, 80) + 70e^{-0.03} - 70e^{-0.06}$

(Hint: You may consider working out this problem both algebraically and geometrically. The put-call parity for gap options can be useful.)

Problem 8.9.10. (Probability that a gap option has a negative payoff) Assume the Black-Scholes framework. For a stock which pays dividends continuously at a rate proportional to its price, you are given:

(i) The continuously compounded expected rate of stock-price *appreciation* is 7%.

(ii) The continuously compounded risk-free interest rate is 2%.

(iii) The 95% lognormal prediction interval for the 2-year stock price is $(45.7693, 241.4658)$.

Calculate the probability that the payoff of a 3-year 96-strike 103-trigger European gap put option is negative.

Problem 8.9.11. [HARDER!] (What can you deduce from a gap option exercise probability and a lognormal prediction interval?) Assume the Black-Scholes framework. For a stock which pays dividends continuously at a rate proportional to its price, you are given:

(i) The probability that a 3-month 70-strike 75-trigger European gap put option on the stock has a positive payoff is 0.3669.

(ii) The 90% lognormal prediction interval for the 6-month stock price is $(57.7734, 92.0014)$.

(iii) The continuously compounded risk-free interest rate is 1.5%.

Calculate the variance of the 9-month stock price.

Problem 8.9.12. (How should the strike price be set so that the gap option is free?) Assume the Black-Scholes framework. For $t \geq 0$, let $S(t)$ be the time-t price of a stock. You are given:

(i) $S(0) = \$48$.

(ii) The stock pays dividends continuously at a rate proportional to its price. The dividend yield is 3%.

(iii) $\text{Var}[\ln S(t)] = 0.04t$ for all $t \geq 0$.

(iv) The continuously compounded risk-free interest rate is 4%.

Consider a 3-month European gap option. If the 3-month stock price is less than $48, the payoff of the option is
$$K - S(0.25);$$
otherwise, the payoff is zero.

The value of K is set so that the gap option is free (i.e., its price is zero). Calculate the current delta of the gap option.

Problem 8.9.13. (Delta of a "truncated" call option) Assume the Black-Scholes framework. For $t \geq 0$, let $S(t)$ be the time-t price of a stock. You are given:

(i) $S(0) = 65$.

(ii) The stock pays dividends continuously at a rate proportional to its price. The dividend yield is 3%.

(iii) The stock's volatility is 25%.

(iv) The continuously compounded risk-free interest rate is 6%.

Consider a special 1-year European "truncated" call option with payoff given by

$$\text{Payoff} = \begin{cases} S(1) - 60, & \text{if } 60 \leq S(1) \leq 80, \\ 0, & \text{otherwise.} \end{cases}$$

Calculate the time-0 delta of this special call option.
    (Warning: The "truncated" call option is NOT a call bull spread!)

Problem 8.9.14. (Elasticity of a gap put given information about the corresponding plain vanilla) Assume the Black-Scholes framework. For $t \geq 0$, let $S(t)$ be the time-t price of a stock that pays dividends continuously at a rate proportional to its price.

Consider a 1-year European gap option. If the 1-year stock price is less than $150, the payoff is

$$120 - S(1);$$

otherwise, the payoff is zero.

You are given:

(i) $S(0) = \$120$.

(ii) The stock's volatility is 35%.

(iii) The price of a 1-year 150-strike European put option on the stock is $34.022.

(iv) The delta of the put option in (iii) is -0.638.

(v) The continuously compounded risk-free interest rate is 5%.

Calculate the elasticity of the gap option.

Problem 8.9.15. (Holding profit for a delta-hedged gap option) Assume the Black-Scholes framework. You are given:

(i) The current stock price is 100.

(ii) The stock pays dividends continuously at a rate proportional to its price. The dividend yield is 2%.

(iii) The volatility of the stock is 20%.

(iv) The price of a 6-month European gap call option on the above stock with a strike price of 90 and a payment trigger of 110 is 7.3528.

(v) The price of a 6-month European gap put option on the above stock with a strike price of 90 and a payment trigger of 110 is -3.4343.

Ryan has just written 10 gap call options in (iv), and he delta-hedged his position immediately with shares of the stock.

After one month, the stock price drops to 95 and the price of the gap call option decreases to 4.2437.

Calculate the one-month holding profit for Ryan.

Problem 8.9.16. (Gamma of a gap option) Assume the Black-Scholes framework. For a 5-month European gap put option on a nondividend-paying stock, you are given:

(i) The current price of the stock is 120.

(ii) The stock's volatility is 20%.

(iii) The gap put option has strike price 110 and payment trigger 130.

(iv) The continuously compounded risk-free interest rate is 4%.

Calculate the current gamma of the gap put option.

Exchange options

Problem 8.9.17. (Conceptual question: What arguments to put in the Black-Scholes function?) You are given the following generic Black-Scholes-type pricing function:

$$BS(s_1, \delta_1; s_2, \delta_2; \sigma, T) := s_1 e^{-\delta_1 T} N(d_1) - s_2 e^{-\delta_2 T} N(d_2),$$

where

$$d_1 = \frac{\ln(s_1/s_2) + (\delta_2 - \delta_1 + \sigma^2/2)T}{\sigma\sqrt{T}} \quad \text{and} \quad d_2 = d_1 - \sigma\sqrt{T},$$

and all variables are positive.

Your boss, who knows nothing about option pricing, has asked you to analyze the following T-year European options. Allergic to formulas, he wants you to simplify notation and express the Black-Scholes pricing formulas of these options in terms of the BS function above: (r is the continuously compounded risk-free interest rate)

Option	Specifics	Pricing Formula
A	A K-strike call option on a stock that pays dividends continuously at a rate proportional to its price	$BS(F^P_{0,T}(S), \boxed{1}; K, r; \sigma, T)$
B	A K-strike put option on a stock that pays discrete dividends of known amounts at known times	$BS(\boxed{2}, 0; \boxed{3}, \boxed{4}; \sigma, T)$
C	A K-strike call option on a futures contract maturing at time T_f ($\geq T$)	$BS(\boxed{5}, r; K, r; \sigma, \boxed{6})$
D	An option to exchange two units of Stock 1 for one unit of Stock 2	$BS(\boxed{7}, 0; \boxed{8}, 0; \sigma, T)$

(a) Using standard notation in this book, fill in $\boxed{1}$ to $\boxed{8}$ in the above table.

(b) Explain to your boss the verbal meaning of σ for options B and D.

Problem 8.9.18. (Exercise probability for an exchange option) Assume the Black-Scholes framework. You are given the following information about two stocks:

(i)

	April	Samlung
Current share price	100	125
Expected rate of return	10%	8%
Dividend yield	1%	2%
Volatility	25%	15%

(ii) The correlation between the continuously compounded returns on the two stocks is -0.3.

(iii) The continuously compounded risk-free interest rate is 6%.

Consider a 3-year European exchange option giving its holder the right to exchange one share of Samlung for one share of April after three years.

Calculate the probability that this exchange option will be exercised.

(Hints: We do not have a formula for the exercise probability for an exchange option in the main text. Work out this problem from first principles by considering the (physical) distribution of the difference between the natural logarithms of the two stock prices.)

Problem 8.9.19. (Swapping parameters: Black-Scholes version) Assume the Black-Scholes framework. Two actuaries, A and B, are computing the prices of a European call and a European put using different parameters.

You are given:

Actuary	Option	Underlying Stock Price	Strike Price	Dividend Yield	Risk-free Interest Rate	Stock Volatility	Option Maturity
A	Call	190	200	5%	3%	30%	1
B	Put	200	190	3%	5%	30%	1

Describe the relationship between the call price computed by Actuary A and the put price computed by Actuary B.

(Hint: View the two options from the perspective of an exchange option and consider the generic Black-Scholes pricing function.)

Problem 8.9.20. (What happens if $\sigma = 0$?) Assume the Black-Scholes framework. Consider two nondividend-paying stocks whose time-t prices are denoted by $S_1(t)$ and $S_2(t)$, respectively.

You are given:

(i) $S_1(0) = S_2(0) = 10$.

(ii) Stock 1's volatility is 25%.

(iii) Stock 2's volatility is 25%.

(iv) The correlation between the continuously compounded returns on the two stocks is 1.

(v) The continuously compounded risk-free interest rate is 5%.

Consider a 1-year European option to exchange Stock 2 for Stock 1 at time 1.
    Calculate the price of the exchange option.
    (Hint: Is it possible for the exchange option will be exercised?)

Problem 8.9.21. (Qualitative question: Effects of model parameters on the exchange option price) Assume the Black-Scholes framework. Consider a European option to exchange Stock 2 for Stock 1 at a certain future date. Each stock pays dividends continuously at a rate proportional to its price.

Determine whether each of the following increases, decreases, or does not affect the price of the exchange option, holding everything else constant.

(a) An increase in the dividend yield of Stock 1

(b) An increase in the current price of Stock 1

(c) An increase in the current price of Stock 2

(d) An increase in the correlation between the continuously compounded returns on the two stocks

(e) An increase in the continuously compounded risk-free interest rate

Problem 8.9.22. (Exchanging unequal volumes of assets) Assume the Black-Scholes framework. For $j = 1, 2$, and $t \geq 0$, let $S_j(t)$ denote the price of one share of stock j at time t (in years). Both stocks pay no dividends.

Let π be the current price of a 4-year European exchange option that provides the right to obtain $S_2(0)/S_1(0)$ shares of stock 1 in exchange for one share of stock 2. The continuously compounded risk-free interest rate is 15%. The volatility of stock 1 is 20%. The volatility of stock 2 is 10%.

The correlation between the continuously compounded returns on the two stocks is 0.4. Calculate the ratio $\pi/S_2(0)$.

Problem 8.9.23. (Ranking the prices of three exchange options) Assume the Black-Scholes framework. You are given:

(i) The current prices of Stock 1 and Stock 2 are 100 and 200, respectively.

(ii) Stocks 1 and 2 pay dividends continuously at a rate proportional to their prices. The dividend yield of Stock 1 is 8%. The dividend yield of Stock 2 is 5%.

(iii) Stocks 1 and 2 have the same volatility.

(iv) The continuously compounded risk-free interest rate is 8%.

Consider the following European options with the same time to maturity:

I. An at-the-money call option on one share of Stock 2

II. An option to exchange two shares of Stock 1 for one share of Stock 2, assuming that the correlation between the continuously compounded returns on the two stocks is 0.7.

III. An option to exchange two shares of Stock 1 for one share of Stock 2, assuming that the correlation between the continuously compounded returns on the two stocks is -0.25.

Rank the prices of the three options in ascending order. Explain your order.
(Hint: How do the volatility parameters of the three options compare with one another?)

Problem 8.9.24. (An exchange straddle) Assume the Black-Scholes framework. Consider a model with two nondividend-paying stocks, Stock 1 and Stock 2, and a special 5-year European straddle on Stock 1, with a strike price given by the 5-year price of Stock 2.
You are given:

(i) The current prices of Stock 1 and Stock 2 are both 60.

(ii) Stock 1's volatility is 40%.

(iii) Stock 2's volatility is 50%.

(iv) The continuously compounded risk-free interest rate is 6%.

(v) The current price of the straddle is 72.

Calculate the correlation coefficient between the continuously compounded returns of the two stocks.

Problem 8.9.25. (General explicit pricing formulas for maximum and minimum claims) Assume the Black-Scholes framework. For $j = 1, 2$ and $t \geq 0$, let $S_j(t)$ denote the time-t price of Stock j.

(a) Consider a T-year European contingent claim whose payoff is the maximum of the two stocks,

$$\max(S_1(T), S_2(T)).$$

Show that the time-0 price of such a claim can be written as

$$V^{\max} = F_{0,T}^P(S_1)N\left(\frac{\ln[F_{0,T}^P(S_1)/F_{0,T}^P(S_2)] + (\sigma^2/2)T}{\sigma\sqrt{T}}\right) + \bullet,$$

where $\sigma = \sqrt{\mathrm{Var}[\ln(F_{t,T}^P(S_1)/F_{t,T}^P(S_2))]/t}$ for $t \in (0, T]$. Give the second term of the pricing formula. How does it compare with the first term?

(b) Repeat part (a) for the T-year European contingent claim whose payoff is the minimum of the two stocks,

$$\min(S_1(T), S_2(T)).$$

Problem 8.9.26. (Correlation coefficient implied from the price of a maximum claim) Assume the Black-Scholes framework. Consider two nondividend-paying stocks whose time-t prices are denoted by $S_1(t)$ and $S_2(t)$, respectively.
　　You are given:

(i) $S_1(0) = \$100$ and $S_2(0) = \$150$.

(ii) Stock 1's volatility is 30%.

(iii) Stock 2's volatility is 40%.

(iv) The continuously compounded risk-free interest rate is 4%.

(v) The price of a 2-year European contingent claim which pays $\max(1.5S_1(2), S_2(2))$ is 201.

Calculate the correlation coefficient between the continuously compounded returns of the two stocks.

Problem 8.9.27. (Valuing a minimum claim) Assume the Black-Scholes framework. For $t \geq 0$, let $S_1(t)$ and $S_2(t)$ be the time-t prices of Stock 1 and Stock 2, respectively. You are given:

(i) $S_1(0) = \$100$ and $S_2(0) = \$120$.

(ii) The volatility of the 4-year prepaid forward on Stock 1 is 12%.

(iii) The volatility of the 4-year prepaid forward on Stock 2 is 15%.

(iv) Stock 1 pays dividends of $0.03S_1(t)\,dt$ between time t and time $t + dt$.

(v) Stock 2 pays a dividend of $\$5$ in two years.

(vi) The correlation between the natural logarithms of the 4-year prepaid forward prices on the two stocks is -0.25.

(vii) The continuously compounded risk-free interest rate is 4%.

　　Consider a 4-year contingent claim which pays

$$\min(S_1(4), S_2(4)).$$

Calculate the current price of the claim.

Compound options

Problem 8.9.28. (Compound option parity) You are given:

(i) Stock XYZ pays no dividends.

(ii) Derivative A gives its holder the right, but not the obligation, to buy an at-the-money European call option for $6 at the end of 6 months. The call option is on one share of stock XYZ and will mature one year from now.

(iii) The price of Derivative A is 2.5944.

(iv) Derivative B is identical to Derivative A, except that it gives its holder the right to sell the same call option in (ii) for $6 at the end of 6 months.

(v) The continuously compounded risk-free interest rate is positive.

(vi) The following prices of at-the-money European options on stock XYZ for various maturities:

Type of Option	Time to Expiration	Price
Call	6 months	4.2581
Call	1 year	6.4225
Put	6 months	2.7804
Put	1 year	3.5107

Calculate the price of Derivative B.

Problem 8.9.29. (A conceptual problem on Greeks of exotic options) Which of the following statements about exotic call options is/are correct?

(A) The gamma of a European cash-or-nothing call option must be positive.

(B) The vega of a European cash-or-nothing call option must be positive.

(C) The delta of a European gap call option must be positive.

(D) If the price of a stock increases, then the price of a CallOnCall (with the underlying call written on the stock) decreases.

(E) If the price of a stock increases, then the price of a CallOnPut (with the underlying put written on the stock) decreases.

(Note: This problem shows that the Greeks for exotic options can differ drastically from those of plain vanilla options.)

Asian options

Problem 8.9.30. (Pricing an Asian option using a binomial tree – I) For a three-period binomial stock price model, you are given:

(i) The length of each period is one year.

(ii) The current price of a nondividend-paying stock is 100.

(iii) $u = 1.1$, where u is one plus the percentage change in the stock price per period if the price goes up.

(iv) $d = 1/1.1$, where d is one plus the percentage change in the stock price per period if the price goes down.

(v) The continuously compounded risk-free interest rate is 5%.

Calculate the current price of a 3-year 105-strike arithmetic average price European call option, with the average calculated based on annual closing stock prices.

Problem 8.9.31. (Pricing Asian options using a binomial tree – II) You use the following information to construct a binomial forward tree for modeling the price movements of a stock:

(i) The length of each period is 6 months.

(ii) The current stock price is 100.

(iii) The stock's volatility is 30%.

(iv) The stock pays dividends continuously at a rate proportional to its price. The dividend yield is 2%.

(v) The continuously compounded risk-free interest rate is 10%.

Calculate the price of:

(a) A 1-year arithmetic average strike Asian European put option on the stock

(b) A 1-year geometric average strike Asian European put option on the stock

For both options, the stock price averages are computed by averaging the 6-month and 1-year stock prices.

Problem 8.9.32. (Replicating portfolio of an Asian option) You use the following information to construct a binomial forward tree for modeling the price movements of a nondividend-paying stock:

(i) The length of each period is 6 months.

(ii) The current stock price is 80.

(iii) The stock's volatility is 30%.

(iv) The continuously compounded risk-free interest rate is 8%.

Determine the replicating portfolio at the initial node of a 1-year at-the-money arithmetic average price Asian European call option on the stock, with the average calculated based on the stock prices in 6 months and in 1 year.

Problem 8.9.33. [HARDER!] (Put-call parity for Asian options) For $t \geq 0$, let $S(t)$ be the time-t price of a stock. You are given:

(i) $S(0) = 15$.

(ii) The stock pays dividends continuously at a rate proportional to its price. The dividend yield is 3%.

(iii) The continuously compounded risk-free interest rate is 6%.

(iv) The price of an arithmetic average price 3-year 15-strike European call option is 1.5. The payoff of the call option is based on the ending stock price every year, i.e.,

$$\text{Payoff} = \left(\frac{S(1) + S(2) + S(3)}{3} - 15 \right)_+ .$$

Calculate the price of an arithmetic average price 3-year 15-strike European put option.

Lookback options

Problem 8.9.34. (Expected rate of return on a lookback call) You use the following information to construct a binomial forward tree for modeling the price movements of a stock:

(i) The length of each period is 6 months.

(ii) The current stock price is 100.

(iii) The stock pays dividends continuously at a rate proportional to its price. The dividend yield is 3%.

(iv) The stock's volatility is 30%.

(v) The continuously compounded expected rate of return on the stock is 10%.

(vi) The continuously compounded risk-free interest rate is 8%.

Calculate the continuously compounded expected rate of return on a 1-year European standard lookback put option over the entire 1-year period.

Problem 8.9.35. (Arbitraging a currency lookback option) You use the following information to construct a binomial forward tree for modeling the movements of the dollar/euro exchange rate:

(i) The length of each period is 3 months.

(ii) The current dollar/euro exchange rate is $1.50/€.

(iii) The volatility of the exchange rate is 20%.

(iv) The continuously compounded risk-free interest rate on dollars is 4%.

(v) The continuously compounded risk-free interest rate on euros is 5%.

(vi) The price of a 6-month standard dollar-denominated *American* lookback call option on euro is $0.1.

Describe transactions that should be entered into *currently* to exploit an arbitrage opportunity (if one exists).

(Note: You need not describe how to update your transactions after 3 months.)

Problem 8.9.36. (Replicating portfolio for an American lookback futures put option) You use the following information to construct a binomial forward tree for modeling the price movements of a futures contract on the S&V 150:

(i) The length of each period is 6 months.

(ii) The initial futures price is 500.

(iii) $u = 1.2363$, where u is one plus the percentage change in the futures price per period if the price goes up.

(iv) $d = 0.8089$, where d is one plus the percentage change in the futures price per period if the price goes down.

(v) S&V 150 pays dividends continuously at a rate proportional to its price. The dividend yield is 2%.

(vi) The continuously compounded risk-free interest rate is 6%.

Consider a 1-year *American* lookback put option on the above S&P 500 futures contract. Determine the replicating portfolio of the lookback put option *at the initial node*.

Shout options

Problem 8.9.37. (Ranking the prices of several exotic options in this chapter)
Consider the following European options having the same strike price, time to maturity,
and underlying stock:

I. A plain vanilla call option

II. A gap call option

III. An extrema lookback call option

IV. A shout call option

Rank the prices of the four options in ascending order.

Problem 8.9.38. (Pricing a shout option in a three-period binomial tree) For a
binomial model for the price of a nondividend-paying stock, you are given:

(i) The length of each period is 1 year.

(ii) The current price of the stock is 120.

(iii) $u = 1.15$, where u is one plus the percentage change in the stock price per period if the
price goes up.

(iv) $d = 0.85$, where d is one plus the percentage change in the stock price per period if the
price goes down.

(v) The continuously compounded risk-free interest rate is 6%.

Calculate the price of a 3-year at-the-money European shout call option.
  (Hint: Should you shout at the u node or wait until the uu node?)

Problem 8.9.39. (How much is the right to shout worth?) For a binomial tree
modeling the price movements of a nondividend-paying stock, you are given:

(i) The length of each period is 4 months.

(ii) The current stock price is 120.

(iii) $u = 1.2212$, where u is one plus the percentage change in the stock price per period if
the price goes up.

(iv) $d = 0.8637$, where d is one plus the percentage change in the stock price per period if
the price goes down.

(v) The continuously compounded risk-free interest rate is 8%.

Let P_I be the price of a 1-year at-the-money European put option and P_{II} be the price of
an otherwise identical shout put option.
  Calculate $P_{II} - P_I$.
  (Hint: You can save some time by *not* calculating P_I and P_{II} individually.)

Barrier options

Problem 8.9.40. (Sometimes a barrier option is just a plain vanilla option) Assume the Black-Scholes framework. You are given:

(i) The current price of a nondividend-paying stock is 80.

(ii) The stock's volatility is 30%.

(iii) The continuously compounded risk-free interest rate is 8%.

(a) Calculate the price of a 1-year knock-out European put option with a barrier of 75 and a strike of 70.

(b) Calculate the price of a 1-year knock-in European put option with a barrier of 75 and a strike of 70.

Problem 8.9.41. (Ranking the prices of barrier options with different barriers) Consider the following up-and-in 60-strike European call options on the same underlying stock with the same time to expiration:

Option	Barrier
I	55
II	60
III	65
IV	70

The current price of the stock is 50.
Rank the four barrier options, from highest to lowest price.

(A) $I \geq II \geq III \geq IV$

(B) $IV \geq III \geq II \geq I$

(C) $I = II \geq III \geq IV$

(D) $IV \geq III \geq II = I$

(E) The correct answer is not given by (A), (B), (C), or (D)

Problem 8.9.42. (Pricing a barrier option in the Black-Scholes framework) Assume the Black-Scholes framework. You are given:

(i) The current stock price is 40.

(ii) The stock pays no dividends.

(iii) The stock's volatility is 30%.

(iv) The continuously compounded risk-free interest rate is 8%.

(v) The price of a 3-month 40-strike down-and-in European call option with a barrier of 35 is 0.08.

Calculate the price of a 3-month 40-strike down-and-out European call option with a barrier of 35.

Problem 8.9.43. (Pricing a barrier option in a three-period binomial tree model)
For a binomial stock price model, you are given:

(i) The length of each period is 1 year.

(ii) The current price of a nondividend-paying stock is 100.

(iii) $u = 1.15$, where u is one plus the percentage change in the stock price per period if the price goes up.

(iv) $d = 0.95$, where d is one plus the percentage change in the stock price per period if the price goes down.

(v) The continuously compounded risk-free interest rate is 10%.

Calculate the price of an up-and-in 100-strike 6-month European call option with a barrier of 110.

Problem 8.9.44. (Pricing a double-barrier option in a binomial tree model) For a four-period binomial stock price model, you are given:

(i) The length of each period is 3 months.

(ii) The current price of a nondividend-paying stock is 100.

(iii) $u = 1.11$.

(iv) $d = 0.90$.

(v) The continuously compounded risk-free interest rate is 8%.

Calculate the price of a "double-barrier" 80-strike 1-year European call option which gets knocked out if either the upper barrier of 120 or the lower barrier of 80 is crossed.
(Hint: Make sure that you determine the number of stock price paths leading to a positive payoff correctly.)

Problem 8.9.45. (Valuing an Asian-barrier collar in a binomial model) You use the following information to construct a binomial forward tree for modeling the price movements of a stock:

(i) The length of each period is 1 year.

(ii) The current stock price is 200.

(iii) The stock's volatility is 30%.

(iv) The stock pays dividends continuously at a rate proportional to its price. The dividend yield is 3%.

(v) The continuously compounded risk-free interest rate is 5%.

Consider a special 3-year geometric average price European 180-250 collar with the average calculated based on yearly closing stock prices and a double barrier feature. The collar is knocked out whenever either the upper barrier of 300 or the lower barrier of 150 is crossed over the life of the collar.
Calculate the price of this special collar.

Other exotic options

Problem 8.9.46. **[HARDER!]** **(Arbitraging a mispriced chooser option)** Consider a chooser option (also known as an as-you-like-it option) on a nondividend-paying stock. At time 1, its holder will choose whether it becomes a European call option or a European put option, each of which will expire at time 3 with a strike price of $18.

The chooser option price is $6 at time $t = 0$. The stock price is $20 at time $t = 0$.

Let $C(T)$ and $P(T)$ denote, respectively, the prices of a European call option and a European put option at time $t = 0$ on the stock expiring at time T, $T > 0$, with a strike price of $18.

You are given:

(i) The continuously compounded risk-free interest rate is 0.

(ii) $P(1) = \$1.04$.

(iii) $C(3) = \$4.33$.

Describe actions you could take at times $t = 0$ and $t = 1$ only using the chooser option and the options in (ii) and (iii) to exploit an arbitrage opportunity. Verify carefully that your trading strategy is indeed an arbitrage, i.e., your cash flows at all times are always non-negative and sometimes strictly positive.

Problem 8.9.47. **(Chooser option on a stock paying discrete dividends)** Consider a chooser option (also known as an as-you-like-it option) on stock ABC. At time τ (in years) with $0 < \tau \le 6$, its holder will choose whether it becomes a European call option or a European put option, both of which will expire at time 6 (in years) with a strike price of $30.

You are given:

(i) The current price of stock ABC is 32.

(ii) Dividends of 1.5 are paid at times 1, 3, 5 and 7.

(iii) The continuously compounded risk-free rate of interest is 5%.

(iv) The following prices of European call options on stock ABC with various strike prices and maturity times:

Strike Price	Maturity Time (In Years)	Call Price
21.84	2	11.23
21.84	6	13.97
27.28	2	7.60
27.28	6	10.29
30.00	2	6.45
30.00	6	9.48

(a) If $\tau = 6$, name, with specification of the contractual details, the option strategy to which the chooser option is financially equivalent, and calculate the price of the chooser option.

(b) If $\tau = 2$, calculate the price of the chooser option.

Problem 8.9.48. (Choosing between two gap options) Assume the Black-Scholes framework. Consider a special chooser option (also known as an as-you-like-it option) on a nondividend-paying stock. One year from now, its holder will choose whether it becomes a European gap call option or a European gap put option, each of which will expire three years from now with a strike price of $80 and a payment trigger of $90.
   You are given:

(i) The current stock price is $80.

(ii) The risk-free interest rate is 0.

(iii) The following information about European call options on the stock with different strike prices and times to maturity:

	Maturity (in Years)	
Strike Price	1	3
80	Price = 8.749	Price = 15.059
	Delta = 0.555	Delta = 0.594
90	Price = 5.129	Price = 11.477
	Delta = 0.386	Delta = 0.496

Calculate the price of the special chooser option.

Problem 8.9.49. [HARDER!] (Chooser option with an exchange option flavor) Consider a chooser option (also known as an as-you-like-it option) on two stocks. At time 1, its holder will choose whether it becomes a European option to exchange two units of Stock B for one unit of Stock A, or a European option to exchange one unit of Stock A for two units of Stock B. Both European exchange options will expire at time 3.
   You are given:

(i) The time-0 price of Stock A is 205.

(ii) The time-0 price of Stock B is 100.

(iii) Both Stock A and Stock B pay dividends continuously at a rate proportional to their prices. They share the same dividend yield, which is 2%.

(iv) The time-0 price of a European option to exchange two units of Stock B for one unit of Stock A at time 1 is 5.

(v) The time-0 price of a European option to exchange two units of Stock B for one unit of Stock A at time 3 is 8.

Calculate the time-0 price of the chooser option.

Problem 8.9.50. (Delta-hedging a chooser option) The current time is $t = 0$ (in years). Assume the Black-Scholes framework. You are given:

(i) The current stock price is 40.

(ii) The stock's volatility is 40%.

(iii) The stock pays dividends continuously at a rate proportional to its price. The dividend yield is 2%.

(iv) The continuously compounded risk-free interest rate is 8%.

A market-maker has just sold 100 chooser options (also known as as-you-like-it options) on the above stock. At $t = 0.25$, the holder of each chooser option will choose whether it becomes a European call option or a European put option, each of which will expire at $t = 1$ with a strike price of $42. The market-maker delta-hedges his/her position with shares of the stock.

Calculate the initial number of shares in the market-maker's delta-hedging program.

Problem 8.9.51. (Two relatives of forward start option) Assume the Black-Scholes framework. Let $S(t)$ be the time-t price of a stock. Consider a special 3-year European contingent claim which pays a certain amount three years from now, provided that $S(3)$ is higher than $S(1)$; otherwise, the contingent claim will pay nothing.

You are given:

(i) $S(0) = 100$.

(ii) The stock pays dividends continuously at a rate proportional to its price. The dividend yield is 2%.

(iii) The stock's volatility is 30%.

(iv) The continuously compounded risk-free interest rate is 8%.

Determine the price today of the contingent claim if it pays, at time 3:

(a) $S(1)$

(b) $S(3)$

(Hint: The concept of all-or-nothing options in Subsection 8.1.2 may be useful.)

Problem 8.9.52. (A contingent forward start option) Assume the Black-Scholes framework.

Consider a special forward start option which, 1 year from today, will give its owner a 1-year European put option with a strike price equal to the one-year stock price, *provided that the one-year stock price is higher than the current stock price*; otherwise, the owner will receive nothing.

You are given:

(i) The European put option is on a stock that pays dividends continuously at a rate proportional to its price. The dividend yield is 2%.

(ii) The current stock price is 100.

(iii) The stock's volatility is 30%.

(iv) The continuously compounded risk-free interest rate is 8%.

For this special forward start option, calculate:

(a) Its current price

(b) Its current delta

Problem 8.9.53. (Valuing a forward start gap option) Assume the Black-Scholes framework. Consider a special forward start option which, 2 years from today, will give its owner a 1-year 100-strike European *gap* call option whose payment trigger is equal to the stock price at that time.

You are given:

(i) The gap call option is on a stock that pays dividends continuously at a rate proportional to its price. The dividend yield is 2.5%.

(ii) The current price of the stock is 100.

(iii) The stock's volatility is 24%.

(iv) The continuously compounded risk-free interest rate is 6%.

Calculate the current price of the special forward start option.

Problem 8.9.54. (Put-call parity for forward start options) You are given:

(i) The current price of a stock is 100.

(ii) The stock pays dividends continuously at a rate proportional to its price. The dividend yield is 1%.

(iii) The continuously compounded risk-free interest rate is 2%.

(iv) The price of a forward start option which, 2 years from today, will give its owner a 1-year European call option with a strike price equal to the stock price at that time is 6.89.

Calculate the price of a forward start option which, 2 years from today, will give its owner a 1-year European put option with a strike price equal to the stock price at that time.

(Hint: Apply the identity $x_+ - (-x)_+ = x$ to time 3 to develop a time-3 payoff identity relating the two forward start options. Then translate the time-3 payoff identity "carefully" into a time-0 price identity. Be extremely cautious!)

Problem 8.9.55. (A special minimum derivative) Assume the Black-Scholes framework. For $t \geq 0$, let $S(t)$ be the time-t price of a stock.

You are given:

(i) $S(0) = 20$.

(ii) The stock's volatility is 25%.

(iii) The stock pays dividends continuously at a rate proportional to its price. The dividend yield is 1.5%.

(iv) The continuously compounded risk-free interest rate is 4%.

(v) The payoff of a 3-year European contingent claim is

$$\min(S(2), S(3)).$$

Calculate the time-0 price of the contingent claim.

(Hint: Express the payoff of the contingent claim in terms of the payoff(s) of exotic option(s) covered in this chapter. The techniques you have seen in Sections 8.2 and 8.8 may be useful.)

Part III

Epilogue

9

General Properties of Option Prices

Chapter overview. In Part II of this book, we deal with the pricing and hedging of options by prescribing different assumptions on the probabilistic behavior of the underlying asset. The price of the asset can evolve either binomially or continuously in accordance with a lognormal process. Because any option pricing model can be criticized for being subjective and unrealistic to some extent, this begs the question: is there anything definite that one can say about option prices regardless of how the price of the underlying asset develops over time? In such paucity of information, the answer turns out to be affirmative, with a surprisingly rich theory describing how option prices should behave to preclude arbitrage opportunities.

Rounding off our study of derivatives, the final chapter of this book is dedicated to studying this theory of general no-arbitrage properties satisfied by option prices in a model-free framework. Unlike earlier parts of this book, where European options are the predominant object of interest, both European and American options are analyzed in this chapter. In Section 9.1, we revisit the familiar notion of put-call parity and generalize it to the context of exchange options introduced in Section 8.2. The introduction of exchange options blurs the conventional distinction between calls and puts and gives rise to a new result known as put-call duality, which we illustrate for currency options. In Section 9.2, we derive bounds on the prices of European and American calls and puts by means of the no-arbitrage argument we first presented back in Chapter 2. The exchange option mindset we foster in Section 9.1 will continue to be useful. In Section 9.3, we view option prices as a function of the strike price and the time to maturity, and investigate how the prices behave with respect to these two variables. It is shown that option prices are always monotonic, Lipschitz, and convex in the strike price, and American prices always increase with the time to maturity. We conclude in Section 9.4 with a quantitative discussion on the factors that affect the early exercise decision surrounding an American option. By putting the associated costs and benefits in perspective, we formulate useful sufficient conditions guaranteeing the non-optimality of early exercise.

9.1 Put-Call Parity and Duality

9.1.1 Generalized Parity

Parity relations for exchange options.

In this book, we have seen several versions of put-call parity governing the prices of otherwise identical European calls and European puts. Among its many uses, put-call parity suggests that we can understand calls if we understand puts, and the other way round. Armed with the concept of exchange options seen in Section 8.2, we can extend put-call parity to a *generalized parity* relating the prices of European exchange options. Following Section 8.2, we let

$S_i(t)$ be the time-t price of Asset i, for $i = 1, 2$,

$F_{0,T}^P(S_i)$ be the time-0 price of a T-year prepaid forward on Asset i, for $i = 1, 2$,

$V(S_1(0), S_2(0), T)$ be the time-0 price of a European exchange option which allows its holder to exchange one unit of Asset 2 for one unit of Asset 1 at time T,

$V(S_2(0), S_1(0), T)$ be the time-0 price of a European exchange option which allows its holder to exchange one unit of Asset 1 for one unit of Asset 2 at time T.

In this general setting, the parity equation reads

$$V(S_1(0), S_2(0), T) - V(S_2(0), S_1(0), T) = F_{0,T}^P(S_1) - F_{0,T}^P(S_2). \tag{9.1.1}$$

In the same spirit as Subsection 3.2.2, a swift proof of (9.1.1) starts with the time-T payoff identity

$$(S_1(T) - S_2(T))_+ - (S_2(T) - S_1(T))_+ = S_1(T) - S_2(T),$$

followed by taking the "prepaid forward price operator" on both sides, immediately leading to the time-0 price identity (9.1.1).

Parenthetically, a direct consequence of (9.1.1) is that the two exchange options share the same price if and only if the two prepaid forward prices, $F_{0,T}^P(S_1)$ and $F_{0,T}^P(S_2)$, are the same.

Special cases of (9.1.1).

The generalized parity (9.1.1) provides a unifying framework for many put-call parity relations in seemingly disparate settings. Some important and more subtle specializations of (9.1.1) are as follows:

- *Ordinary stock options:* The most well-known version of (9.1.1) is when Asset 1 is a risky stock and Asset 2 represents cash of K. In this case, the exchange option that exchanges Asset 2 for Asset 1 (resp. Asset 1 for Asset 2) is a plain vanilla K-strike call option (resp. put option). With $S_1(T) = S(T)$, the time-T price of the risky stock, and $S_2(T) = K$, the strike price, (9.1.1) reduces to the familiar put-call parity

$$
\begin{aligned}
C - P &= F_{0,T}^P(S) - F_{0,T}^P(K) \\
&= \begin{cases} S(0)e^{-\delta T} - Ke^{-rT}, & \text{if the stock pays continuous proportional} \\ & \text{dividends at a dividend yield of } \delta, \\ S(0) - \sum_{i=1}^{n} D(t_i)e^{-rt_i} & \text{if the stock pays deterministic dividends} \\ -Ke^{-rT}, & \text{of } D(t_i) \text{ at time } t_i \text{ for } i = 1, 2, \ldots, n, \end{cases}
\end{aligned}
$$

we saw in Subsection 3.2.2.

Example 9.1.1. (SOA Exam IFM Introductory Derivatives Sample Question 2: No dividends) You are given the following:

- The current price to buy one share of XYZ stock is 500.

- The stock does not pay dividends.

- The annual risk-free interest rate, compounded continuously, is 6%.

- A European call option on one share of XYZ stock with a strike price of K that expires in one year costs 66.59.

- A European put option on one share of XYZ stock with a strike price of K that expires in one year costs 18.64.

Using put-call parity, calculate the strike price, K.

(A) 449

(B) 452

(C) 480

(D) 559

(E) 582

Solution. By put-call parity, we solve

$$66.59 - 18.64 = 500 - Ke^{-0.06},$$

which yields $K = \boxed{480}$. (**Answer: (C)**) □

Example 9.1.2. (SOA Exam MFE Spring 2007 Question 1: Discrete dividends) On April 30, 2007, a common stock is priced at \$52.00. You are given the following:

(i) Dividends of equal amounts will be paid on June 30, 2007 and September 30, 2007.

(ii) A European call option on the stock with strike price of \$50.00 expiring in six months sells for \$4.50.

(iii) A European put option on the stock with strike price of \$50.00 expiring in six months sells for \$2.45.

(iv) The continuously compounded risk-free interest rate is 6%.

Calculate the amount of each dividend.

(A) \$0.51

(B) \$0.73

(C) \$1.01

(D) \$1.23

(E) \$1.45

Solution. Let D be the amount of each quarterly dividend. By put-call parity,

$$4.50 - 2.45 = \underbrace{[52 - D(e^{-0.06/6} + e^{-0.06\times 5/12})]}_{\text{Prepaid forward price}} - 50e^{-0.06/2},$$

which gives $D = \boxed{0.7264}$. (**Answer: (B)**) □

- *Currency options:* When Assets 1 and 2 are both currencies, say euro and dollar, with respective interest rates $r_{\text{€}}$ and $r_{\$}$, then the currency counterpart of (9.1.1) for dollar-denominated options on euro is

$$\boxed{C(X(0), K, T) - P(X(0), K, T) = X(0)e^{-r_{\text{€}}T} - Ke^{-r_{\$}T},}$$
(9.1.2)

where both sides are expressed in terms of dollar, and $X(0)$ is the time-0 dollar/euro exchange rate.

Example 9.1.3. (CAS Exam 3 Fall 2007 Question 15: Application of currency put-call parity) A nine-month dollar-denominated call option on euros with a strike price of $1.30 is valued at $0.06. A nine-month dollar-denominated put option on euros with the same strike price is valued at 0.18. The current exchange rate is $1.2/euro and the continuously compounded risk-free rate on dollars is 7%.

What is the continuously compounded risk-free rate on euros?

(A) Less than 7.5%

(B) At least 7.5%, but less than 8.5%

(C) At least 8.5%, but less than 9.5%

(D) At least 9.5%, but less than 10.5%

(E) At least 10.5%

Solution. A direct application of (9.1.2) yields

$$0.06 - 0.18 = 1.2e^{-r_{\text{€}} \times 3/4} - 1.3e^{-0.07 \times 3/4},$$

resulting in $r_{\text{€}} = \boxed{0.0997}$. **(Answer: (C))** □

9.1.2 Currency Put-call Duality

Exchange option duality.

In the presentation of (9.1.1), the usage of the terms "call" and "put" is deliberately avoided. As pointed out in Section 8.2, this is because whether an exchange option is a call or a put critically depends on which asset is labeled as the underlying asset and which asset is labeled as the strike asset. Viewing Asset 1 as the underlying asset and Asset 2 as the strike asset, we can regard the exchange option with payoff $(S_1(T) - S_2(T))_+$ as a call on Asset 1 because the payoff resembles the plain vanilla call payoff formula $(S(T) - K)_+$. Alternatively, the exchange option can also be viewed as a put on Asset 2 with Asset 1 being the strike asset if one associates $(S_1(T) - S_2(T))_+$ with the plain vanilla put payoff formula $(K - S(T))_+$. Such blurred distinction between a call or a put is termed *put-call duality.*

Currency option put-call duality.

The most prominent manifestation of put-call duality lies in currency markets, where European calls and puts present different ways of viewing the essentially same contract. To see this concretely, consider the following two European currency options expiring at a generic time T:

Option 1:   This is a dollar-denominated European call on euros giving you the right to use $\$K$ to buy €1 at expiration. The time-0 call price, in dollars, is denoted by $C_\$(X(0), K)$, where the subscript "$\$$" emphasizes the fact that the call price and strike price are expressed in dollars.

Option 2:   This is a euro-denominated European put on dollars giving you the right to sell $\$1$ for €$(1/K)$ at expiration. The time-0 put price, in euros, is denoted by $P_\text{€}(1/X(0), 1/K)$, where the subscript "€" again emphasizes the euro denomination.

Comparing Options 1 and 2, we can see that the underlying currency (resp. strike currency) of Option 1 is the strike currency (resp. underlying currency) of Option 2:

Option	Underlying Currency	Strike Currency
1	Euros	Dollars
2	Dollars	Euros

Despite this, the two options are exercised simultaneously, because Option 1 is exercised if and only if

$$X(T) > K,$$

and Option 2 is exercised if and only if

$$\frac{1}{K} > \frac{1}{X(T)},$$

which is the same exercise condition as Option 1. The similarity between the two currency options will become even more conspicuous if we rewrite the defining properties of Options 1 and 2 in the language of exchange options as

"*Option 1 gives you the right to give up $\$K$ for €1*", and

"*Option 2 gives you the right to give up $\$1$ for €(1/K)*".

This shows that Options 1 and 2 are essentially the same option in the sense that exercising either of them results in the loss of dollars and the possession of euros; only the scales of transaction are different.

On the basis of the above discussions, we can perform the following two steps to link the price of Option 1 (a dollar-denominated call on euros) to that of Option 2 (a euro-denominated put on dollars):

Step 1.   From the above discussions, we see that Option 1 is equivalent to K units of Option 2.[i] Hence

$$\$C_\$(X(0), K) = \text{€}K \times P_\text{€}\left(\frac{1}{X(0)}, \frac{1}{K}\right).$$

This adjusts for the difference in transaction scales of Options 1 and 2.

---

[i]This can be formalized mathematically by writing the payoff of Option 1 in euros as

$$(\text{€}1 - \$K)_+ = \left(\text{€}1 - \text{€}\frac{K}{X(T)}\right)_+ = K \times \text{€}\left(\frac{1}{K} - \frac{1}{X(T)}\right)_+,$$

which is K times the payoff of Option 2.

Step 2.   To express the two option prices on the basis of the same denomination, we use the time-0 dollar/euro exchange rate (recall that the two option prices are paid at time 0) and obtain

$$\text{€}K \times P_{\text{€}}\left(\frac{1}{X(0)}, \frac{1}{K}\right) = \$X(0) \times K \times P_{\text{€}}\left(\frac{1}{X(0)}, \frac{1}{K}\right).$$

In conclusion, we have obtained the following *put-call duality*, which shows that a call on one currency can be relabeled as a scaled put on another currency, and their prices are related via

$$C_{\$}(X(0), K) = X(0) \times K \times P_{\text{€}}\left(\frac{1}{X(0)}, \frac{1}{K}\right). \tag{9.1.3}$$

To better remember this formula, you may observe that the two factors, $X(0)$ and K, when multiplied by the corresponding argument of the option on the right-hand side, give unity.

Example 9.1.4. (SOA Exam MFE Spring 2009 Question 9: Put-call parity plus put-call duality) You are given:

(i) The current exchange rate is $0.011\$/\text{¥}$.

(ii) A four-year dollar-denominated European put option on yen with a strike price of $0.008 sells for $0.0005.

(iii) The continuously compounded risk-free interest rate on dollars is 3%.

(iv) The continuously compounded risk-free interest rate on yen is 1.5%.

Calculate the price of a four-year yen-denominated European put option on dollars with a strike price of ¥125.

(A) 35

(B) 37

(C) 39

(D) 41

(E) 43

Solution. Symbolically, we are given that the current dollar/yen exchange rate is $X(0) = 0.011$ and $P_{\$}(0.011, 0.008) = 0.0005$. We are asked to find $P_{\text{¥}}(1/0.011, 125)$. For this purpose, we may pursue two different routes:

1. Apply put-call parity to get $C_{\$}(0.011, 0.008)$, then use put-call duality for $P_{\text{¥}}(1/0.011, 125)$.

2. Apply put-call duality to get $C_{\text{¥}}(1/0.011, 125)$, then use put-call parity for $P_{\text{¥}}(1/0.011, 125)$.

For definiteness, we illustrate the first route here. By currency put-call parity (9.1.2), the price of a 4-year dollar-denominated \$0.008-strike European call option on yen is

$$
\begin{aligned}
C_\$(0.011, 0.008) &= P_\$(0.011, 0.008) + X(0)e^{-r_f T} - Ke^{-rT} \\
&= 0.0005 + 0.011e^{-0.015(4)} - 0.008e^{-0.03(4)} \\
&= 0.003764.
\end{aligned}
$$

Then by currency put-call duality (9.1.3), we have

$$
P_¥(1/0.011, 125) = \frac{1}{0.011} \times \frac{1}{0.008} \times C_\$(0.011, 0.008) = \boxed{42.77}. \quad \textbf{(Answer: (E))}
$$

$\square$

Remark. The following flowchart depicts the two possible routes to get $P_¥(1/0.011, 125)$ from $P_\$(0.011, 0.008)$:

$$
\begin{array}{ccc}
P_\$(0.011, 0.008) & \xrightarrow{\text{put-call parity}} & C_\$(0.011, 0.008) \\[2pt]
\Big\downarrow{\text{put-call duality}} & & \Big\downarrow{\text{put-call duality}} \\[2pt]
C_¥(1/0.011, 125) & \xrightarrow{\text{put-call parity}} & P_¥(1/0.011, 125)
\end{array}
$$

9.2 Upper and Lower Bounds on Option Prices

Even in the absence of a concrete option pricing model, the prices of options, be they European or American, cannot behave too irregularly. In this section, we investigate the maximum and minimum values in terms of (prepaid forward) prices of the stock and strike satisfied by the fair option prices of European and American calls and puts by means of different arguments, which enjoy relative merits. Intuitive explanations, though not rigorous, help us make sense of these no-arbitrage restrictions, sharpen our understanding of the properties of option prices, and, most importantly, allow us to formulate these constraints in the first place. No-arbitrage proofs, arguably the most important argument you should master, enable us to validate these no-arbitrage bounds systematically and provide us with a recipe to create an arbitrage strategy to reap riskless profits if the options are mispriced. Moreover, they can be applied even when intuitions fall apart. Algebraic proofs are simple to formulate, but they have limited applicability and are not particularly informative.

European call prices.

In a no-arbitrage market, the price of a K-strike T-year European call, C^E (the superscript "E" denotes "European"), must be bounded as follows:

$$
\boxed{\left(F^P_{0,T}(S) - F^P_{0,T}(K)\right)_+ \leq C^E \leq F^P_{0,T}(S).} \tag{9.2.1}
$$

We establish these bounds in three ways:

- *Intuitive explanations:* The upper bound on the call price is obvious because the best the call can offer is the ownership of the underlying stock at maturity, and one will not pay more than the price of a prepaid forward, which gives its holder a unit of the stock at maturity for sure, just to acquire a right.

  Similarly, the European call, which represents the value of the right to buy the stock for K at time T, should be more valuable than an obligation to buy the stock for the same price K at time T. The latter is precisely the T-year synthetic forward with a forward price of K and a time-0 price of $F_{0,T}^P(S) - F_{0,T}^P(K)$.

- *No-arbitrage proof:* To develop bounds on the price of a general European derivative, we can first look at the corresponding bounds on its payoff. To this end, we observe that the time-T payoff of a K-strike T-year European call is bounded above and below as follows: (Exercise: Draw a diagram to see why.)

$$S(T) - K \leq (S(T) - K)_+ \leq S(T), \tag{9.2.2}$$

  where the upper bound is the payoff of a T-year prepaid forward contract on the stock, and the lower bound is the payoff of a T-year forward contract on the stock with a forward price of K. With time-T payoffs translated to time-0 prices, it follows from (9.2.2) that the time-0 prices of the three derivatives are ordered in the same way:

$$F_{0,T}^P(S) - F_{0,T}^P(K) \leq C^E \leq F_{0,T}^P(S).$$

  Furthermore, as the call price must be non-negative, the lower bound can be sharpened to $\left(F_{0,T}^P(S) - F_{0,T}^P(K)\right)_+$.

  The payoff inequalities in (9.2.2) can be used to show how to effect an arbitrage strategy if at least one of the two bounds in (9.2.1) does not hold. Suppose that $F_{0,T}^P(S) - F_{0,T}^P(K) > C^E$. Following the "buy low, sell high" principle, we buy the European call, sell a prepaid forward on the stock and lend $F_{0,T}^P(K)$. Immediately, we receive $F_{0,T}^P(S) - F_{0,T}^P(K) - C^E$, which is strictly positive by hypothesis. At time T, our payoff is

$$(S(T) - K)_+ - (S(T) - K)$$

  which is always non-negative because of (9.2.2).

- *Algebraic proof of the lower bound:* Algebraically, the lower bound in (9.2.1) can also be obtained by means of put-call parity, which is only applicable to European options, along with the fact that a put price must be non-negative:

$$C^E = P^E + F_{0,T}^P(S) - F_{0,T}^P(K) \geq F_{0,T}^P(S) - F_{0,T}^P(K).$$

Example 9.2.1. (When does the upper bound in (9.2.1) become tight?) Explain intuitively when you will expect that $C^E \approx F_{0,T}^P(S)$.

Solution. As the strike price decreases, the call price naturally increases (see Subsection 9.3.1) and the upper bound $F_{0,T}^P(S)$ becomes tighter. We expect that as the strike price approaches zero, we have $C^E \approx F_{0,T}^P(S)$. Intuitively, if the strike price is zero, then possessing the call option means that we can get the stock for free at maturity. In other words, the call option is essentially equivalent to a derivative which delivers one unit of the stock at maturity, or the prepaid forward contract on the stock. Algebraically, we can also see the same conclusion from (9.2.1), where the upper and lower bounds coincide when $K = 0$. $\qquad\square$

American call prices.

For American options, the payoff inequalities in (9.2.2) are not useful since the options can be exercised prior to maturity. It turns out that for American call prices, the lower bound in 9.2.1 is tightened to take into account the payoff as a result of immediate exercise, while the upper bound is relaxed to the current stock price:

$$\max \left\{ (S(0) - K)_+ , \left(F_{0,T}^P(S) - F_{0,T}^P(K) \right)_+ \right\} \leq C^A \leq S(0).$$ (9.2.3)

If the underlying stock pays no dividends, then the lower bound simplifies to

$$\max \left\{ (S(0) - K)_+ , \left(S(0) - F_{0,T}^P(K) \right)_+ \right\} = \left(S(0) - F_{0,T}^P(K) \right)_+ ,$$

because $F_{0,T}^P(K) \leq K$. In the presence of dividends, there is generally no definite order between $(S(0) - K)_+$ and $\left(F_{0,T}^P(S) - F_{0,T}^P(K) \right)_+$, and both lower bounds are of use.

We again use no-arbitrage arguments to establish the bounds in (9.2.3). Because an American call is always at least as much as an otherwise identical European call, we have $C^A \geq C^E \geq \left(F_{0,T}^P(S) - F_{0,T}^P(K) \right)_+$, so we only need to prove $(S(0) - K)_+ \leq C^A \leq S(0)$. The lower bound is easy: If $S(0) > K$ and $S(0) - K = (S(0) - K)_+ > C^A$, we can buy the American call, exercise it immediately, and realize an immediate cash inflow of $S(0) - K - C^A > 0$ without being exposed to any loss in the future.

We now turn to the upper bound $C^A \leq S(0)$. Assume on the contrary that $C^A > S(0)$. Then we buy one unit of the stock for $S(0)$ and sell the American call option for C^A, for a positive net cash inflow of $C^A - S(0)$ at time 0. Depending on the behavior of the American call option holder, two cases can arise:

Case 1. If the call holder exercises the option at some intermediate time $\tau \in [0, T]$, then at the same time we sell one unit of the stock we are holding, for an overall payoff of $-(S(\tau) - K) + S(\tau) = K$. If the stock pays no dividends, we have closed all of our positions. If the stock pays dividends (in cash or shares), we still have these dividends at hand.

Case 2. If the call holder never exercises the option, then at time T the call expires worthless while we still hold the stock, possibly with dividends.

In both cases, our cash flows in the future must be non-negative. This constructs an arbitrage strategy. Consequently, we must have $C^A \leq S(0)$ in a no-arbitrage market.

Put prices.

To derive analogous bounds on put prices, we can repeat the above derivations for puts (see Problem 9.5.5). Alternatively and more instructively, we may view a call (resp. put) as an exchange option which entails giving up the strike price (resp. the underlying stock) in return for the underlying stock (resp. the strike price) in return for the strike price (resp. the underlying stock). With this exchange option perspective, we can readily adapt the bounds for call prices by reversing the roles of the stock and the strike, resulting almost effortlessly in

$$\left(F_{0,T}^P(K) - F_{0,T}^P(S) \right)_+ \leq P^E \leq F_{0,T}^P(K)$$ (9.2.4)

and

$$\max \left\{ (K - S(0))_+ , \left(F_{0,T}^P(K) - F_{0,T}^P(S) \right)_+ \right\} \leq P^A \leq K.$$ (9.2.5)

Inequalities (9.2.1), (9.2.3), (9.2.4), and (9.2.5) together define different upper and lower bounds on American and European option prices as a function of the current stock price $S(0)$.

Example 9.2.2. (SOA Exam IFM Advanced Derivatives Sample Question 26: Graphical problem) Consider European and American options on a nondividend-paying stock.

You are given:

(i) All options have the same strike price of 100.

(ii) All options expire in six months.

(iii) The continuously compounded risk-free interest rate is 10%.

You are interested in the graph for the price of an option as a function of the current stock price. In each of the following four charts I–IV, the horizontal axis, S, represents the current stock price, and the vertical axis, π, represents the price of an option.

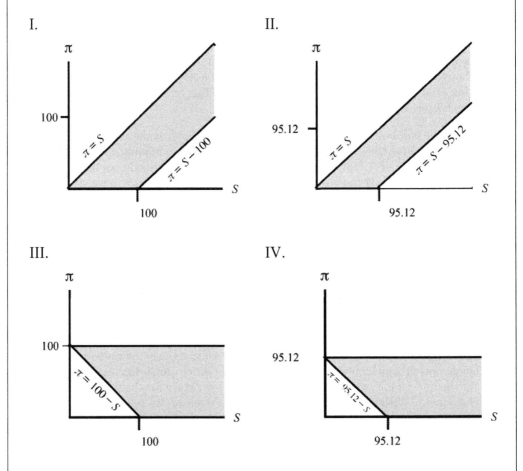

Match the option with the shaded region in which its graph lies. If there are two or more possibilities, choose the chart with the smallest shaded region.

	European Call	American Call	European Put	American Put
(A)	I	I	III	III
(B)	II	I	IV	III
(C)	II	I	III	III
(D)	II	II	IV	III
(E)	II	II	IV	IV

Solution. By (9.2.1) and (9.2.3), we have

$$(S - 95.12)_+ = \left(F^P_{0,0.5}(S) - Ke^{-0.1(0.5)}\right)_+ \leq C^E \leq F^P_{0,0.5}(S) = S$$

and

$$S - 95.12 = \max\{(S - 100)_+, (S - 95.12)_+\} \leq C^A \leq S.$$

Thus, the shaded region in II contains C^A and C^E. (Although the shaded region in I does, it is a larger region)

For put prices, we use (9.2.4) and (9.2.5), yielding

$$(95.12 - S)_+ \leq P^E \leq F^P_{0,T}(K) = 95.12$$

and

$$100 - S = \max\{(100 - S)_+, (95.12 - S)_+\} \leq P^A \leq 100.$$

Therefore, the shaded regions in III and IV contain P^A and P^E, respectively. **(Answer: (D))** □

Remark. In fact, in the absence of dividends the European call and American call share the same price (see Section 9.4).

9.3 Comparing Options with Respect to Contract Characteristics

While the preceding section look at bounds fulfilled by no-arbitrage option prices, in this section, we treat option prices as functions of the strike price and time to maturity by varying one of the contract characteristics at a time while keeping everything else unchanged. This allows us to study the effects of these characteristics on option prices on a fair basis.

9.3.1 Strike Price

It is easy to see that option prices are monotonic functions of the strike price. Common sense suggests and our no-arbitrage arguments confirm that call prices are decreasing in and put prices are increasing in the strike price. It turns out that much more about the geometry of these prices as a function of the strike price can be said even when no option pricing model is prescribed. Throughout this subsection, we consider options identical in every aspect except for their strike prices. To emphasize their dependence on K, we write $C^E(K), C^A(K), P^E(K)$, and $P^A(K)$, respectively, for the prices of a European call, an American call, a European put, and an American put, all of which are K-strike. The symbols

$C(K)$ and $P(K)$ will be used when the concerned option price property is true for both European and American options.

Property 1: Call prices (resp. put prices) are decreasing (resp. increasing) functions of the strike price.

If $K_1 \le K_2$, then[ii]

$$C(K_1) \ge C(K_2) \quad \text{and} \quad P(K_1) \le P(K_2).$$

These inequalities are true for both European and American option prices.

- *Intuitive justification:* These monotonicity relations are intuitively obvious as the higher the strike price, the lower (resp. higher) the payoff of a call option (resp. put option) when exercised, and hence the less valuable (resp. more valuable) the right to buy (resp. sell) the underlying asset for the strike price.

- *No-arbitrage proof:* To avoid repetition, we only prove the inequality for call option prices; the proof for put prices proceeds along the same line. First, consider the European case. Suppose by way of contradiction that $C^E(K_1) < C^E(K_2)$. Following the "buy low, sell high" principle, we buy the K_1-strike call and sell the K_2-strike call, in effect buying a K_1-K_2 call bull spread. At time 0, we receive $C^E(K_2) - C^E(K_1) > 0$. At maturity, the payoff of a call bull spread is always non-negative (recall Subsection 3.3.1). This creates an arbitrage strategy.

  We now turn to the more interesting American case. Again suppose that $C^A(K_1) < C^A(K_2)$, and we buy the K_1-strike American call and sell the K_2-strike American call. At time 0, we receive $C^A(K_2) - C^A(K_1) > 0$. Two cases need to be distinguished depending on the behavior of the holder of the K_2-strike American call.

  Case 1. If, for whatever reason, the holder of the K_2-strike American call decides to exercise his/her option at some intermediate time $\tau \in [0, T]$, then at the same time we exercise our K_1-strike American call, with the effect being that we realize $[S(\tau) - K_1] - [S(\tau) - K_2] = K_2 - K_1$, which is non-negative, at time τ.

  Case 2. If the holder of the K_2-strike American does not exercise his/her call at all, then at time T we are still entitled to the non-negative payoff of $(S(T) - K_1)_+$.

  In both cases, we successfully realize risk-free profits.

Example 9.3.1. (Profit of an arbitrageur exploiting put price monotonicity)
You are given:

(i) The following 1-year European put option prices on the same stock:

Strike Price	Put Price
40	6
45	5

(ii) The continuously compounded risk-free interest rate is 6%.

[ii]The inequalities can be sharp. Consider the pathological case when the future stock price is a constant less than K_1 and K_2. Since both call options do not pay off for certain, we have $C(K_1) = C(K_2) = 0$.

You take advantage of any possible mispricing by means of an appropriate spread position.

Calculate your 1-year profit when the 1-year stock price is 42.

Solution. Since $P(40) > P(45)$, the increasing monotonicity of put prices is violated. To pursue an arbitrage strategy, we buy the 45-strike put and sell the 40-strike put, creating a put bear spread and resulting in a net cash inflow of $6 - 5 = 1$ at time 0. The 1-year profit when $S(1) = 42$ is

$$(45 - 42)_+ - (40 - 42)_+ - e^{0.06} = 3 - e^{0.06} = \boxed{1.9382}.$$

□

The following Lipschitz property of option prices complements the monotonicity property by describing how much less expensive the call and more expensive the put will be when one raises the strike price.

Property 2: Option prices are Lipschitz with respect to the strike price.

If $K_1 \le K_2$, then

$$\boxed{C^E(K_1) - C^E(K_2) \le e^{-rT}(K_2 - K_1) \quad \text{and} \quad P^E(K_2) - P^E(K_1) \le e^{-rT}(K_2 - K_1),}$$

and

$$\boxed{C^A(K_1) - C^A(K_2) \le K_2 - K_1 \quad \text{and} \quad P^A(K_2) - P^A(K_1) \le K_2 - K_1.}$$

If the option prices are differentiable in the strike price,[iii] then the Lipschitz property implies that the partial derivatives $\partial C(K)/\partial K$ and $\partial P(K)/\partial K$ are all bounded with

$$-e^{-rT} \le \frac{\partial C^E(K)}{\partial K} \le 0 \le \frac{\partial P^E(K)}{\partial K} \le e^{-rT} \quad \text{and} \quad -1 \le \frac{\partial C^A(K)}{\partial K} \le 0 \le \frac{\partial P^A(K)}{\partial K} \le 1.$$

- *Intuitive justification:* Consider the case of a European put. The extra payoff as a result of exercising the K_2-strike put as compared to the K_1-strike put is at most $K_2 - K_1$; it is exactly $K_2 - K_1$ when both puts are in-the-money and less otherwise. For a European put, this extra payoff is realized at maturity only. The difference in the two otherwise identical put prices must be less than the present value of $K_2 - K_1$, i.e., $e^{-rT}(K_2 - K_1)$. The case of a European call admits similar explanations.

  For American options, the extra payoff can be realized any time at or before maturity. In some cases, the options may be exercised immediately. This relaxes the upper bound from the present value of $K_2 - K_1$ to just $K_2 - K_1$.

- *Simple algebraic proof for European prices:* For *European* options, the Lipschitz property is simply a restatement of the monotonicity property. For call prices, this can be seen from the following chain of equivalences with the aid of put-call parity (which applies

---

[iii]This is indeed true as a consequence of Lipschitz continuity.

only to European options):

$$C^E(K_1) - C^E(K_2) \quad \leq \quad e^{-rT}(K_2 - K_1)$$

$$\updownarrow$$

$$[P^E(K_1) + F_{0,T}^P(S) - K_1 e^{-rT}] - [P^E(K_2) + F_{0,T}^P(S) - K_2 e^{-rT}] \quad \leq \quad e^{-rT}(K_2 - K_1)$$

$$\updownarrow$$

$$P^E(K_1) - P^E(K_2) + e^{-rT}(K_2 - K_1) \quad \leq \quad e^{-rT}(K_2 - K_1)$$

$$\updownarrow$$

$$P^E(K_1) \quad \leq \quad P^E(K_2).$$

Analogously, the Lipschitz property for European put prices is an equivalence of the monotonicity property of European call prices. The implication is that the Lipschitz property for European call prices (resp. put prices) brings no new information[iv] about these prices compared to the monotonicity property for otherwise identical European put prices (resp. call prices).

- *No-arbitrage proof for American prices:* To avoid repetition, we only prove the Lipschitz inequality for American put prices. Assume on the contrary that $P^A(K_2) - P^A(K_1) > K_2 - K_1$. Let us buy a K_1-strike American put, sell a K_2-strike American put, and invest the proceeds of $K_2 - K_1$ into the risk-free account. If the short K_2-strike put is exercised at some intermediate time $\tau \in [0, T]$, then at the same time we exercise our K_1-strike put and collect the accumulated proceeds from the risk-free account. Our time-τ payoff is

$$-(K_2 - S(\tau))_+ + (K_1 - S(\tau))_+ + (K_2 - K_1)e^{r\tau}$$

$$= \begin{cases} (K_1 - K_2) + (K_2 - K_1)e^{r\tau} = (K_2 - K_1)(e^{r\tau} - 1), & \text{if } S(\tau) < K_1, \\ -[K_2 - S(\tau)] + (K_2 - K_1)e^{r\tau} \geq (K_2 - K_1)(e^{r\tau} - 1), & \text{if } K_1 \leq S(\tau) < K_2, \\ (K_2 - K_1)(e^{r\tau} - 1), & \text{if } K_2 \leq S(\tau), \end{cases}$$

which is always positive. If the short K_2-strike put is never exercised, then we still have the accumulated proceeds from the risk-free account, leading to an arbitrage strategy.

Example 9.3.2. (SOA Exam MFE Spring 2009 Question 12: Strike-related inequalities) You are given:

- $C(K, T)$ denotes the current price of a K-strike T-year European call option on a nondividend-paying stock.

- $P(K, T)$ denotes the current price of a K-strike T-year European put option on the same stock.

- S denotes the current price of the stock.

- The continuously compounded risk-free interest rate is r.

Which of the following is (are) correct?

$$\text{(I)} \quad 0 \leq C(50, T) - C(55, T) \leq 5e^{-rT}$$

$$\text{(II)} \quad 50e^{-rT} \leq P(45, T) - C(50, T) + S \leq 55e^{-rT}$$

$$\text{(III)} \quad 45e^{-rT} \leq P(45, T) - C(50, T) + S \leq 50e^{-rT}$$

[iv]The Lipschitz property for European call prices, however, is not equivalent to the Lipschitz property for European put prices.

(A) (I) only

(B) (II) only

(C) (III) only

(D) (I) and (II) only

(E) (I) and (III) only

Solution. • (I) is true. Call price is a decreasing function of K, so $C(50, T) \geq C(55, T)$. The second inequality follows from the Lipschitz property.

• (II) is incorrect, but (III) is true. By put-call parity,

$$
\begin{aligned}
P(45, T) - C(50, T) + S &= [C(45, T) - S + 45e^{-rT}] - C(50, T) + S \\
&= C(45, T) - C(50, T) + 45e^{-rT}.
\end{aligned}
$$

Adapting (I), we have $0 \leq C(45, T) - C(50, T) \leq (50 - 45)e^{-rT}$, which is equivalent to

$$
45e^{-rT} \leq C(45, T) - C(50, T) + 45e^{-rT} \leq 50e^{-rT}.
$$

(Answer: (E))   □

The monotonicity and Lipschitz properties both pertain to a pair of strike prices of interest. A comparison involving a triplet of strike prices reveals the convexity property, which is by far the most technical no-arbitrage property.

Property 3: Option prices are convex with respect to the strike price.

If $K_1 < K_2$ and λ is a number between 0 and 1, then

$$
\boxed{C[\lambda K_1 + (1 - \lambda)K_2] \leq \lambda C(K_1) + (1 - \lambda)C(K_2)}
$$

and

$$
\boxed{P[\lambda K_1 + (1 - \lambda)K_2] \leq \lambda P(K_1) + (1 - \lambda)P(K_2).}
$$

As a function of K, $C(K)$ and $P(K)$ are both convex.[v] Again, these inequalities apply to both European and American options.

• *Intuitive justification:* Not easily available for the convexity property, unfortunately! ☺

• *No-arbitrage proof.* For compactness, let $K_\lambda = \lambda K_1 + (1 - \lambda)K_2$. We only prove $C^A(K_\lambda) \leq \lambda C^A(K_1) + (1 - \lambda)C^A(K_2)$; the case for American put prices is similar and European prices are even simpler. Suppose by way of contradiction that

---

[v]The convexity condition is equivalent to the *chord inequalities*: For $K_1 < K_2 < K_3$,

$$
\frac{C(K_2) - C(K_1)}{K_2 - K_1} \leq \frac{C(K_3) - C(K_2)}{K_3 - K_2}
$$

and

$$
\frac{P(K_2) - P(K_1)}{K_2 - K_1} \leq \frac{P(K_3) - P(K_2)}{K_3 - K_2}.
$$

In words, the rate of change of the option prices increases as the strike price increases (for calls, the rate of change becomes less negative).

$C^A(K_\lambda) > \lambda C^A(K_1) + (1 - \lambda)C^A(K_2)$, which means that the K_λ-strike American call is overpriced. Then we buy λ units of the K_1-strike American call, $1 - \lambda$ units of the K_2-strike American call, and sell a K_λ-strike American call[vi]. This results in a strictly positive cash inflow at time 0. If the holder of the K_λ-strike American call exercises his/her option at some intermediate time $\tau \in [0, T]$, then at the same time we exercise our K_1-strike calls and K_2-strike American calls. Our time-τ payoff equals that of a τ-year *asymmetric call butterfly spread*, which is always non-negative. Thus $C^A(K_\lambda) \le \lambda C^A(K_1) + (1 - \lambda)C^A(K_2)$ has to be true to preclude arbitrage opportunities. □

To inspect the convexity property for any given strike prices K_1, K_2 and K_3 with $K_1 < K_2 < K_3$, we need to express the sandwiched strike price K_2 as a weighted average of K_1 and K_3. By linear interpolation, the precise representation is

$$K_2 = \left(\frac{K_3 - K_2}{K_3 - K_1}\right) K_1 + \left(\frac{K_2 - K_1}{K_3 - K_1}\right) K_3.$$

Then inspect whether

$$C(K_2) \le \left(\frac{K_3 - K_2}{K_3 - K_1}\right) C(K_1) + \left(\frac{K_2 - K_1}{K_3 - K_1}\right) C(K_3).$$

Example 9.3.3. (Calculating the maximum and minimum profits of an arbitrage strategy) You are given:

(i) The following prices of 3-year European call options on the same stock:

Strike Price ($)	Call Price ($)
100	36
110	28
120	19

(ii) The continuously compounded risk-free interest rate is 2%.

To earn arbitrage profit, you buy two 100-strike call options, two 120-strike call options, sell some 110-strike call options, and invest the proceeds (or finance the cost) at the risk-free rate.

Calculate your maximum and minimum profit at the end of 3 years.

Solution. Observe that $110 = 0.5(100 + 120)$ and that

$$0.5C(100) + 0.5C(120) = 27 < C(110) = 28,$$

violating the convexity condition of call prices. We then buy two 100-strike call options, two 120-strike call options, short four 110-strike call options, and invest the proceeds of $4(28) - 2(36) - 2(19) = 2$ into the risk-free account.

At the end of 3 years, the maximum payoff and minimum payoff are $10(2) = 20$ (realized at $S(3) = 110$) and 0 (realized when $S(3) < 100$ or $S(3) > 120$), respectively. It follows that the maximum profit and minimum profit are, respectively, $20 + 2e^{0.02(3)} = \boxed{22.1237}$ and $2e^{0.02(3)} = \boxed{2.1237}$. □

[vi]If trading of fractional units of options is not allowed in the market, then we multiply both sides of the inequality $C^A(K_\lambda) > \lambda C^A(K_1) + (1 - \lambda)C^A(K_2)$ by some "large" integer so that all coefficients are integers.

Example 9.3.4. [HARDER!] (SOA Exam IFM Advanced Derivatives Sample Question 2: Different sets of arbitrage strategies for the same set of price data) Near market closing time on a given day, you lose access to stock prices, but some European call and put prices for a stock are available as follows:

Strike Price	Call Price	Put Price
$40	$11	$3
$50	$6	$8
$55	$3	$11

All six options have the same expiration date.

After reviewing the information above, John tells Mary and Peter that no arbitrage opportunities can arise from these prices.

Mary disagrees with John. She argues that one could use the following portfolio to obtain arbitrage profit: Long one call option with strike price 40; short three call options with strike price 50; lend $1; and long some calls with strike price 55.

Peter also disagrees with John. He claims that the following portfolio, which is different from Mary's, can produce arbitrage profit: Long 2 calls and short 2 puts with strike price 55; long 1 call and short 1 put with strike price 40; lend $2; and short some calls and long the same number of puts with strike price 50.

Which of the following statements is true?

(A) Only John is correct.

(B) Only Mary is correct.

(C) Only Peter is correct.

(D) Both Mary and Peter are correct.

(E) None of them is correct.

Solution. Let's analyze Mary's and Peter's strategies separately and explore the motivations that underlie their strategies.

Mary.
- *Specifying the strategy:* Let x be the number of 55-strike calls to buy. Mary's initial investment is

$$[C(40) - 3C(50) + C(55)x] + 1 = [11 - 3(6) + 3x] + 1$$
$$= 3x - 6.$$

A non-positive initial investment requires that $x \leq 7/3$. To investigate Mary's terminal payoff, consider its shape: (go from left to right because calls are used)

	Stock Price			
	$S(T) < 40$	$40 < S(T) < 50$	$50 < S(T) < 55$	$55 < S(T)$
Slope	0	+1	−2	$x - 2$

The slope when $S(T) > 55$ must be non-negative, or else the terminal payoff will be negative when $S(T)$ is large enough. This implies that $x \geq 2$. The allowable range of values of x is therefore $2 \leq x \leq 7/3$. If one restricts x to be an integer, then the only possible value of x is 2.

- *Rationale:* Since $50 = \frac{1}{3}(40) + \frac{2}{3}(55)$ but $C(50) = 6 \not\leq \frac{1}{3}C(40) + \frac{2}{3}C(55) = 17/3$, the call prices (but not the put prices) violate the convexity condition. Mary is therefore buying a 40-50-55 call butterfly spread, which requires buying a 40-strike call, buying two($= x$) 55-strike calls and selling three 50-strike calls.

- *Why is it an arbitrage strategy?* Mary pays nothing at time 0 and receives a non-negative payoff together with the future value of \$1 at expiration.

Peter.

- *Specifying the strategy:* Let y be the number of 50-strike puts to buy and 50-strike calls to sell. Peter's initial investment is

$$[C(40) - C(50)y + 2C(55)] - [P(40) - P(50)y + 2P(55)] + 2$$
$$= [11 - 6y + 2(3)] - [3 - 8y + 2(11)] + 2$$
$$= 2y - 6.$$

A non-positive initial investment requires that $y \leq 4$. Since buying a call and selling an otherwise identical put creates a synthetic forward, Peter's terminal payoff is

$$[S(T) - 40] - y[S(T) - 50] + 2[S(T) - 55] = (3 - y)S(T) + (50y - 150).$$

For this linear function to be non-negative for all $S(T) \geq 0$, it is necessary and sufficient that its slope and intercept are both non-negative. This is equivalent to $3 - y \geq 0$, or $y \leq 3$ and $50y - 150 \geq 0$, or $y \geq 3$. Therefore, the only possible value of y is 3.

- *Rationale:* Peter is buying a 40-50-55 call butterfly spread and selling a 40-50-55 put butterfly spread. The idea behind Peter's strategy is that an asymmetric butterfly spread constructed from calls shares the same *payoff and profit* as an asymmetric butterfly spread constructed from otherwise identical puts (see Problem 3.5.31). In this case, the call butterfly spread costs $11 + 2(3) - 3(6) = -1$ whereas the put butterfly spread costs $3 + 2(11) - 3(8) = 1$. Then one should buy the call butterfly spread (i.e., buy one 40-strike call, buy two 55-strike calls, and sell three 50-strike calls) and sell the put butterfly spread (i.e., sell one 40-strike put, sell two 55-strike puts, and buy three 50-strike puts).

- *Why is it an arbitrage strategy?* Just like Mary, Peter need not pay anything at time 0, but is entitled to the future value of \$2 at expiration.

Both Mary and Peter have set up an arbitrage portfolio. (**Answer: (D)**)   □

Remark. The SOA's official solution does not explain why $x = 2$ and $y = 3$ are the only possibilities, although these two values are probably the most intuitive choices that align with standard arbitrage strategies.

We remark that the Lipschitz property and the convex property do not imply each other. It is possible for a set of option prices to violate the Lipschitz property, but not the convex property (see Problem 9.5.10), or the other way round.

9.3.2 Maturity

American prices are increasing in the time to maturity.

Given $0 \leq T_1 \leq T_2$, consider a T_1-year American call and a T_2-year American call, which are otherwise identical, with respective prices denoted by $C^A(T_1)$ and $C^A(T_2)$. Then the prices of these two American calls are ordered as

$$\boxed{C^A(T_1) \leq C^A(T_2).}$$

- *Intuitive explanations:* The T_2-year American call provides a greater degree of flexibility to its holder, who can make the early exercise decision any time until time T_2, than the T_1-year American call. The former call can always be exercised at the same time as the latter call.

- *No-arbitrage proof:* Suppose on the contrary that $C^A(T_1) > C^A(T_2)$. Then we buy the T_2-year American call and sell the T_1-year American call for a positive net cash inflow of $C^A(T_1) - C^A(T_2) > 0$. In the time period $[0, T_1]$, there are two possibilities:

  Case 1. If the holder of the T_1-year American call exercises the option at some time $\tau \in [0, T_1]$. At the same time, we exercise our T_2-year American call. The overall cash flow is

  $$\underbrace{[S(\tau) - K]}_{\text{long } T_2\text{-year call}} \quad \underbrace{-[S(\tau) - K]}_{\text{short } T_1\text{-year call}} = 0.$$

  In other words, what we gain from our T_2-year call exactly offsets what we have to pay to the T_1-year call holder.

  Case 2. Suppose that the holder of the T_1-year American call does not exercise his/her option, which expires at time T_1. We are still holding the T_2-year American call option and hence entitled to a non-negative payoff in the future without the need for paying anything.

  In both cases, there is no chance of loss after time 0, but we earn risk-free money at time 0.

By the same token, a T_2-year American put must be at least as valuable as a T_1-year American put, i.e.,

$$\boxed{P^A(T_1) \leq P^A(T_2).}$$

Example 9.3.5. (CAS Exam 3 Fall 2007 Question 13: Ranking option prices – I) Given the following chart about call options on a particular dividend-paying stock, which option has the highest value?

Option	Option Style	Time Until Expiration	Strike Price	Stock Price
A	European	1 year	50	42
B	American	1 year	50	42
C	European	2 years	50	42
D	American	2 years	50	42
E	American	2 years	55	42

(A) Option A

(B) Option B

(C) Option C

(D) Option D

(E) Option E

Solution. An American call must be at least as valuable as its otherwise identical European counterpart, so $C_B \geq C_A$ and $C_D \geq C_C$. Meanwhile, American call prices are increasing in the time to maturity and decreasing in the strike price, so $C_D \geq C_B$ and $C_D \geq C_E$. In conclusion, call D has the highest value **(Answer: (D))** ☐

Warning: The same cannot be said for European prices.

You should be cautioned that European option prices are generally not increasing in the time to maturity. It is possible for a longer-lived European call to be worth less than an otherwise identical shorter-lived call. As an extreme case, consider a stock which will pay a liquidating dividend in three years, i.e., the entire value of the stock will be paid to shareholders in three years. Then a 1-year European call in general has a positive price because it can be exercised at the time (i.e., one year from now) when the stock still has a positive value, but the stock will always be worthless after three years, so will be the 3-year European call. We then have $0 = C(3) < C(1)$. In general, the dividends between time T_1 and time T_2, where $T_1 < T_2$, lower the price of the T_2-year call but leave the price of the T_1-year call unaffected.

Likewise, it is not necessary that a longer-lived European put costs more than an otherwise identical shorter-lived European put. For instance, a longer-lived European put on a bankrupt company (i.e., $S(t) = 0$ for all $t \geq 0$) is less valuable than a short-lived European put because in this case the put price is simply the present value of the strike price. The longer the time to maturity, the lower the present value.

Exceptions: No dividends or interest.

If the underlying asset pays no dividends between time T_1 and time T_2, however, then it is indeed true that $C^E(T_1) \leq C^E(T_2)$. To see this, we expand our notation and write $C^E(t, T)$ for the time-t price of a generic T-year European call with $0 \leq t \leq T$. In the absence of dividends in the period $[T_1, T_2]$, we have $F^P_{T_1,T_2}(S) = S(T_1)$. It follows from the lower bound in (9.2.1) that

$$
\begin{aligned}
C^E(T_1, T_2) &\geq \left(F^P_{T_1,T_2}(S) - Ke^{-r(T_2-T_1)} \right)_+ \\
&= \left(S(T_1) - Ke^{-r(T_2-T_1)} \right)_+ \\
&\geq (S(T_1) - K)_+ \\
&= C^E(T_1, T_1).
\end{aligned}
$$

This time-T_1 price/payoff inequality then results in the time-0 price inequality

$$
C^E(T_2) = C^E(\boxed{0}, T_2) \geq C^E(\boxed{0}, T_1) = C^E(T_1).
$$

If we further assume that there are no dividends between time 0 and time T_2, then the proof can be simplified. This is because both European calls are identical to their American counterparts (see Section 9.4) and they obey the increasing relationship with respect to the time to maturity:

$$C^E(T_1) = C^A(T_1) \le C^A(T_2) = C^E(T_2).$$

By reversing the roles of the stock and the strike, we can deduce that $P^E(T_1) \le P^E(T_2)$ if the risk-free interest rate is zero between time T_1 and time T_2, although such a situation is less conceivable in practice.

9.4 Early Exercise Decisions for American Options

In Section 4.3, we studied the valuation of American options in the context of the binomial tree model. We saw that whether to exercise an American option early boils down to a series of comparisons between its holding value and its exercise value at different nodes of the tree. In the absence of an option pricing model, the explicit determination of holding value is almost impossible, making this kind of comparisons not of much practical value. In this section, we analyze the early exercise decision for American options in a general context by studying the economic factors that encourage or discourage early exercise. While devising the early exercise strategy (i.e., to determine for what range of stock price we exercise the option early) for an American option requires an option pricing model, which is not the linchpin of this section, our general analysis allows us to formulate easily verifiable *sufficient* conditions for the early exercise right to be of no value and, by extension, for the price of an American option to coincide with the price of an otherwise identical European option.

> ### IMPORTANT PROPOSITION
>
> $C^E = C^A$ for nondividend-paying stocks.

A very important but somewhat surprising fact in the theory of American options is that it is never optimal to exercise an American call on a *nondividend-paying stock* early. Informally speaking, an American call in this case is worth more "alive" than "dead." Even if you are holding a deep in-the-money American call at some time point before maturity (i.e., $S(\tau) \gg K$ for some $\tau \in [0, T]$), and have a strong "urge" to exercise it early, you should subdue your urge and pursue a better decision (what is it?). There are several ways to show this proposition; we shall supply two proofs.

9.4.1 Proof 1: A Proof Based on No-arbitrage Bounds

The most typical proof of the equality of C^E and C^A results from an application of the lower bound in (9.2.1), so that

$$C^A \ge C^E \ge F_{0,T}^P(S) - Ke^{-rT} \stackrel{\text{(no dividends)}}{=\joinrel=} S(0) - Ke^{-rT} > S(0) - K,$$

where the last term is the payoff arising from immediate exercise. Hence *selling the American call* to the market at the price of C^A will always bring us a higher payoff than exercising the American call early, however deep in-the-money the call is. This implies that any rational American call holder will not exploit the early exercise right, which becomes worthless and does not cause the American call price to outstrip the European call price.

Note that if the underlying stock pays dividends, then we only have

$$C^A \geq C^E \geq \left(F_{0,T}^P(S) - Ke^{-rT}\right)_+,$$

which may or may not be greater than the immediate exercise value of $(S(0) - K)_+$. Hence no definite conclusion about the optimality of early exercise can be drawn.

9.4.2   Proof 2: A Cost-benefit Dissection Proof

Insurance value.

A considerably more informative proof is based on decomposing the American call price into several pieces, each of which impacts on the early exercise decision differently. We start by introducing the *insurance value* of the American call defined as

$$\mathrm{IV}(C^A) := C^A - [F_{0,T}^P(S) - F_{0,T}^P(K)],$$

which is the difference between the current price for a *right* (the call) to buy the underlying asset for K anytime on or before the maturity date, and the current price of an *obligation* (the synthetic forward contract) to buy the underlying asset for K at maturity. Such a difference between the price of a right and the price of an obligation to buy the asset returns the value of the insurance against a *drop* in the asset price in the future.

Note that by (9.2.1), we have $C^A \geq C^E \geq F_{0,T}^P(S) - F_{0,T}^P(K)$, so the insurance value of an option must be non-negative.

Call price decomposition.

Armed with the notion of the insurance value of an American call introduced above, we proceed to decompose its price into four distinct components:

$$
\begin{aligned}
C^A &= [F_{0,T}^P(S) - F_{0,T}^P(K)] + \mathrm{IV}(C^A) \\
&= [S(0) - \mathrm{PV}_{0,T}(\mathrm{Div})] - Ke^{-rT} + \mathrm{IV}(C^A) \\
&= \underbrace{[S(0) - K]}_{\text{intrinsic value}} + \underbrace{K(1 - e^{-rT})}_{\text{time value}} + \underbrace{\mathrm{IV}(C^A)}_{\text{insurance value}} - \underbrace{\mathrm{PV}_{0,T}(\mathrm{Div})}_{\text{dividends}}, \qquad (9.4.1)
\end{aligned}
$$

where

$S(0) - K$ is the *intrinsic value*, or the immediate exercise value, of the call,

$K(1 - e^{-rT})$ is the *time value* of the call, or the difference between what we pay now (i.e., K) and the present value of what we pay at time T (i.e., Ke^{-rT}),

$\mathrm{IV}(C^A)$ is the insurance value of the call, and

$\mathrm{PV}_{0,T}(\mathrm{Div})$ is the present value of the dividend payouts from the underlying stock (if any) over the life of the call.

Because $C^A \geq S(0) - K$ due to the lower bound in (9.2.3), the early exercise decision boils down to whether:

- $C^A > S(0) - K$, in which case the call holder would be better off selling the call to the market than exercising the call early, and early exercise is not optimal. By (9.4.1), an equivalent condition for the non-optimality of early exercise is

$$\underbrace{\text{time value} + \mathrm{IV}(C^A)}_{\text{costs of early exercise}} > \underbrace{\text{dividends}}_{\text{benefits of early exercise}}. \qquad (9.4.2)$$

- $C^A = S(0) - K$, in which case early exercise is optimal.

Cost-benefit interpretation.

Albeit longish compared to Proof 1, the current proof based on decomposition (9.4.1) sheds important light on the factors influencing the early exercise decision. Economically, it is a manifestation of the common-sense economic principle that whether an action should be taken involves weighing the associated costs and the benefits. From a cost-benefit perspective, the early exercise of a call achieves two effects:

- *Benefits:* It accelerates the receipt of dividends, if any, on the stock. This line of reasoning also shows that the only incentive for the American call holder to exercise the option early stems from the early receipt of dividends.

- *Costs:* Meanwhile, doing so expedites the payment of the strike price K, resulting in an interest cost, and throws away the implicit insurance.

In view of these cost-benefit considerations, (9.4.2) can be interpreted as follows:

When the costs of early exercise strictly outweigh the benefits, it is not judicious to exercise the American call early.

As a simple application of (9.4.1), note that inequality (9.4.2), in the absence of dividends, is trivially satisfied for its right-hand side is zero. Therefore, we can conclude that $C^A = C^E$ when the stock pays no dividends.

A useful sufficient condition for non-optimality of early exercise.

In the presence of dividends, inequality (9.4.2) is mainly of theoretical interest. This is because the insurance value of an American call (more generally, an American option) is difficult to determine in general. For a European call, however, an application of put-call parity shows that its insurance value is simply the price of an otherwise identical put. Using the insurance value of an otherwise identical call as a proxy for the insurance value of an American call, we replace (9.4.2) by the approximate condition

$$\underbrace{\text{time value} + P^E}_{\text{approximate costs of early exercise}} > \underbrace{\text{dividends}}_{\text{benefits of early exercise}}. \tag{9.4.3}$$

Since $\text{IV}(C^A) \geq \text{IV}(C^E)$, (9.4.3) implies (9.4.2), which in turn implies the non-optimality of early exercise. Mathematically speaking, (9.4.3) provides an easily verifiable *sufficient* condition for the non-optimality of early exercise. However, the fact that (9.4.3) is violated does *not* mean that early exercise is optimal. In order that early exercise is optimal, it is *necessary* that the present value of the dividends is large enough so that condition (9.4.3) is refuted.

Example 9.4.1. (Continuation of Example 4.2.5: 10-period binomial tree for an American call!?) Consider a 10-period binomial model. The length of each period is a year. The time-0 stock price is $1,000. Every year the stock price will either increase by 5% or decrease by 5%. Assume that the continuously compounded risk-free interest rate is $i = 2\%$. The stock pays dividends continuously at rate $\delta = 1\%$.

Calculate the time-0 price of an American 10-year 1,400-strike call option on the stock.

Ambrose's comments:

Because of the existence of dividends, one cannot immediately rule out the possibility of early exercise. With some work, one will see that the American call holder will never early exercise.

Solution. It turns out that the right to early exercise is never used. To see this, we show that the time value alone (not including the insurance value) is always greater than the present value of the dividends, i.e., $1,400(1 - e^{-0.02(10-t)}) > S(t)(1 - e^{-0.01(10-t)})$, or

$$S(t) < \frac{1,400(1 - e^{-0.02(10-t)})}{1 - e^{-0.01(10-t)}} = 1,400(1 + e^{-0.01(10-t)}) \quad \text{for all } t = 0, 1, \ldots, 10.$$

$$(9.4.4)$$

Note that the highest value of the stock price over the 10-year horizon is $S_{u^{10}} = 1,000(1.05)^{10} = 1,628.8946$. Because $1,400(1 + e^{-0.01(10-t)})$ is increasing in t, its lowest value is achieved at $t = 0$, or $1,400(1 + e^{-0.01(10)}) = 2,666.7724$. Thus (9.4.4) is always satisfied. It follows that the American call price is equal to the call price of an otherwise identical European call. Standard binomial tree calculations then yield $C^E = \boxed{3.5924}$. □

9.4.3   Early Exercise Criterion for American Puts

To investigate the economic determinants of the early exercise for American puts, one may develop a put price decomposition analogous to (9.4.1). More efficiently and instructively, we may view a put as an exchange option which entails giving up the underlying stock in return for the strike price. From such an exchange option perspective, we can readily adapt the results for an American call, which is an exchange option giving up the strike price in return for the underlying stock, and obtain the following useful sufficient condition for the non-optimality of the early exercise of an American put:

$$\underbrace{\text{dividends} + C^E}_{\text{approximate costs of early exercise}} > \underbrace{\text{time value}}_{\text{benefits of early exercise}}. \quad (9.4.5)$$

In particular, when the risk-free interest rate is zero, early exercising an American put is never a wise decision.

Example 9.4.2. (SOA Exam MFE Spring 2007 Question 4: Early exercise for a put) For a stock, you are given:

(i) The current stock price is $50.00.

(ii) $\delta = 0.08$.

(iii) The continuously compounded risk-free interest rate is $r = 0.04$.

(iv) The prices for one-year European calls (C) under various strike prices (K) are shown below:

K	C
$40	$9.12
$50	$4.91
$60	$0.71
$70	$0.00

You own four special put options, each with one of the strike prices listed in (iv). Each of these put options can only be exercised immediately or one year from now.

Determine the lowest strike price for which it is optimal to exercise these special put option(s) immediately.

(A) $40

(B) $50

(C) $60

(D) $70

(E) It is not optimal to exercise any of these put options.

Solution 1. We show that (9.4.5) is true for all $K = 40, 50, 60$, and 70:

K	Dividends $+ C^E$		Time Value
40	$50(1 - e^{-0.08}) + 9.12 = 12.9642$	>	$40(1 - e^{-0.04}) = 1.5684$
50	$50(1 - e^{-0.08}) + 4.91 = 8.7542$	>	$50(1 - e^{-0.04}) = 1.9605$
60	$50(1 - e^{-0.08}) + 0.71 = 4.5542$	>	$60(1 - e^{-0.04}) = 2.3526$
70	$50(1 - e^{-0.08}) + 0.00 = 3.8842$	>	$70(1 - e^{-0.04}) = 2.7447$

As the costs of early exercising the special puts are always greater than the benefits, it is not optimal to exercise any of these put options immediately. (**Answer: (E)**) □

Solution 2. Due to the special nature of the special puts (they are essentially Bermudan options), it is possible to determine their exact holding values easily. This requires the observation that if these puts are not exercised immediately, they can only be exercised in one year, making them European in nature. By put-call parity, their holding values and exercise values are:

K	Holding Value ($P^E = Ke^{-r} - Se^{-\delta} + C$)	Exercise Value ($K - 50$)
40	$40e^{-0.04} - 50e^{-0.08} + 9.12 = 1.3958$	-10
50	$50e^{-0.04} - 50e^{-0.08} + 4.91 = 6.7937$	0
60	$60e^{-0.04} - 50e^{-0.08} + 0.71 = 12.2015$	10
70	$70e^{-0.04} - 50e^{-0.08} + 0.00 = 21.0994$	20

Since the holding value is greater than the exercise value at each strike, it is not optimal to exercise any of these put options immediately. (**Answer: (E)**) □

9.5   Problems

Put-call parity and duality

Problem 9.5.1. (SOA Exam IFM Advanced Derivatives Sample Question 1: No dividends) Consider a European call option and a European put option on a nondividend-paying stock. You are given:

(i) The current price of the stock is 60.

(ii) The call option currently sells for 0.15 more than the put option.

(iii) Both the call option and put option will expire in 4 years.

(iv) Both the call option and put option have a strike price of 70.

Calculate the continuously compounded risk-free interest rate.

(A) 0.039

(B) 0.049

(C) 0.059

(D) 0.069

(E) 0.079

Problem 9.5.2. (Put-call "inequalities" in the presence of market frictions) You are given:

(i) The quoted ask (resp. bid) prices of a K-strike T-year call option and a K-strike T-year put option on the same stock are denoted by $C^a(K,T)$ and $P^a(K,T)$ (resp. $C^b(K,T)$ and $P^b(K,T)$), respectively.

(ii) The current ask and bid prices for the stock are $S^a(0)$ and $S^b(0)$, respectively.

(iii) You can borrow at the rate r^b and lend at the rate r^l.

(iv) The stock pays dividends continuously at a rate proportional to its price. The dividend yield is δ.

Use no-arbitrage arguments to derive condition(s), in terms of the above symbols, under which you cannot profitably perform a parity arbitrage.

Problem 9.5.3. [HARDER!] (Given the prices of four call options) You are given:

(i) The current dollar-euro exchange rate is $1.25/€.

(ii) The price of a 3-year dollar-denominated European call option on euros with a strike price of $1.20 is $0.06545.

(iii) The price of a 3-year dollar-denominated European call option on euros with a strike price of $1.50 is $0.05842.

(iv) The price of a 3-year euro-denominated European call option on dollars with a strike price of €0.6667 is €0.1641.

(v) The price of a 3-year euro-denominated European call option on dollars with a strike price of €0.8333 is €0.03315.

Calculate the continuously compounded risk-free interest rate on dollars.

Problem 9.5.4. [HARDER!] (Currency parity arbitrage) You are given:

(i) The current euro/dollar exchange rate is 0.72.

(ii) The price of a 1-year euro-denominated European call option on dollars with a strike price of €0.70 is €0.09.

(iii) The price of a 1-year euro-denominated European put option on dollars with a strike price of €0.70 is €0.05.

(iv) The continuously compounded risk-free interest rate on euros is 2%.

(v) The continuously compounded risk-free interest rate on dollars is 1%.

Describe actions you could take to earn arbitrage profits at time 0, and calculate the arbitrage profits in euros.

Upper and lower bounds on option prices

Problem 9.5.5. (Upper bound on European put prices) Justify $P^E \leq F^P_{0,T}(K)$ by means of:

(a) Intuitive explanations

(b) A no-arbitrage proof

(c) An algebraic proof

Problem 9.5.6. (Upper and lower bounds on option prices for dividend-paying stocks) Consider European and American call options on the same underlying stock. You are given:

(i) Both options have the same strike price of 100.

(ii) Both options expire in six months.

(iii) The stock pays dividends continuously at a rate proportional to its price. The dividend yield is 6%.

(iv) The continuously compounded risk-free interest rate is 10%.

Sketch two separate graphs to show the possible prices of European and American calls against the current stock price $S(0)$. The bounds in your graphs should be as tight as possible. Label the equations of the boundaries and the key points in your graphs clearly.

Problem 9.5.7. (Upper and lower bounds on currency option prices) To settle an urgent debt payable in US dollars in one year, Jeff has decided to (reluctantly!) sell his favorite Rolls Royce car in exchange for a fixed sum payable in British pounds in one year. Because the British pound may lose value relative to the US dollar, Jeff decides to buy appropriate 1-year dollar-denominated $1.6-strike currency options on pounds to hedge against the exchange rate risk, but he is not sure about whether he should buy European or American options.

You are given:

(i) The continuously compounded risk-free interest rate on dollars is 6%.

(ii) The continuously compounded risk-free interest rate on pounds is 8%.

Sketch two separate graphs to show the possible (dollar) prices of each European and American currency option that Jeff should buy against the current dollar/pound exchange rate, $X(0)$. Shade the relevant regions and clearly indicate the equations of the relevant boundaries.

Comparing options with respect to contract parameters

Problem 9.5.8. (Yet another strike-related inequality) Suppose the current time is 0. Consider two European put options on the same underlying stock and the same maturity date T, but with different strike prices K_1 and K_2, where $K_1 \leq K_2$. The prices of the above options are denoted by $P(K_1)$ and $P(K_2)$, respectively.

Use no-arbitrage arguments to show that

$$\frac{P(K_1)}{P(K_2)} \leq \frac{K_1}{K_2}.$$

(Note: The same inequality does *not* hold for call option prices.)

Problem 9.5.9. (Linear combinations fulfilling or violating convexity) For $K \geq 0$, let $C(K)$ denote the price of a K-strike European call option on a stock. Let $0 < K_1 < K_2$. Define $K_\lambda = \lambda K_1 + (1 - \lambda)K_2$, where λ is a real number, *not necessarily between 0 and 1*.

(a) Determine all values of λ for which

$$C(K_\lambda) \leq \lambda C(K_1) + (1 - \lambda)C(K_2).$$

(b) Determine all values of λ for which

$$C(K_\lambda) \geq \lambda C(K_1) + (1 - \lambda)C(K_2).$$

Explain your reasons.

(Hints: For part (b), you may pursue an algebraic solution or a graphical solution. Regardless, you should explain your reasons clearly. The answer is of the form "$\lambda \leq \odot$ or $\odot \leq \lambda \leq \bullet$". Find the three faces. Also observe that strike prices should not be negative.)

Problem 9.5.10. (Convexity or something else violated?) You are given:

(i) The following prices of 1-year European call options on the same stock for various strikes:

Strike Price	Call Price
$20	$50.0
$25	$44.5
$30	$42.0

(ii) The continuously compounded risk-free interest rate is 6%.

Describe transactions you can enter into to exploit an arbitrage opportunity (if one exists).

Problem 9.5.11. (Constructing two arbitrage strategies based on different principles) You are given the following European call and put prices on the same stock:

Strike Price	Call Price	Put Price
$20	$5.16	$1.35
$25	$3.59	$2.64
$35	$2.45	$4.36

All six options have the same expiration date.
Propose two sets of arbitrage strategies based on different principles. For each strategy:

• Identify the principle upon which your strategy is based.

• Describe clearly the transactions you should enter into.

• Check that your strategy is indeed an arbitrage strategy.

Problem 9.5.12. (Ranking option prices) You are given the following European and American call options written on the same stock:

Option	European/American	Strike Price	Time to Expiration
I	European	100	1 years
II	European	100	2 years
III	American	100	1 years
IV	American	100	2 years
V	American	200	2 years

Rank, as far as possible, the prices of these five options.

Appendix A

Standard Normal Distribution Table

Entries below represent the area under the standard normal distribution function from $-\infty$ to z, $N(z) = \mathbb{P}(Z \leq z)$. The value of z to the first decimal is given in the left column. The second decimal is given in the top row. Areas for negative values of z can be obtained by symmetry, i.e., $N(z) = 1 - N(-z)$.

z	0.00	0.01	0.02	0.03	0.04	0.05	0.06	0.07	0.08	0.09
0.0	0.5000	0.5040	0.5080	0.5120	0.5160	0.5199	0.5239	0.5279	0.5319	0.5359
0.1	0.5398	0.5438	0.5478	0.5517	0.5557	0.5596	0.5636	0.5675	0.5714	0.5753
0.2	0.5793	0.5832	0.5871	0.5910	0.5948	0.5987	0.6026	0.6064	0.6103	0.6141
0.3	0.6179	0.6217	0.6255	0.6293	0.6331	0.6368	0.6406	0.6443	0.6480	0.6517
0.4	0.6554	0.6591	0.6628	0.6664	0.6700	0.6736	0.6772	0.6808	0.6844	0.6879
0.5	0.6915	0.6950	0.6985	0.7019	0.7054	0.7088	0.7123	0.7157	0.7190	0.7224
0.6	0.7257	0.7291	0.7324	0.7357	0.7389	0.7422	0.7454	0.7486	0.7517	0.7549
0.7	0.7580	0.7611	0.7642	0.7673	0.7703	0.7734	0.7764	0.7794	0.7823	0.7852
0.8	0.7881	0.7910	0.7939	0.7967	0.7995	0.8023	0.8051	0.8078	0.8106	0.8133
0.9	0.8159	0.8186	0.8212	0.8238	0.8264	0.8289	0.8315	0.8340	0.8365	0.8389
1.0	0.8413	0.8438	0.8461	0.8485	0.8508	0.8531	0.8554	0.8577	0.8599	0.8621
1.1	0.8643	0.8665	0.8686	0.8708	0.8729	0.8749	0.8770	0.8790	0.8810	0.8830
1.2	0.8849	0.8869	0.8888	0.8907	0.8925	0.8944	0.8962	0.8980	0.8997	0.9015
1.3	0.9032	0.9049	0.9066	0.9082	0.9099	0.9115	0.9131	0.9147	0.9162	0.9177
1.4	0.9192	0.9207	0.9222	0.9236	0.9251	0.9265	0.9279	0.9292	0.9306	0.9319
1.5	0.9332	0.9345	0.9357	0.9370	0.9382	0.9394	0.9406	0.9418	0.9429	0.9441
1.6	0.9452	0.9463	0.9474	0.9484	0.9495	0.9505	0.9515	0.9525	0.9535	0.9545
1.7	0.9554	0.9564	0.9573	0.9582	0.9591	0.9599	0.9608	0.9616	0.9625	0.9633
1.8	0.9641	0.9649	0.9656	0.9664	0.9671	0.9678	0.9686	0.9693	0.9699	0.9706
1.9	0.9713	0.9719	0.9726	0.9732	0.9738	0.9744	0.9750	0.9756	0.9761	0.9767
2.0	0.9772	0.9778	0.9783	0.9788	0.9793	0.9798	0.9803	0.9808	0.9812	0.9817
2.1	0.9821	0.9826	0.9830	0.9834	0.9838	0.9842	0.9846	0.9850	0.9854	0.9857
2.2	0.9861	0.9864	0.9868	0.9871	0.9875	0.9878	0.9881	0.9884	0.9887	0.9890
2.3	0.9893	0.9896	0.9898	0.9901	0.9904	0.9906	0.9909	0.9911	0.9913	0.9916
2.4	0.9918	0.9920	0.9922	0.9925	0.9927	0.9929	0.9931	0.9932	0.9934	0.9936
2.5	0.9938	0.9940	0.9941	0.9943	0.9945	0.9946	0.9948	0.9949	0.9951	0.9952
2.6	0.9953	0.9955	0.9956	0.9957	0.9959	0.9960	0.9961	0.9962	0.9963	0.9964
2.7	0.9965	0.9966	0.9967	0.9968	0.9969	0.9970	0.9971	0.9972	0.9973	0.9974
2.8	0.9974	0.9975	0.9976	0.9977	0.9977	0.9978	0.9979	0.9979	0.9980	0.9981
2.9	0.9981	0.9982	0.9982	0.9983	0.9984	0.9984	0.9985	0.9985	0.9986	0.9986

Values of z for selected values of $\mathbb{P}(Z \leq z)$							
z	0.842	1.036	1.282	1.645	1.960	2.326	2.576
$\mathbb{P}(Z \leq z)$	0.800	0.850	0.900	0.950	0.975	0.990	0.995

Appendix B

Solutions to Odd-Numbered End-of-Chapter Problems

B.1 Chapter 1

1. *Solution.* Solving the two payoff equations

$$\begin{cases} S - F_{0,T} = -5 \\ 1.1S - F_{0,T} = 1 \end{cases}$$

we get $S = 60$ and $F_{0,T} = 65$. If the spot price at expiration were 20% higher, then Aaron's profit, which is the same as his payoff, would be $1.2S - F_{0,T} = 1.2(60) - 65 = \boxed{7}$. $\square$

3. *Solution.* Because Rose purchases a put and makes a profit, the put must be in-the-money at expiration, i.e., it must be the case that $S(0.5) < 50$. With this information available, Jack's loss is

$$-\text{Profit} = 8e^{0.04(0.5)} - \underbrace{(S(0.5) - 50)_+}_{0} = 8.1616,$$

and Rose's profit is

$$(50 - S(0.5))_+ - 6e^{0.04(0.5)} = (50 - S(0.5)) - 6e^{0.02} = 43.8788 - S(0.5).$$

As Rose's profit is twice as large as Jack's loss, we have $43.8788 - S(0.5) = 2(8.1616)$, which implies that $S(0.5) = \boxed{27.56}$. $\square$

5. *Solution.* We begin by computing the future value of the put premiums:

Strike	FV of Put Premium
35	$0.44 \times 1.08 = 0.48$
40	$1.99 \times 1.08 = 2.15$
45	$5.08 \times 1.08 = 5.49$

The profit functions of the three short puts are sketched in Figure B.1.1. The 35-strike line crosses the 40-strike and 45-strike lines, respectively, at $40 - (2.15 - 0.48) = 38.33$ and $45 - (5.49 - 0.48) = 39.99$. Visually inspecting the profit diagram, we conclude that the 35-strike put produces a higher profit than the 45-strike put, but a lower profit than the 40-strike put, when $\boxed{38.33 < S(1) < 39.99}$. $\square$

7. *Solution.* Only I and II are correct. For III, the short European put, being long in nature, results in an obligation to buy the underlying asset if the long put holder chooses to exercise his/her put. (**Answer: (B)**) $\square$

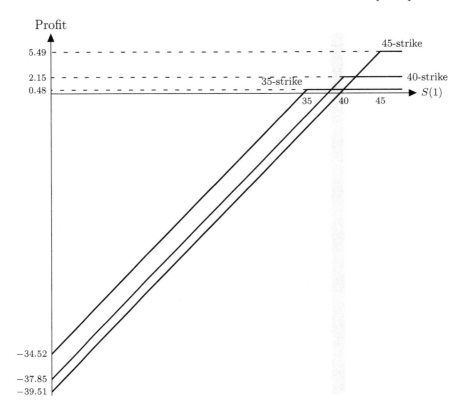

FIGURE B.1.1
Profit diagrams of the three puts in Problem 1.4.5.

B.2   Chapter 2

1. *Solution.* Using (2.2.1), we have

$$
\begin{aligned}
F_{0,1}^P &= 50 - \mathrm{PV}_{0,1}(\mathrm{Div}) \\
&= 50 - (e^{-0.06(0.25)} + e^{-0.06(0.5)} + e^{-0.06(0.75)} + e^{-0.06}) \\
&= \boxed{46.15}
\end{aligned}
$$

$\square$

3. *Solution.* Let i be the effective annual risk-free rate of interest. By the first formula in (2.3.2), the 1-year forward price satisfies the equation

$$
\mathrm{FV}_{0,1}(S(0)) - \mathrm{FV}_{0,1}(\mathrm{Div}) = F_{0,1},
$$

or

$$
80(1 + i) - 2(1 + i)^{1/2} - 3 = 84. \tag{B.2.1}
$$

Let $x = (1 + i)^{1/2}$. Then (B.2.1) becomes the quadratic equation (in x)

$$
80x^2 - 2x - 87 = 0,
$$

whose positive solution is

$$x = \frac{2 + \sqrt{2^2 - 4(80)(-87)}}{2(80)} = 1.0554.$$

Then $i = x^2 - 1 = \boxed{0.1139}$. □

5. *Solution.* By the first formula in (2.3.2), we have

$$\begin{aligned}
\Gamma_{0,3} \quad &- \quad S(0)e^{3r} - 1.5[e^{2.75r} + e^{2.5r}(1.01) + \cdots + (1.01)^{11}] \\
&= \quad 200e^{3(0.04)} - 1.5e^{2.75(0.04)}\left[\frac{1 - (1.01e^{-0.25(0.04)})^{12}}{1 - 1.01e^{-0.25(0.04)}}\right] \\
&= \quad \boxed{205.4119}. \qquad \textbf{(Answer: (B))}
\end{aligned}$$

□

Remark. In the second equality, we use the geometric series formula $a + ar + ar^2 + \cdots + ar^{n-1} = a(1 - r^n)/(1 - r)$ for $r \neq 1$.

7. *Solution.* By the second formula in (2.3.2), the fair 1-year forward price is

$$F_{0,1}^{\text{fair}} = S(0)e^{(r-\delta)T} = 120e^{(0.06-0.04)(1)} = 122.4242,$$

which is higher than the observed price of $F_{0,1}^{\text{obs}} = 121$. In other words, the observed forward in the market is underpriced. To exploit the arbitrage opportunity, we should buy the observed forward in the market and sell the synthetic forward. The table below shows the details and the associated cash flows:

	Cash Flows	
Transaction	**Time 0**	**Time 1**
Buy observed forward	0	$S(1) - 121$
Short sell $e^{-0.04} = 0.9608$ units of Stock XYZ	$120e^{-0.04} = 115.2947$	$-S(1)$
Lend 115.2947	-115.2947	122.4242
Total	0	$122.4242 - 121 = 1.4242$

□

9. *Solution.* There are two cash flows between November 1, 3016 and November 1, 3018:

- On November 1, 3016, you short sold 100 shares of stock X at the *bid* price of 95.00 per share, together with the commission, for a total cash inflow of

$$95.00 \times 100 \times 0.98 = 9,310.$$

- On November 1, 3018, due to the reinvestment of dividends you have to buy back $100e^{0.04(2)} = 108.328707$ shares at the *ask* price of 100.50 per share, along with the commission, for a total cash outflow of

$$108.328707 \times 100.50 \times 1.02 = 11,104.77573.$$

Therefore, your profit measured as of November 1, 3018 equals

$$-11,104.77573 + 9,310e^{0.05(2)} = \boxed{-815.63}.$$

□

11. *Solution.* In symbols, we have

$$
\begin{array}{rcl}
\text{(A)} & = & S(0), \\
\text{(B)} & = & F_{0,1} = S(0)e^{r-\delta}, \\
\text{(C)} & = & F_{0,2} = S(0)e^{2(r-\delta)}, \\
\text{(D)} & = & S(0)e^{-2\delta}, \\
\text{(E)} & = & \mathbb{E}[S(2)].
\end{array}
$$

Because $\delta < r$, the correct ranking is:

$$\boxed{(D) < (A) < (B) < (C) < (E)}.$$

□

13. *Solution.* The no-arbitrage interval of the 3-year forward price is

$$
\left[(40 \underbrace{-1}_{\text{paid at time 0}})e^{(0.06-0.03)(3)} \underbrace{-2}_{\text{paid at time 3}} , (41+1)e^{(0.07-0.03)(3)} + 2 \right]
$$
$$= [40.6728, 49.3549];$$

remember that market frictions enter the interval in such a way to make it as wide as possible. Since the observed forward price of 38 is lower than the lower end-point of the no-arbitrage interval, we engage in a reverse cash-and-carry arbitrage by performing the following actions:

Transaction	Cash Flows	
	Time 0	Time 3
Buy observed forward	0	$[S(3) - 38] - 2 = S(3) - 40$
Short sell $e^{-0.09} = 0.9139$ units of Stock Y	$(40-1)e^{-0.03(3)} = 35.6433$	$-S(3)$
Lend 35.6433	-35.6433	$35.6433e^{0.06(3)} = 42.6728$
Total	0	$42.6728 - 40 = 2.6728$

□

15. *Solution.* The notional value of Peter's futures contracts is

$$8 \times 250 \times 1,629 = 3,258,000.$$

The initial margin is 10% of the notional value, or $10\% \times 3,258,000 = 325,800$.

The maintenance margin is 80% of the initial margin, or $70\% \times 325,800 = 228,060$.

There will be a margin call one month from now if the new margin balance is less than the maintenance margin, i.e.

$$\overbrace{325,800e^{0.06/12} + 8 \times 250 \times (F_{\underbrace{1,3}_{\text{in months}}}}^{\text{Mark-to-market proceed}} - 1,629) \leq 228,060,$$

which results in $F_{1,3} \leq 1,579.31346$.

Solving

$$S(1)e^{(0.06-0.02)(2/12)} = F_{1,3} \leq 1,579.31346,$$

we get $(0 \leq) \boxed{S(1) \leq 1,568.82}$. □

B.3 Chapter 3

1. *Solution.* (A) No

   (B) Yes

   (C) No (the answer would be "Yes" for a long collar.)

   (D) Yes

   (E) No

   □

3. *Solution.* The final payoff as a result of using a floor is $\max(S(T), K)$, which increases with the floor level K. The higher the strike price, the higher the payoff and the more expensive the put is. (**Answer: (A)**) □

5. *Solution.* You can combine the payoff diagrams of a floor and cap and see that the resulting diagram is that of a long straddle. Alternatively, recall that a floor consists of a long asset plus a long put, and that a cap comprises a short asset plus a long call. Adding a floor and a cap cancels the asset and leaves only the long put and the long call. This is simply a $\boxed{\text{long straddle}}$. (**Answer: (C)**) □

7. *Solution.* The positions that Richard should take in the call and put options are:

   - $\boxed{\text{Short}}$ call
   - $\boxed{\text{Long}}$ put

   By put-call parity, the cost required to establish these two positions equals (recall that $F_{0,T}^P(S) = S(0)$ for a nondividend-paying stock)

   $$P - C = \text{PV}_{0,0.5}(1,020) - F_{0,1}^P = e^{-0.05(0.5)}(1,020) - 1,000 = \boxed{-5.1839}.$$

   □

9. *Solution.* • Actuary A, being short the 70-strike call, has a profit of

   $$\text{Pr}_A = 1.5e^{0.06} - (S(1) - 70)_+.$$

- Actuary B, being long the synthetic forward, has the same profit as a genuine long forward, whose profit is (recall that the stock pays no dividends, so $F_{0,T} = S(0)e^{rT}$)

$$\mathrm{Pr}_B = S(1) - F_{0,1} = S(1) - 60e^{0.06}.$$

Since $\mathrm{Pr}_B = 2\,\mathrm{Pr}_A$, we solve

$$S(1) - 60e^{0.06} = 2[1.5e^{0.06} - (S(1) - 70)_+]. \tag{B.3.1}$$

Case 1.   If $S(1) < 70$, then (B.3.1) implies that $S(1) - 60e^{0.06} = 2(1.5e^{0.06})$, or $S(1) = 66.90$.

Case 2.   If $S(1) \geq 70$, then (B.3.1) results in

$$S(1) - 60e^{0.06} = 2[1.5e^{0.06} - (S(1) - 70)],$$

which can be solved to yield $S(1) = 68.97$, contradicting $S(1) \geq 70$.

The only possible 1-year stock price is $S(1) = \boxed{66.90}$. $\square$

11. *Solution.* Two applications of put-call parity give

$$\begin{cases} 14.3782 - 0.4394 = S(0)e^{-0.02(2)} - 98e^{-2r} \\ 12.7575 - 0.6975 = S(0)e^{-0.02(2)} - 100e^{-2r} \end{cases}.$$

Solving the two simultaneous equations, we have $e^{-2r} = 0.9394$ and $S(0) = \boxed{110.33}$ $\square$

13. *Solution.* Since $C(K)$ is decreasing in K, we have $|C(60) - C(65)| = C(60) - C(65) = 3$. By put-call parity,

$$\begin{cases} C(60) - P(60) = F_{0,0.25}^P - 60e^{-0.05(0.25)} \\ C(65) - P(65) = F_{0,0.25}^P - 65e^{-0.05(0.25)} \end{cases}.$$

Subtracting the second equation from the first one yields

$$[C(60) - C(65)] + [P(65) - P(60)] = 5e^{-0.05(0.25)},$$

and because $P(K)$ is increasing in K, we have $|P(60) - P(65)| = P(65) - P(60) = \boxed{1.9379}$. $\square$

15. *Solution.* By definition, a bear spread is set up by selling a low-strike option and buying a high-strike otherwise identical option. Both (A) and (B) give rise to a bull spread instead. For (C) and (D), note that $C(K)$ is non-increasing in K whereas $P(K)$ is non-decreasing in K. It follows that the call (resp. put) with a premium of 6 has a higher (resp. lower) strike than the call (resp. put) with a premium of 10.

(A) No

(B) No

(C) Yes

(D) No

     $\square$

17. *Solution 1 (Preferred).* Note that a 70-80 bear spread constructed using puts has the same profit as one constructed using calls. The latter is constructed by selling a 70-strike call and buying a 80-strike call. The investment required is

$$C(80) - C(70) = 2.7 - 8.3 = -5.6.$$

For the profit to be 4, the payoff has to be

$$\text{Profit} + FV_{0,1}(\text{Investment}) = 4 + (-5.6e^{0.06}) = -1.9463,$$

which happens when the 1-year stock price is $S(1) - 70 + 1.9403 = \boxed{71.95}$. ⊔

Solution 2 (Not preferred). A 70-80 put bear spread is constructed by constructed selling a 70-strike put and buying a 80-strike put. By put-call parity,

$$\begin{cases} C(70) - P(70) = PV_{0,1}(F_{0,1}) - PV_{0,1}(70) \\ C(80) - P(80) = PV_{0,1}(F_{0,1}) - PV_{0,1}(80) \end{cases}.$$

Hence

$$\begin{aligned} [C(70) - C(80)] + [P(80) - P(70)] &= PV_{0,1}(10) \\ (8.3 - 2.7) + [P(80) - P(70)] &= 10e^{-0.06} \\ P(80) - P(70) &= 3.817645. \end{aligned}$$

Setting $4 = \text{Profit} = \text{Payoff} - FV_{0,1}(\text{Investment})$, we get

$$\text{Payoff} = 4 + 3.817645e^{0.06} = 8.053715,$$

which is realized at $S(1) = 80 - 8.053715 = \boxed{71.95}$. □

19. *Solution.* From the given table of option premiums, it is possible to deduce the T-year present value factor $PV_{0,T}$. Specifically, put-call parity implies (recall that $F_{0,T}^P = S(0)$ if the stock pays no dividends)

$$\begin{cases} 8.4 - 0.8 = S(0) - 50PV_{0,T} \\ 2.6 - 4.7 = S(0) - 60PV_{0,T} \end{cases},$$

which gives $PV_{0,T} = 0.97$ and $S(0) = 56.1$. Since a bear spread constructed by calls has the same profit as one constructed by puts, with respect to profit we may assume without loss of generality that the bear spread is constructed by puts, or more precisely, by buying the 60-strike put and selling the 50-strike put. The initial investment in this case is

$$P(60) - P(50) = 4.7 - 0.8 = 3.9,$$

which accumulates to $3.9/0.97 = 4.0206$ at expiration. For the profit of the bear spread to be zero, the stock price at expiration has to be $S(T) = 60 - 4.0206 = 55.9794$, meaning that the stock price will move by $55.9794 - 56.1 = \boxed{-0.1206}$. □

21. *Solution.* A (long) box spread is created by buying a 25-strike call, selling a 25-strike put, selling a 35-strike call and buying a 35-strike put. The investment required is

$$C(25) - P(25) - C(35) + P(25) = 9.51,$$

and the payoff at expiration is $35 - 25 = 10$. The implicit 1-year accumulation factor is $10/9.51 = 1.0515$, whereas the 1-year accumulation factor in the market is $e^{0.06} = 1.0618$. In other words, the long box spread is worse than a risk-free investment in the market and should be short:

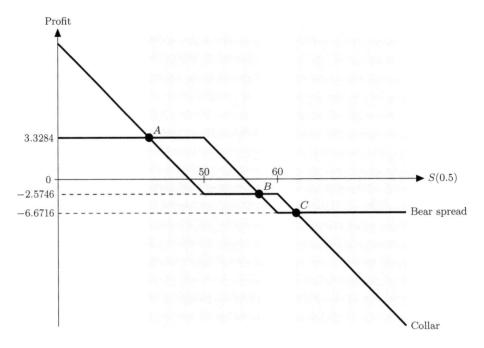

FIGURE B.3.1
The profit diagrams of the 50-60 bear spread and the 50-60 collar in Problem 3.5.23.

Strike	Position in Call	Position in Put
25	Short	Long
35	Long	Short

In the language of bull/bear spread(s), we should:

- Buy a 25-35 call bear spread (or equivalently sell a 25-35 call bull spread)
- Buy a 25-35 put bull spread (or equivalently sell a 25-35 put bear spread)

Together with a long 1-year zero-coupon bond with a face value of 10, the resulting profit at time 0 is $9.51 - 10e^{-0.06} = \boxed{0.0924}$. (At time 1, our payoff is constant at $\underbrace{-10}_{\text{from box spread}} \quad \underbrace{+10}_{\text{from bond}} = 0$) $\qquad\qquad\square$

23. *Solution.* Without loss of generality (why?), we may assume that the 50-60 bear spread is constructed by calls, i.e., a short 50-strike call coupled with a long 60-strike call. The initial investment is $C(60) - C(50) = 1.96 - 5.19 = -3.23$, which grows to $-3.23e^{0.06(0.5)} = -3.3284$ after 6 months. The 50-60 collar is set up by a long 50-strike put and a short 60-strike call. By put-call parity, the put price is

$$P(50) = 5.19 - 50 + 0.75e^{-0.06/12} + 50e^{-0.06(0.5)} = 4.4585.$$

The initial investment of the collar is $P(50) - C(60) = 4.4585 - 1.96 = 2.4985$, which accumulates to $2.4985e^{0.06(0.5)} = 2.5746$ after 6 months.

The profit functions of the 50-60 bear spread and the 50-60 collar are sketched in Figure B.3.1 (not drawn to scale). The horizontal coordinates of points A, B, and C are 44.0970, 55.9030, and 64.097, respectively. It follows that the bear spread's profit outweighs the collar's profit when and only when $\boxed{44.10 < S(0.5) < 55.90}$ and $\boxed{S(0.5) > 64.10}$. $\qquad\square$

25. *Solution.* (A) False. It is a short with respect to the underlying stock and is a bet on the stock price going down in the future.

(B) True. Simply set $K_1 = K_2 = F_{0,T}$.

(C) False. There are infinitely many possible zero-cost collars.

(D) True. Recall that K_1 and K_2 are restricted by $K_1 \leq F_{0,T} \leq K_2$. If $F_{0,T} \geq S(0)$, then K_1 can possibly take the value of $S(0)$.

(E) False. If $F_{0,T} < S(0)$, which happens when the dividends are sufficiently large, then K_2 can possibly be less than $S(0)$, in which case the call is in-the-money.

□

27. *Solution.* By inspection (starting from the right because of put options), the payoff of the derivative equals

$$\text{Payoff} = 3(5 - S(1))_+ + 2(10 - S(1))_+ - 6(15 - S(1))_+ - 2S(1) + 80,$$

which shows that the derivative can be replicated by:

- 3 long 5-strike puts
- 2 long 10-strike puts
- 6 short 15-strike puts
- 2 short stocks
- A long zero-coupon bond with a face value of 80 (i.e. lend PV(80))

Therefore, the fair price of the derivative is

$$\begin{aligned}
& 3P(5) + 2P(10) - 6P(15) - 2S(0) + 80e^{-0.04} \\
= \ & 3(0.11) + 2(1.75) - 6(9.52) - 2(10) + 80e^{-0.04} \\
= \ & 3.5732,
\end{aligned}$$

which is lower than the observed price by 0.02684.

To exploit an arbitrage opportunity, we sell the observed derivative and buy the synthetic derivative by:

- 3 long 5-strike puts
- 2 long 10-strike puts
- 6 short 15-strike puts
- 2 short stocks
- A long zero-coupon bond with a face value of 80 (i.e. lend PV(80))

The amount of arbitrage profit per unit of stock is $\boxed{0.02684}$.

□

29. *Solution.* Note that we should buy the 70-strike straddle. By put-call parity, the put price is

$$P(70) = C(70) - F_{0,T}^P + Ke^{-rT} = 7 - 70e^{-0.02(4/12)} + 70e^{-0.08(4/12)} = 5.6231.$$

Then the stock price has to move by $FV_{0,4/12}[C(70) + P(70)] = 12.6231e^{0.08(4/12)} = \boxed{12.96}$ in either direction after 4 months to result in a positive profit.

□

31. *Solution.* (A) False.

     Explanations: The holder of a long butterfly spread benefits the most when the spot price at expiration is very close to the current price. A long butterfly spread is therefore a bet on the volatility of the underlying asset being *lower* than that perceived by the market.

    (B) True.

     Explanations: Adding the payoff functions of a long K_1-K_2 bear spread and a long K_2-K_3 bull spread yields a payoff structure having the same general shape (horizontal, down, up, horizontal, from left to right) as a short K_1-K_2-K_3 butterfly spread.

    (C) True.

     Explanations: Algebraically, this is a consequence of put-call parity. Write the middle strike price as a weighted average of the two extreme strike prices, i.e., $K_2 = \lambda K_1 + (1 - \lambda)K_3$ for some $\lambda \in (0,1)$. Then the cost of a call (generally asymmetric) butterfly spread is

$$
\begin{aligned}
&\lambda C(K_1) - C(K_2) + (1 - \lambda)C(K_3) \\
={} &\lambda[P(K_1) + \mathrm{PV}_{0,T}(F_{0,T} - K_1)] - [P(K_2) + \mathrm{PV}_{0,T}(F_{0,T} - K_2)] \\
&+ (1 - \lambda)[P(K_3) + \mathrm{PV}_{0,T}(F_{0,T} - K_3)] \\
={} &\lambda P(K_1) - P(K_2) + (1 - \lambda)P(K_3),
\end{aligned}
$$

     which is the cost of the otherwise identical put butterfly spread.

     Geometrically, you can also employ the slope adjustment technique and see that exactly the same payoff structure can be produced by call or put options. For example, try to construct Figure 3.4.5 by put options.

    (D) True.

     Explanations: Scaling up the constituent options of a butterfly spread by a common factor of k leads to the maximum profit multiplied by k, which can be any arbitrary positive number. □

33. *Solution.* From (i) and (ii), the asymmetric butterfly spread can be created by buying two 20-strike calls, selling five 23-strike calls and buying three 25-strike calls. From the given table, the net investment is

$$
I = 2C(20) + 3C(25) - 5C(23) = 2(3.59) + 3(1.89) - 5(2.45) = 0.6.
$$

To find the present value factor, we apply put-call parity at $K = 20$ and $K = 23$ (or any other pair):

$$
\begin{cases}
3.59 - 2.64 = F^P_{0,T} - 20\mathrm{PV}_{0,T} \\
2.45 - 4.36 = F^P_{0,T} - 23\mathrm{PV}_{0,T}
\end{cases},
$$

which gives $\mathrm{PV}_{0,T} = 143/150$ (and $F^P_{0,T} = 1,201/60$). If $S(T) = 21$, then the payoff is 2 (by slope considerations), so the profit equals $2 - 0.6(150/143) = \boxed{1.37}$. □

B.4   Chapter 4

1. *Solution.* It turns out that the strangle pays $(50 - 70)_+ + (70 - 65)_+ = 5$ in the up state and $(50 - 45)_+ + (45 - 65)_+ = 5$ in the down state as well. It is the same as a risk-free

bond with a face value of 5. The current price of the strangle is therefore the present value of 5, or

$$V_0 = \frac{5}{1.1} = \boxed{4.5455}.$$

□

Remark. There is no need to determine the tree parameters and the risk-neutral probability of an up move. For your information, they are $u = 70/55 = 14/11$, $d = 45/55 = 9/11$ and

$$p^* = \frac{(1+0.1) - 9/11}{14/11 - 9/11} = 0.62.$$

3. *Solution.* The payoff of the special derivative is $V_u = (38 - 32)^2 = 36$ in the up-state and $V_d = (28 - 25)^3 = 27$ in the down-state. With the risk-neutral probability of an up move being

$$p^* = \frac{30(1.1) - 25}{38 - 25} = \frac{8}{13},$$

the current price of the derivative is

$$V_0 = \frac{1}{1.1}[36p^* + 27(1 - p^*)] = \boxed{29.5804}.$$

□

Remark. For your information, the replicating portfolio consists of $\Delta = 9/13$ shares and $B = 1{,}260/143$ in the risk-free bond.

5. This is a hard but interesting question on replication. Two solutions are provided, both of which are instructive.

Solution 1 (Replication). Call option A pays $(12-9)_+ = 3$ in Outcome 1, and $(8-9)_+ = 0$ in Outcome 2. To determine the value of Security 2 in Outcome 1, let's use α units of Security 1 and β units of call option A to replicate Security 2. Matching the time-0 prices and the payoffs in Outcome 2, we are prompted to solve the equations (see the remark below)

$$\begin{cases} 10.4\alpha + 1.8\beta = 10 & \text{(time-0 price)} \\ 8\alpha + 0(\beta) = 2.5 & \text{(payoff in Outcome 2)} \end{cases}.$$

They imply that $\alpha = 0.3125$ and $\beta = 3.75$. Then the value of Security 2 in Outcome 1 must be $12\alpha + 3\beta = 15$.

Now that the price evolution of Security 2 is known, we can deduce that call option B pays $(15 - 11)_+ = 4$ in Outcome 1, and $(2.5 - 11)_+ = 0$ in Outcome 2. Call option B is therefore equivalent to $4/3$ units of call option A and should cost

$$\frac{4}{3} \times \text{time-0 price of call option A} = \frac{4}{3} \times 1.8 = \boxed{2.4}.$$

□

Remark. (i) Notice carefully that having the same time-0 price does not mean having the same payoff at expiration. However, having the same time-0 price and the same payoff at one future outcome necessitates having the same payoff at the other outcome, or else arbitrage opportunities will exist (the question says that this is an "arbitrage-free binomial model").

(ii) You can replicate call option B by something else, but this is unnecessary.

Solution 2 (Risk-neutral pricing). To identify the risk-free rate implicit in this model, we replicate the risk-free bond using a portfolio of α' units of Security 1 and β' units of Call option A, so that the portfolio has the same payoff in Outcome 1 and in Outcome 2. Specifically, we require

$$12\alpha' + 3\beta' = 8\alpha' + 0(\beta') = 8\alpha',$$

which implies that α' and β' are related via

$$4\alpha' + 3\beta' = 0.$$

With this relation, the time-0 cost of the portfolio is

$$10.4\alpha' + 1.8\beta' = 10.4\alpha' + 1.8(-\tfrac{4}{3}\alpha') = 8\alpha'.$$

As $8\alpha'$ at time 0 grows to the same amount of $8\alpha'$ at the end of the binomial model, we deduce that the risk-free interest rate is $r = 0$.

Now we proceed to determine the risk-neutral probability of moving to Outcome 1. To this end, we apply the risk-neutral pricing formula to Security 1 and get

$$10.4 = e^{-r}[12p^* + 8(1 - p^*)] = 12p^* + 8(1 - p^*),$$

meaning that $p^* = 0.6$. Then applying the risk-neutral pricing formula to Security 2 leads to

$$10 = ? \times p^* + 2.5(1 - p^*),$$

so that $? = 15$. A final application of the risk-neutral pricing formula shows that the current price of call option B is

$$p^*(15 - 11)_+ + (1 - p^*)(2.5 - 11)_+ = \boxed{2.4}.$$

$\square$

7. *Solution.* Since $\mathrm{Var}[\ln[S(t)]] = \sigma^2 t$ for all $t > 0$, we have $\sigma = 0.5$. For a forward tree, the risk-neutral probability of an up move is

$$p^* = \frac{1}{1 + e^{\sigma\sqrt{h}}} = \frac{1}{1 + e^{0.5\sqrt{0.25}}} = \boxed{0.4378}.$$

$\square$

9. *Solution.* In a forward tree,

$$
\begin{aligned}
u &= \exp[(r - \delta)h + \sigma\sqrt{h}] = \exp[(0.03 - 0.05)(1) + 0.3\sqrt{1}] = 1.323130, \\
d &= \exp[(r - \delta)h - \sigma\sqrt{h}] = \exp[(0.03 - 0.05)(1) - 0.3\sqrt{1}] = 0.726149.
\end{aligned}
$$

The risk-neutral probability of an up move is

$$p^* = \frac{e^{(r-\delta)h} - d}{u - d} = \frac{e^{(0.03-0.05)(1)} - 0.726149}{1.323130 - 0.726149} = 0.425557,$$

or, more simply,

$$p^* = \frac{1}{1 + e^{\sigma\sqrt{h}}} = \frac{1}{1 + e^{0.3\sqrt{1}}} = 0.425557.$$

Then the four possible 1-year stock prices are

$$S_{uuu} = 440.1097, \quad S_{uud} = 241.5373, \quad S_{udd} = 132.5585, \quad S_{ddd} = 72.7496$$

and the corresponding payoffs of the 150-250 strangle (being $(S(3) - 250)_+ + (150 - S(3))_+$) are

$$V_{uuu} = 190.1097, \quad V_{uud} = 0, \quad V_{udd} = 17.4415, \quad V_{ddd} = 77.2504.$$

By risk-neutral pricing, the current price of the strangle is

$$
\begin{aligned}
V_0 &= e^{-rT}[(p^*)^3 V_{uuu} + 3(p^*)^2(1 - p^*)V_{uud} + 3p^*(1 - p^*)^2 V_{udd} + (1 - p^*)^3 V_{ddd}] \\
&= e^{-0.03(3)}[(0.425557)^3(190.1097) + 3(0.425557)(1 - 0.425557)^2(17.4415) \\
&\quad + (1 - 0.425557)^3(77.2504)] \\
&= \boxed{33.4888}.
\end{aligned}
$$

$\square$

11. *Solution.* The 3-period binomial tree is constructed as follows:

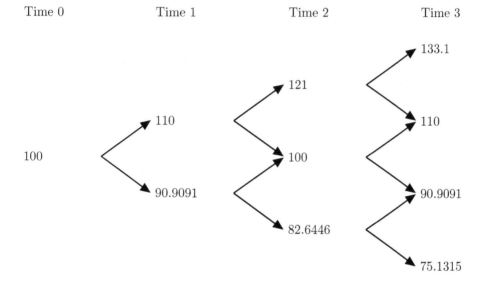

The risk-neutral probability of an up move is

$$p^* = \frac{e^{0.05} - 1/1.1}{1.1 - 1/1.1} = 0.744753.$$

The possible payoffs at expiration are

$$
\begin{aligned}
V_{uuu} &= 21 + 33.1 = 54.1, \\
V_{uud} &= 21 + 10 = 31, \\
V_{udu} = V_{duu} &= 0 + 10 = 10,
\end{aligned}
$$

and zero otherwise. By risk-neutral pricing, the price of the special derivative is

$$V_0 = e^{-0.05(3)}[(p^*)^3 V_{uuu} + \underbrace{(p^*)^2(1 - p^*)(V_{uud} + V_{udu} + V_{duu})}_{\text{not } 3(p^*)^2(1-p^*)V_{uud}!}] = \boxed{25.4495}.$$

$\square$

13. This is a challenging question in which several financial instruments (stock, put, Derivative X, Derivative Y), some with different expiration dates, co-exist. To answer this question well, one needs to identify the nature and underlying asset of each derivative clearly.

Solution. (a) Note that Derivative X is a 1-year European *call* option on a European *put* option which matures two years from now. To value the call, we need the possible values of the put option (underlying asset) at the end of the first year. We first start with the evolution of the stock price as shown below:

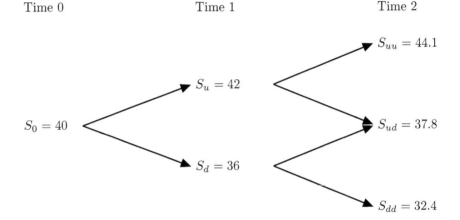

Time 0   Time 1   Time 2

$S_0 = 40$

$S_u = 42$

$S_d = 36$

$S_{uu} = 44.1$

$S_{ud} = 37.8$

$S_{dd} = 32.4$

The risk-neutral probability of an up move is

$$p^* = \frac{e^{0.03} - 0.9}{1.05 - 0.9} = 0.869697.$$

The possible time-2 payoffs of the 2-year put option are

$$P_{uu} = (38 - 44.1)_+ = 0, \quad P_{ud} = (38 - 37.8)_+ = 0.2, \quad P_{dd} = (38 - 32.4)_+ = 5.6.$$

By risk-neutral pricing, the possible time-1 values of this put option are

$$P_u = e^{-0.03}(1 - p^*)(0.2) = 0.025290,$$
$$P_d = e^{-0.03}[p^*(0.2) + (1 - p^*)(5.6)] = 0.876930.$$

Here is the evolution of the put price:

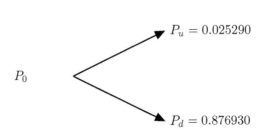

Time 0   Time 1

P_0

$P_u = 0.025290$

$P_d = 0.876930$

As a call on the above put, Derivative X pays off only at the d node with a non-zero payoff of

$$V_d^X = \underbrace{(P_d - 0.5)_+}_{\text{like the usual } (S(T)-K)_+ \text{ formula}} = 0.376930.$$

By risk-neutral pricing again, the time-0 price of Derivative X is

$$V_0^X = e^{-0.03}(1 - p^*)V_d^X = \boxed{0.0477}.$$

(b) This time Derivative Y is a 1-year European *put* option on the *put* option which matures two years from now. Derivatives X and Y therefore form a call-put pair. To apply put-call parity, we need the time-0 price of the underlying asset, which is the 2-year put option:

$$P_0 = e^{-0.03}[0.025290p^* + 0.876930(1 - p^*)] = 0.13223.$$

By put-call parity for nondividend-paying underlying assets (the 2-year put in this case), we have

$$V_0^X - V_0^Y = \underbrace{P_0}_{\text{not } S(0)} - \underbrace{0.5}_{\substack{\text{common} \\ \text{strike price}}} \times \underbrace{e^{-r}}_{\substack{\text{discount for} \\ \text{1 year only}}}$$

$$0.04766 - V_0^Y = 0.13223 - 0.5e^{-0.03}$$

$$V_0^Y = \boxed{0.4007}.$$

$\square$

Remark. (i) Note that when applying the put-call parity, we discount for one year only because the time to maturity of Derivatives X and Y is one year, although their underlying asset, which is the 2-year put, matures in two years.

(ii) There are many good reasons for forcing you to use put-call parity to answer part (b). First, it challenges you with this compound option version of put-call parity. Second and more importantly, in the continuous-time Black-Scholes framework it is hard (at least beyond the scope of this book) to compute the price of a compound option directly. Given the price of Derivative X, one can first calculate the price of the underlying put by the Black-Scholes formula, then determine the price of Derivative Y by the compound option put-call parity.

15. This challenging problem illustrates the essence of *dynamic* replication in a multi-period binomial model and its applications to the synthetic construction of a derivative as well as to arbitraging a mispriced derivative. The most interesting feature is that your strategies vary with time and with the prevailing stock price. You cannot just sit on the sofa and watch TV forever!

Solution. Let's begin by determining the (one-period) replicating portfolios at the initial node, u node, and d node. The stock price evolution is described as follows:

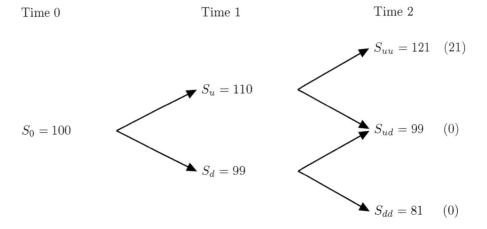

Time 0   Time 1   Time 2

$S_0 = 100$ $S_u = 110$ $S_{uu} = 121$ (21)

$S_d = 99$ $S_{ud} = 99$ (0)

$S_{dd} = 81$ (0)

- *Replicating portfolio:* With $C_{uu} = 21$ and $C_{ud} = C_{dd} = 0$, we have

$$\Delta_u = \frac{C_{uu} - C_{ud}}{S_u - S_d} = \frac{21 - 0}{121 - 99} = 0.954545,$$

$$B_u = \frac{uC_{ud} - dC_{uu}}{(1+i)(u-d)} = \frac{0 - 0.9(21)}{1.02(1.1 - 0.9)} = -92.647059,$$

$$C_u = \Delta_u S_u + B_u = 12.352891$$

and

$$\Delta_d = \frac{C_{ud} - C_{dd}}{S_u - S_d} = 0,$$

$$B_d = \frac{uC_{dd} - dC_{ud}}{(1+i)(u-d)} = 0.$$

$$C_d = \Delta_d S_d + B_d = 0.$$

(Note: If the stock price hits the d node, then the call will never pay off and becomes worthless.) At the initial node,

$$\Delta_0 = \frac{C_u - C_d}{S_u - S_d} = \frac{12.352891 - 0}{110 - 90} = 0.617645,$$

$$B_0 = \frac{uC_d - dC_u}{(1+i)(u-d)} = \frac{0 - 0.9(12.352891)}{1.02(1.1 - 0.9)} = -54.498049,$$

$$C_0 = \Delta_0 S_0 + B_0 = 7.266451. \quad (< 7.5)$$

In other words, the observed call is overpriced.

- *Arbitrage strategies:* To exploit the arbitrage opportunity, Peter should, *at time 0*, sell the observed call for \$7.5, buy $\Delta_0 = 0.617645$ shares and borrow \$54.498049 for a cost of $C_0 = \$7.266451$, thereby immediately realizing a positive cash inflow of \$0.233549.

  His actions at time 1 depend on the stock price that prevails at that time:

  Case 1. If $S(1) = S_u = 110$, then Peter should buy $\Delta_u - \Delta_0 = 0.3369$ *additional* shares and borrow an *additional* amount of $\$|B_u - B_0(1.02)| = \$37.0590$. No additional investment is required because the cost of \$0.3369(110) = \$37.0590 is the same (except for minor rounding errors) as the amount of borrowing.

(Alternatively, Peter can liquidate his position by selling $\Delta_0 = 0.617645$ shares for \$110 each, or for \$67.9410 in total and repaying the loan (with interest) of \$54.4980(1.02) = \$55.5880. Subsequently, he can use the \$12.353 he receives to buy $\Delta_u = 0.954545$ shares and borrow \$$|B_u|$ = \$92.6471 for one year.)

Case 2. If $S(1) = S_d = 90$, then Peter should completely liquidate his position by selling $\Delta_0 = 0.617645$ shares for \$90 each, or for \$55.5881 in total, and repaying the loan (with interest) of \$54.4980(1.02) = \$55.5880. Again, no cash inflow and outflow arises

In both cases, Peter will have identically zero payoff at time 2 because the payoff of his long synthetic call exactly offsets that of the short observed call. An arbitrage strategy is thus constructed due to the risk-less cash inflow of \$0.233549 at time 0. □

Remark. Peter cannot do nothing at time 0 and wait until time 1 for the exact actions to take, depending on whether the u node or d node is reached. The reason is that we are not sure of the observed call price, which can be greater than, smaller than or equal to the fair call price, at time 1.

17. *Solution.* In a forward tree,

$$u = \exp[(r - \delta)h + \sigma\sqrt{h}] = \exp[(0.06)(0.5) + 0.2\sqrt{0.5}] = 1.186991,$$
$$d = \exp[(r - \delta)h - \sigma\sqrt{h}] = \exp[(0.06)(0.5) - 0.2\sqrt{0.5}] = 0.894562.$$

The risk-neutral probability of an up move is

$$p^* = \frac{e^{(r-\delta)h} - d}{u - d} = \frac{e^{(0.06)(0.5)} - 0.894562}{1.186991 - 0.894562} = 0.464703,$$

or

$$p^* = \frac{1}{1 + e^{\sigma\sqrt{h}}} = \frac{1}{1 + e^{0.2\sqrt{0.5}}} = 0.464703.$$

The stock prices are depicted below:

Time 0	Time 0.5	Time 1

$S_{uu} = 140.8947$
$P_{uu} = \boxed{0}$

$S_u = 118.6991$
$P_u = \boxed{7.1772}$

$S_0 = 100$
$P_{II} = \boxed{20}$
$P_I = \boxed{17.2612}$
(Early exercise)

$S_{ud} = 106.1837$
$P_{ud} = \boxed{13.8163}$

$S_d = 89.4562$
$P_d^A = \boxed{30.5438}$
$P_d^E = \boxed{26.9973}$
(Early exercise)

$S_{dd} = 80.0241$
$P_{dd} = \boxed{39.9759}$

Note that early exercise is optimal at the d node because

$$\begin{aligned} \text{Holding value} \quad &= \quad e^{-0.06(0.5)}[p^* P_{ud} + (1-p^*)P_{dd}] \\ &= \quad 26.9973 \\ &< \quad (120 - S_d)_+ \\ &= \quad 30.5438 = \text{Exercise value,} \end{aligned}$$

so that $P_d^A = 30.5438$, as well as at the initial node, because

$$\begin{aligned} \text{Holding value} \quad &= \quad e^{-0.06(0.5)}[p^* P_u + (1-p^*)P_d] \\ &= \quad 19.1035 \\ &< \quad (120 - S_0)_+ \\ &= \quad 20 = \text{Exercise value.} \end{aligned}$$

Therefore, it pays to exercise the American put right away, with its price being $P_{II} = 20$. Meanwhile, the price of the corresponding European put is

$$P_I = e^{-0.06(1)}[2p^*(1-p^*)(13.8163) + (1-p^*)^2(39.9759)] = 17.2612.$$

It follows that $P_{II} - P_I = \boxed{2.739}$.   □

19. *Solution.* In a forward tree,

$$\begin{aligned} u \quad &= \quad \exp[(r-\delta)h + \sigma\sqrt{h}] = \exp[(0.05-0.035)/3 + 0.3\sqrt{1/3}] = 1.195070, \\ d \quad &= \quad \exp[(r-\delta)h - \sigma\sqrt{h}] = \exp[(0.05-0.035)/3 - 0.3\sqrt{1/3}] = 0.845180. \end{aligned}$$

The risk-neutral probability of an up move is

$$p^* = \frac{e^{(r-\delta)h} - d}{u - d} = \frac{e^{(0.05-0.035)/3} - 0.845180}{1.195070 - 0.845180} = 0.456807,$$

or

$$p^* = \frac{1}{1 + e^{\sigma\sqrt{h}}} = \frac{1}{1 + e^{0.3/\sqrt{3}}} = 0.456807.$$

The stock prices and put prices are depicted below:

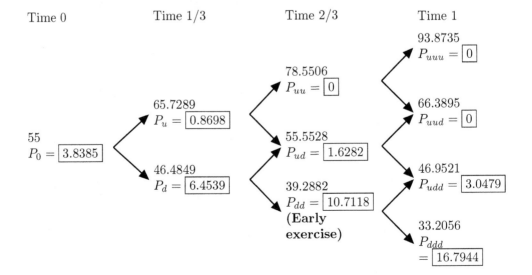

Early exercise is optimal at (and only at) the dd node because

$$
\begin{aligned}
\text{Holding value} \quad &= \quad e^{-0.05}[p^* P_{udd} + (1 - p^*) P_{ddd}] \\
&= \quad 10.3411 \\
&< \quad (50 - S_{dd})_+ \\
&= \quad 10.7118 = \text{Exercise value.}
\end{aligned}
$$

The time-0 price of the American put is $P_0 = \boxed{3.8385}$. $\qquad\qquad\square$

21. *Solution.* The forward tree parameters are

$$
\begin{aligned}
u &= \quad e^{(0.04-0.05)(0.25)+0.2\sqrt{0.25}} = 1.102411, \\
d &= \quad e^{(0.04-0.05)(0.25)-0.2\sqrt{0.25}} = 0.902578.
\end{aligned}
$$

The risk-neutral probability of an up move in the exchange rate is

$$
p^* = \frac{1}{1 + e^{0.2\sqrt{0.25}}} = 0.475021.
$$

The evolution of the dollar-euro exchange rate is shown below:

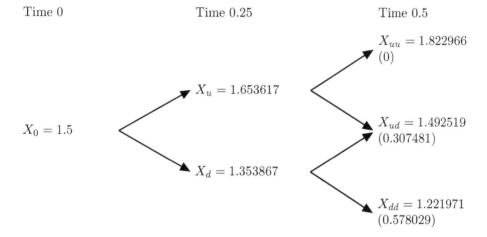

Time 0	Time 0.25	Time 0.5

$X_{uu} = 1.822966$ (0)

$X_u = 1.653617$

$X_0 = 1.5$

$X_{ud} = 1.492519$ (0.307481)

$X_d = 1.353867$

$X_{dd} = 1.221971$ (0.578029)

By risk-neutral pricing, the possible 3-month values of the put are

$$
\begin{aligned}
P_u &= \quad \max\{e^{-0.04(0.25)}[0.475021(0) + (1 - 0.475021)(0.307481)], (1.8 - 1.653617)_+\} \\
&= \quad \max\{0.159815, 0.146383\} \\
&= \quad 0.159815
\end{aligned}
$$

and

$$
\begin{aligned}
P_d &= \quad \max\{e^{-0.04(0.25)}[0.475021(0.307481) + (1 - 0.475021)(0.578029)], (1.8 - 1.353867)_+\} \\
&= \quad \max\{0.445040, 0.446133\} \\
&= \quad 0.446133. \quad \textbf{(early exercise optimal here)}
\end{aligned}
$$

Finally, the current price of the put is

$$
\begin{aligned}
P_0 &= \quad \max\{e^{-0.04(0.25)}[0.475021(0.159815) + (1 - 0.475021)(0.446133)], (1.8 - 1.5)_+\} \\
&= \quad \max\{0.307040, 0.3\} \\
&= \quad \boxed{0.3070}.
\end{aligned}
$$

$\qquad\qquad\square$

23. *Solution.* Given that $u = 1.2363$ and $d = 0.8089$, the risk-neutral probability of an up move is

$$p^* = \frac{1-d}{u-d} = 0.4471.$$

The evolution of the futures prices is shown below:

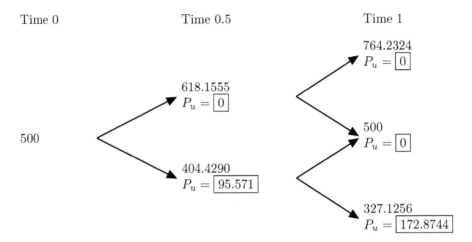

| Time 0 | Time 0.5 | Time 1 |

The possible values of the American put at time $t = 0.5$ are $P_u = 0$ at the u node, and

$$
\begin{aligned}
P_d &= \max\{e^{-0.06(0.5)}[p^*(0) + (1-p^*)(172.8744)], 500 - 404.4290\} \\
&= \max\{92.7465, 95.571\} \\
&= 95.571 \quad \textbf{(early exercise optimal)}
\end{aligned}
$$

at the d node. The replicating portfolio at $t = 0$ for the futures put is defined by

$$\Delta = \frac{P_u - P_d}{F_u - F_d} = \frac{0 - 95.571}{618.1555 - 404.4290} = \boxed{-0.4472}$$

and

$$B = e^{-0.06(0.5)}[p^*(0) + (1-p^*)(95.571)] = \boxed{51.27}.$$

$\square$

25. *Solution.* The forward tree parameters are

$$u = e^{0.08/3+0.3/\sqrt{3}} = 1.221246 \quad \text{and} \quad d = e^{0.08/3-0.3/\sqrt{3}} = 0.863693.$$

The risk-neutral probability of an up move is

$$p^* = \frac{1}{1 + e^{0.3/\sqrt{3}}} = 0.456807.$$

As $C_{uu} = 34.5721$, $C_{ud} = 12.7391$ and $C_{dd} = 0$, we have

$$
\begin{aligned}
C_u &= e^{-0.08/3}[p^*(34.5721) + (1-p^*)(12.7391)] = 22.1149, \\
C_d &= e^{-0.08/3}[p^*(12.7391) + (1-p^*)(0)] = 5.6662.
\end{aligned}
$$

As the true probability of an up move is

$$p = \frac{e^{0.15/3} - 0.863693}{1.221246 - 0.863693} = 0.524616,$$

the continuously compounded true discount rate for the call option during the first time period, γ, satisfies

$$e^{-\gamma/3}[pC_u + (1-p)C_d] = e^{-r/3}[p^*C_u + (1-p^*)C_d],$$

or

$$e^{-\gamma/3}[0.524616(22.1149) + (1 - 0.524616)(5.6662)]$$
$$= e^{-0.08/3}[0.456807(22.1149) + (1 - 0.456807)(5.6662)],$$

leading to $\gamma = \boxed{0.3237}$. $\qquad\square$

B.5 Chapter 5

1. *Solution.* As
$$\hat{d}_2 = \frac{\ln(100/75) + (0.1 - 0.3^2/2)(0.75)}{0.3\sqrt{0.75}} = 1.26606,$$

the probability that a nine-month 75-strike European call option on the stock will be exercised is $N(1.26606) = \boxed{0.89725}$. $\qquad\square$

Remark. Because you are to find the actual probability, the risk-free interest rate is not used.

3. *Solution.* As $N^{-1}(0.975) = 1.95996$, a 95% lognormal prediction interval for the price of the stock in 3 months is

$$S(0)\exp\left[\left(\alpha - \delta - \frac{1}{2}\sigma^2\right)t \pm N^{-1}(0.975)\sigma\sqrt{t}\right]$$
$$= 100\exp\left[\left(0.15 - \frac{1}{2} \times 0.35^2\right)(0.25) \pm 1.95996 \times 0.35\sqrt{0.25}\right]$$
$$= (100e^{-0.32081}, 100e^{0.36518})$$
$$= \boxed{(72.56, 144.08)}.$$

$\qquad\square$

5. *Solution.* The 90% lognormal prediction interval for $S(2)$ is

$$\left(S(0)e^{(\alpha-\delta-\sigma^2/2)(2)-\sigma\sqrt{2}\times1.645}, S(0)e^{(\alpha-\delta-\sigma^2/2)(2)+\sigma\sqrt{2}\times1.645}\right).$$

The ratio of the upper end-point to the lower end-point gives

$$e^{2\sigma\sqrt{2}\times1.645} = \frac{41.9448}{13.1072},$$

which in turn implies that $\sigma = 0.25$. Moreover, the geometric averages of the two endpoints of the prediction intervals for $S(2)$ and $S(4)$ give

$$\begin{cases} S(0)e^{(\alpha-\delta-\sigma^2/2)(2)} = \sqrt{(13.1072)(41.9448)} = 23.447364 \\ S(0)e^{(\alpha-\delta-\sigma^2/2)(4)} = \sqrt{(25.7923)(183.1083)} = 68.722516 \end{cases}$$

resulting in $\alpha - \delta - \sigma^2/2 = 0.537659$ and $S(0) = 8$. As $N^{-1}(0.995) = 2.576$, it follows that the 99% lognormal prediction interval for $S(6)$ is

$$\left(S(0)e^{(\alpha-\delta-\sigma^2/2)(6)-\sigma\sqrt{6}\times 2.576}, S(0)e^{(\alpha-\delta-\sigma^2/2)(6)+\sigma\sqrt{6}\times 2.576} \right)$$

$$= \left(8e^{0.537659(6)-0.25\sqrt{6}\times 2.576}, 8e^{0.537659(6)+0.25\sqrt{6}\times 2.576} \right)$$

$$= (41.5927, 975.4188)$$

with a width of $975.4188 - 41.5927 = \boxed{934}$.   □

7. *Solution.* We are given from (v) that $\mathbb{E}[S(1)] = S(0)e^{(\alpha-\delta)T} = 80e^{\alpha-\delta} = 84.2069$, or $\alpha - \delta = 0.051250$. Then

$$\hat{d}_1 = \frac{\ln[S(0)/K] + (\alpha - \delta + \sigma^2/2)T}{\sigma\sqrt{T}} = \frac{\ln(1) + (0.051250 + 0.25^2/2)(1)}{0.25\sqrt{1}} = 0.33,$$

$$\hat{d}_2 = \hat{d}_1 - \sigma\sqrt{T} = 0.08 - 0.25\sqrt{1} = 0.08,$$

$$N(-\hat{d}_1) = 0.37070,$$

$$N(-\hat{d}_2) = 0.46812,$$

and the expected 1-year stock price, given that $S(1) < 80$, is

$$\mathbb{E}[S(1)|S(1) < 80] = S(0)e^{(\alpha-\delta)T}\frac{N(-\hat{d}_1)}{N(-\hat{d}_2)}$$

$$= 84.2069 \times \frac{0.37070}{0.46812}$$

$$= \boxed{66.6827}.$$

□

Remark. The risk-free interest rate is not used at all.

9. *Solution.* From (i), (ii), and (iii), we have:

(i) $N(-\hat{d}_2) = 0.512$

(ii) $S(0)e^{(\alpha-\delta)T} = 10.134$

(iii) $S(0)e^{(\alpha-\delta)T}\dfrac{N(-\hat{d}_1)}{N(-\hat{d}_2)} = 8.483$

Thus $N(-\hat{d}_1) = 8.483(0.512)/10.134 = 0.4286$. Then

$$\hat{d}_1 = -N^{-1}(0.4286) = -(-0.18) = 0.18,$$

$$\hat{d}_2 = -N^{-1}(0.512) = -0.03,$$

$$\sigma = \frac{\hat{d}_1 - \hat{d}_2}{\sqrt{T}} = \frac{0.18 - (-0.03)}{\sqrt{2/3}} = 0.257196.$$

The 98% lognormal prediction interval for the 8-month stock price $S(2/3)$ is

$$S(0)e^{(\alpha-\delta-\sigma^2/2)T\pm\sigma\sqrt{T}N^{-1}(0.99)} = 10.134e^{-(0.257196^2/2)(2/3)\pm 0.257196\sqrt{2/3}(2.326)}$$

$$= \boxed{(6.08, 16.16)}.$$

□

B.6   Chapter 6

1. *Solution.* To insure against a reduction of more than 40% in the value of the equity index fund at the end of one year, the strike price of the put option should be set to $1,000(1 - 40\%) = 600$. As

$$d_1 = \frac{\ln(1,000/600) + (0.025 - 0.02 + 0.2^2/2)(1)}{0.2\sqrt{1}} = 2.67913,$$

$$d_2 = d_1 - 0.2\sqrt{1} = 2.47913,$$

$$N(-d_1) = 0.00369,$$

$$N(-d_2) = 0.00659,$$

the price of the put option is given by

$$
\begin{aligned}
P &= Ke^{-rT}N(-d_2) - Se^{-\delta T}N(-d_1) \\
&= 600e^{-0.025(1)}(0.00659) - 1,000e^{-0.02}(0.00369) \\
&= \boxed{0.23944}.
\end{aligned}
$$

$\square$

3. *Solution.* Note that the 3-year payoff of the contingent claim can be written as

$$
\text{Payoff} = \begin{cases} 0, & \text{if } S(3) < 45 \\ S(3) - 45, & \text{if } 45 \le S(3) < 75 \\ 2(S(3) - 60), & \text{if } 75 \le S(3) \end{cases}
$$
$$
= (S(3) - 45)_+ + (S(3) - 75)_+,
$$

which is the sum of the payoff of a 45-strike long call and that of a 75-strike long call.

- At $K = 45$,

$$d_1 = \frac{\ln(45/45) + (0.06 - 0.03 + 0.2^2/2)(3)}{0.2\sqrt{3}} = 0.43301,$$

$$d_2 = d_1 - 0.2\sqrt{3} = 0.08660,$$

$$N(d_1) = 0.66750,$$

$$N(d_2) = 0.53451,$$

$$C = 45e^{-0.03(3)}(0.66750) - 45e^{-0.06(3)}(0.53451) = 7.36150.$$

- At $K = 75$,

$$d_1 = \frac{\ln(45/75) + (0.06 - 0.03 + 0.2^2/2)(3)}{0.2\sqrt{3}} = -1.04161,$$

$$d_2 = d_1 - 0.2\sqrt{3} = -1.38802,$$

$$N(d_1) = 0.14880,$$

$$N(d_2) = 0.08257,$$

$$C = 45e^{-0.03(3)}(0.14880) - 75e^{-0.06(3)}(0.08257) = 0.94706.$$

The price of the contingent claim is $V = 7.36150 + 0.94706 = \boxed{8.3086}$. $\square$

5. *Solution.* The 9-month prepaid forward price of the stock is

$$F_{0,0.75}^P(S) = 50 - 2e^{-0.05(0.25)} = 48.0248444.$$

With

$$d_1 = \frac{\ln(48.0248444/50e^{-0.05(0.75)}) + (0.3^2/2)(0.75)}{0.3\sqrt{0.75}} = 0.11911,$$

$$d_2 = d_1 - 0.3\sqrt{0.75} = -0.14070,$$

$$N(-d_1) = 0.45259,$$

$$N(-d_2) = 0.55595,$$

the price of the put option is

$$P = 50e^{-0.05(0.75)}(0.55595) - 48.0248444(0.45259) = \boxed{5.0388}.$$

To price this put option in the Black-Scholes framework, it is assumed that the prepaid forward prices of the stock are lognormally distributed, with σ being the volatility of the prepaid forward price, or the standard deviation per unit time of the natural logarithm of the prepaid forward price. Mathematically,

$$\sigma^2 t = \text{Var}\{\ln[F_{t,0.75}^P(S)]\} \quad \text{or} \quad \sigma = \sqrt{\frac{\text{Var}\{\ln[F_{t,0.75}^P(S)]\}}{t}}, \quad 0 < t \le 0.75.$$

$\square$

7. *Solution.* The 45-55 put bear spread is constructed by combining a short 45-strike put with a long 55-strike put. The 9-month prepaid forward price of the stock is

$$F_{0,0.75}^P(S) = 50 - 2.5(e^{-0.08(2/12)} + e^{-0.08(5/12)}) = 45.115072.$$

At the strike price of 45,

$$d_1 = \frac{\ln(45.115072/45e^{-0.08(0.75)}) + (0.4^2/2)(0.75)}{0.4\sqrt{0.75}} = 0.35378,$$

$$d_2 = d_1 - 0.4\sqrt{0.75} = 0.00737,$$

$$N(-d_1) = 0.36317,$$

$$N(-d_2) = 0.49601,$$

and the price of the 45-strike put is

$$P^{45\text{-strike}} = 45e^{-0.08(0.75)}(0.49601) - 45.115072(0.36317) = 4.63617.$$

At the strike price of 55,

$$d_1 = \frac{\ln(45.115072/55e^{-0.08(0.75)}) + (0.4^2/2)(0.75)}{0.4\sqrt{0.75}} = -0.22550,$$

$$d_2 = d_1 - 0.4\sqrt{0.75} = -0.57191,$$

$$N(-d_1) = 0.58920,$$

$$N(-d_2) = 0.71631,$$

and the price of the 55-strike put is

$$P^{55\text{-strike}} = 55e^{-0.08(0.75)}(0.71631) - 45.115072(0.58920) = 10.52094.$$

Finally, the price of the 45-55 put bear spread is $V = 10.52094 - 4.63617 = \boxed{5.8848}.$ $\square$

9. *Solution.* Observe that the 3-month payoff of the contingent claim, which is a cap, is parallel to that of a long 50-strike put. Therefore, the two derivatives must possess the same profit.

The 3-month prepaid forward price of the stock is

$$F^P_{0,0.25}(S) = 50 - 1.5e^{-0.1(2/12)} = 48.524793.$$

With

$$\begin{aligned}
d_1 &= \frac{\ln(48.524793/50) + (0.1 + 0.3^2/2)(0.25)}{0.3\sqrt{0.25}} = 0.04201, \\
d_2 &= d_1 - 0.3\sqrt{0.25} = -0.10799, \\
N(-d_1) &= 0.4840, \\
N(-d_2) &= 0.5438,
\end{aligned}$$

the price of the 3-month 50-strike put option is

$$P = 50e^{-0.1(0.25)}(0.5438) - 48.524793(0.4840) = 3.0327.$$

When $S(0.25) = 45$, the payoff of the put is $(50 - 45)_+ = 5$. Then the profit of the contingent claim, equal to that of the put, is $5 - 3.0327e^{0.1(0.25)} = \boxed{1.8905}$. □

11. *Solution.* Let $X(t)$ be the time-t exchange rate of US dollars per British pound. In three months, you will pay 200,000 pounds, or $\$200,000X(0.25)$, but you will only have $\$320,000$. You are therefore worried about $X(0.25)$ being too high (to be precise, higher than 1.6). To cover the shortfall of

$$\$[200,000X(0.25) - 320,000]_+ = 200,000 \underbrace{\$[X(0.25) - 1.6]_+}_{\text{payoff of a currency call}},$$

you can buy 200,000 $1.6-strike dollar-denominated currency pound *call* options. As

$$\begin{aligned}
d_1 &= \frac{\ln(1.6/1.6) + (0.01 - 0.02 + 0.2^2/2)(0.25)}{0.2\sqrt{0.25}} = 0.025, \\
d_2 &= d_1 - 0.2\sqrt{0.25} = -0.075, \\
N(d_1) &= 0.50997, \\
N(d_2) &= 0.47011,
\end{aligned}$$

the price of each currency call option is

$$C = 1.6e^{-0.02(0.25)}(0.50997) - 1.6e^{-0.01(0.25)}(0.47011) = 0.061585.$$

The total cost of the currency options is $200,000(0.061585) = \boxed{12,316.90}$. □

Remark. (i) Together with the long pound calls, you have set up a *cap*.

(ii) This problem is dual to Example 6.2.4.

- In Example 6.2.4, you are to *receive* an amount (therefore long) in the foreign currency, and worry about the exchange rate going *down*. A put option can hedge against the downside risk.
- In the current problem, you are to *pay* an amount (therefore short) in the foreign currency, and worry about the exchange rate going *up*. This time a call option can hedge against the upside risk.

13. *Solution.* (a) Note that euro is the foreign (underlying) currency while dollar is the domestic (strike) currency. With $r = 0.04$, $\delta = 0.05$,

$$d_1 = \frac{\ln(1.5/1.5) + (0.04 - 0.05 + 0.2^2/2)(0.5)}{0.2\sqrt{0.5}} = 0.03536,$$

$$d_2 = d_1 - 0.2\sqrt{0.5} = -0.10607,$$

$$N(-d_1) = 0.48590,$$

$$N(-d_2) = 0.54224,$$

the price of the euro put option is

$$P = 1.5e^{-0.04(0.5)}(0.54224) - 1.5e^{-0.05(0.5)}(0.48590) = \boxed{0.08640}.$$

(b) (1) No. Apple will need to pay an amount in euro in 6 months and is therefore short with respect to the 6-month dollar/euro exchange rate. He needs a long call on euros to hedge against his upside exchange rate risk.

(2) Yes. Ambrosio will receive a fixed amount in euro in 6 months and is thus long with respect to the 6-month dollar/euro exchange rate. The currency put in question allows him to hedge against the downside exchange rate risk he faces by setting up a floor.

(c) As

$$\hat{d}_2 = \frac{\ln(1.5/1.5) + (\overbrace{0.015}^{\text{exchange rate appreciation}} - 0.2^2/2)(0.5)}{0.2\sqrt{0.5}} = -0.01768,$$

the true probability that the put option will be exercised is $N(-\hat{d}_2) = \boxed{0.50705}$.

□

15. *Solution.* Note that the underlying futures contract matures three years from now. From the table in (iv), it has a current price of 563.75 and a volatility of 30%. With

$$d_1 = \frac{\ln(563.75/550) + (0.3^2/2)(\overbrace{2}^{\text{lifespan of option}})}{0.3\sqrt{2}} = 0.27033,$$

$$d_2 = d_1 - 0.3\sqrt{2} = -0.15393,$$

$$N(-d_1) = 0.39345,$$

$$N(-d_2) = 0.56117,$$

the current price of the required put option is

$$P = 550e^{-0.06(2)}(0.5596) - 563.75e^{-0.06(2)}(0.3936) = \boxed{77.02}.$$

□

17. *Solution.* We are given in (ii) that $\Delta_P = -e^{-0.03(0.5)}N(-d_1) = -0.3382$, or $N(-d_1) = 0.34331$. Then $d_1 = -N^{-1}(0.34331) = 0.40345$. Because

$$d_1 = \frac{\ln(95/90) + (\sigma^2/2)(0.5)}{\sigma\sqrt{0.5}} = 0.40345,$$

or
$$0.25\sigma^2 - 0.40345\sqrt{0.5}\sigma + \ln(95/90) = 0,$$

we have
$$\sigma = \frac{0.40345\sqrt{0.5} \pm \sqrt{0.40345^2(0.5) - 4(0.25)(\ln(95/90))}}{2(0.25)} = \underbrace{0.901132}_{\text{rejected}} \text{ or } 0.24.$$

Finally, the gamma of the futures put is
$$\begin{aligned}
\Gamma &= e^{-rT} \times N'(d_1) \times \frac{1}{F_0\sigma\sqrt{T}} \\
&= e^{-0.03(0.5)} \times \frac{1}{\sqrt{2\pi}}e^{-0.40345^2/2} \times \frac{1}{95(0.24)\sqrt{0.5}} \\
&= \boxed{0.0225}.
\end{aligned}$$

$\square$

19. The correct graph is **(B)**. Delta of a (plain vanilla) European call is always bounded between 0 and 1. This eliminates (A), (C), and (E).

Solution 1 (Verbal explanations). A short-term deep out-the-money call option will almost surely end up out-of-the-money and is very insensitive to stock price movements, hence a delta of approximately 0. As the time to maturity increases, the call has a higher probability of ending up in-the-money. The increase in the sensitivity to stock price movements translates into an increase in delta (although it remains bounded from above by 1). $\square$

Solution 2 (For math geeks). Mathematically, consider the partial derivative
$$\frac{\partial d_1}{\partial T} = -\frac{\ln(S/K)}{2\sigma T^{3/2}} + \frac{r + \sigma^2/2}{2\sigma\sqrt{T}} \tag{B.6.1}$$
$$= \frac{\overbrace{-\ln(S/K)}^{>0} + (r + \sigma^2/2)T}{2\sigma T^{3/2}}$$
$$> 0, \quad \text{if } S \ll K \text{ (i.e., deep out-of-the-money).}$$

Then
$$\frac{\partial \Delta_C}{\partial T} = \frac{\partial}{\partial T}N(d_1) = N'(d_1)\frac{\partial d_1}{\partial T} > 0,$$
which means that Δ_C as a function of T is increasing. $\square$

21. *Solution.* (a) The price of a European call option will decrease as K increases.

(b) The payoff of a European call option is $(S(T) - K)_+$, which decreases with the strike price K (i.e., the call becomes less attractive). A lower payoff means a lower call price.

(c) Mathematically, we examine the partial derivative of the European call price with respect to the strike price K (there is no Greek defined for this partial derivative):
$$\frac{\partial C}{\partial K} = F^P_{0,T}(S)N'(d_1)\frac{\partial d_1}{\partial K} - e^{-rT}N(d_2) - Ke^{-rT}N'(d_2)\underbrace{\frac{\partial d_2}{\partial K}}_{=\partial d_1/\partial K}$$
$$\overset{(6.3.2)}{=} -e^{-rT}N(d_2),$$

which is negative. Therefore, the call price is a decreasing function of the strike price.

□

23. *Solution.* The gamma of the bull spread equals

$$\Gamma_{\text{bull spread}} = \Gamma^{50\text{-strike}} - \Gamma^{60\text{-strike}}.$$

Initially:

- For the 50-strike call,

$$d_1 = \frac{\ln(50/50) + (0.05 + 0.2^2/2)(0.25)}{0.2\sqrt{0.25}} = 0.175,$$

$$\Gamma^{50\text{-strike}} = e^{-\delta T} \times N'(d_1) \times \frac{1}{S\sigma\sqrt{T}}$$

$$= \frac{1}{\sqrt{2\pi}} e^{-0.175^2/2} \times \frac{1}{50(0.2)\sqrt{0.25}} = 0.07858.$$

- For the 60-strike call,

$$d_1 = \frac{\ln(50/60) + (0.05 + 0.2^2/2)(0.25)}{0.2\sqrt{0.25}} = -1.64822,$$

$$\Gamma^{60\text{-strike}} = \frac{1}{\sqrt{2\pi}} e^{-(-1.64822)^2/2} \times \frac{1}{50(0.2)\sqrt{0.25}} = 0.02051.$$

The original gamma is thus $0.07858 - 0.02051 = 0.05807$.

After 1 month (the remaining time to expiration is 2 months):

- For the 50-strike call,

$$d_1 = \frac{\ln(50/50) + (0.05 + 0.2^2/2)/6}{0.2\sqrt{1/6}} = 0.14289,$$

$$\Gamma^{50\text{-strike}} = \frac{1}{\sqrt{2\pi}} e^{-0.14289^2/2} \times \frac{1}{50(0.2)\sqrt{1/6}} = 0.09673.$$

- For the 60-strike call,

$$d_1 = \frac{\ln(50/60) + (0.05 + 0.2^2/2)/6}{0.2\sqrt{1/6}} = -2.09009,$$

$$\Gamma^{60\text{-strike}} = \frac{1}{\sqrt{2\pi}} e^{-(-2.09009)^2/2} \times \frac{1}{50(0.2)\sqrt{1/6}} = 0.01100.$$

The new gamma is $0.09673 - 0.01100 = 0.08573$. Therefore, the change in gamma is $0.08573 - 0.05807 = \boxed{0.02766}$.

□

25. *Solution.* As

$$d_1 = \frac{\ln(82/80) + (0.08 - 0.03 + 0.3^2/2)(0.25)}{0.3\sqrt{0.25}} = 0.32295,$$

$$d_2 = d_1 - 0.3\sqrt{0.25} = 0.17295,$$

$$N(d_1) = 0.62663,$$

$$N(d_2) = 0.56865,$$

the price of the call option is

$$C = 82e^{-0.03(0.25)}(0.62663) - 80e^{-0.08(0.25)}(0.56865) = 6.40853,$$

so its elasticity is

$$\Omega_C = \frac{S\Delta}{C} = \frac{82(e^{-0.03(0.25)})(0.62663)}{6.40853} = \boxed{7.95811}.$$

□

27. *Solution.* The straddle is constructed by a long 3-month 32-strike European call and a long otherwise identical European put. We first compute

$$
\begin{aligned}
d_1 &= \frac{\ln(30/32) + (0.05 - 0.02 + 0.3^2/2)(0.25)}{0.3\sqrt{0.25}} = -0.30526, \\
d_2 &= d_1 - 0.3\sqrt{0.25} = -0.45526, \\
N(d_1) &= 0.38008, \\
N(-d_1) &= 0.61992, \\
N(d_2) &= 0.32446, \\
N(-d_2) &= 0.67554.
\end{aligned}
$$

The current prices of the call and put are respectively

$$C = 30e^{-0.02(0.25)}(0.38008) - 32e^{-0.05(0.25)}(0.32446) = 1.09179$$

and

$$P = 32e^{-0.05(0.25)}(0.67554) - 30e^{-0.02(0.25)}(0.61992) = 2.84390.$$

Their current deltas are respectively

$$\Delta_C = e^{-0.02(0.25)}(0.38008) = 0.37818$$

and

$$\Delta_P = -e^{-0.02(0.25)}(0.61992) = -0.61683.$$

Therefore, the current elasticity of the straddle is

$$\Omega_{\text{straddle}} = \frac{S\Delta_{\text{straddle}}}{V_{\text{straddle}}} = \frac{S(\Delta_C + \Delta_P)}{C + P} = \frac{30[0.37818 + (-0.61683)]}{1.09179 + 2.84390} = \boxed{-1.81912}.$$

□

29. *Solution.* The contingent claim is composed of a long 9-month zero-coupon bond of face value 20 plus one 9-month at-the-money European call. With

$$
\begin{aligned}
d_1 &= \frac{\ln(1) + (0.06 - 0.02 + 0.35^2/2)(0.75)}{0.35\sqrt{0.75}} = 0.25053, \\
N'(d_1) &= \frac{1}{\sqrt{2\pi}}e^{-d_1^2/2} = 0.38662,
\end{aligned}
$$

the current gamma of the contingent claim equals

$$\Gamma = e^{-\delta T}N'(d_1)\frac{1}{S\sigma\sqrt{T}} = e^{-0.02(0.75)}(0.38662)\frac{1}{S(0.35)\sqrt{0.75}} = 0.0314,$$

which gives $S = 40.0167$. With $N(d_1) = 0.59891$, $d_2 = d_1 - 0.35\sqrt{0.75} = -0.05258$, and $N(d_2) = 0.47903$, the time-0 price and time-0 delta of the contingent claim are, respectively,

$$
\begin{aligned}
V &= 20e^{-rT} + C \\
&= 20e^{-0.06(0.75)} + \underbrace{[40.0167e^{-0.02(0.75)}(0.59891) - 40.0167e^{-0.06(0.75)}(0.47903)]}_{5.28388} \\
&= 24.40383
\end{aligned}
$$

and

$$
\Delta_V = \Delta_C = e^{-0.02(0.75)}(0.59891) = 0.58999.
$$

It follows that the time-0 contingent-claim elasticity is

$$
\Omega_V = \frac{S\Delta_V}{V} = \frac{40(0.58999)}{24.40383} = \boxed{0.9670}.
$$

$\square$

B.7   Chapter 7

1. *Solution.* For simplicity, we let four months ago be time 0 and the current time be time $1/3$.

   (a) We use the current delta (the remaining time to expiration is 8 months) to deduce the volatility of the stock:

$$
d_1 = \frac{\ln(50/45) + (0.05 + \sigma^2/2)\overbrace{(2/3)}^{\text{not 1!}}}{\sigma\sqrt{2/3}} = N^{-1}(0.73507) = 0.62822.
$$

   The solutions to this quadratic equation are $\sigma = 1.19$ and $\sigma = 0.35$. We take $\sigma = \boxed{0.35}$ as it is the only value less than 50%.

   (b) We also need the delta of the call 4 months ago and the current price of the call.

   • Four months ago,

$$
\begin{aligned}
d_1(0) &= \frac{\ln(40/45) + (0.05 + 0.35^2/2)(1)}{0.35\sqrt{1}} = -0.01867, \\
\Delta_C(0) &= N(d_1) = 0.49255.
\end{aligned}
$$

   To delta-hedge his position, four months ago Eric should have *sold* $100(0.49255) = 49.255$ shares of the stock.

   • Currently, with

$$
\begin{aligned}
d_2(1/3) &= d_1(1/3) - 0.35\sqrt{2/3} = 0.34245, \\
N(d_2(1/3)) &= 0.63399,
\end{aligned}
$$

   the call price is

$$C(1/3) = 50(0.73507) - 45e^{-0.05(2/3)}(0.63399) = 9.15926.$$

Then Eric's four-month holding profit is

$$100\{[C(1/3) - \Delta_C(0)S(1/3)] - e^{r/3}[C(0) - \Delta_C(0)S(0)]\}$$
$$= 100\{[9.15926 - 0.49255(50)] - e^{0.05(1/3)}[4.45539 - 0.49255(40)]\}$$
$$= \boxed{3.46}.$$

□

3. *Solution.* We are told in (ii) that $\Delta_C = 750/1{,}000 = 0.75$. Since $\Delta_C = e^{-\delta T}N(d_1) = N(d_1)$, we have

$$d_1 = \frac{\ln(1) + (0.07 + \sigma^2/2)(1)}{\sigma\sqrt{1}} = N^{-1}(0.75) = 0.67449,$$

or

$$0.5\sigma^2 - 0.67449\sigma + 0.07 = 0.$$

Then

$$\sigma = \frac{0.67449 \pm \sqrt{(-0.67449)^2 - 4(0.5)(0.07)}}{2(0.5)} = 1.235682 \text{ or } 0.11330.$$

As $\sigma < 1$, we take $\sigma = 0.11330$. Then $d_2 = d_1 - \sigma\sqrt{T} = 0.67449 - 0.11330\sqrt{1} = 0.56119$ and $N(d_2) = 0.71267$, and the price of each call option is

$$C = 80(0.75) - 80e^{-0.07(1)}(0.71267) = \boxed{6.84}.$$

□

5. *Solution.* By put-call parity, the delta and theta of the put option are, respectively,

$$\Delta_P = \Delta_C - 1 = -0.4579,$$
$$\theta_P = \theta_C + rKe^{-rT} = -2.41519 + 0.02(20)e^{-0.02(0.25)} = -2.01719.$$

Let x be the number of units of the put to *buy*. To maintain delta-neutrality, we have to solve the equation

$$-1000(0.54210) - 0.4579x = 0,$$

resulting in $x = -1{,}183.8829$. In other words, $1{,}183.8829$ units of the put should be *sold*.

Finally, the theta of the overall position is

$$\theta_V = -1000(-2.41519) - 1{,}183.8829(-2.01719) = \boxed{4{,}803}.$$

□

Remark. (i) To relate the call theta to the put theta, we differentiate both sides of put-call parity and *don't forget to negate*:

$$\theta_C - \theta_P = \boxed{-}\frac{\partial}{\partial T}(C - P) = \boxed{-}\frac{\partial}{\partial T}[S(0) - Ke^{-rT}] = -rKe^{-rT}.$$

(ii) That the interest rate is 2% can be deduced from the Black-Scholes equation.

7. *Solution.* For the 60-strike call,

$$d_1 = \frac{\ln(60/60) + (0.05 + 0.30^2/2)(1)}{0.30\sqrt{1}} = 0.31667,$$

$$\Delta_C^{60\text{-strike}} = N(d_1) = 0.62425,$$

$$\Gamma_C^{60\text{-strike}} = \frac{1}{\sqrt{2\pi}} e^{-d_1^2/2} \times \frac{1}{S\sigma\sqrt{T}}$$

$$= \frac{1}{\sqrt{2\pi}} e^{-0.31667^2/2} \times \frac{1}{60(0.30)\sqrt{1}}$$

$$= 0.02108.$$

For the 65-strike put,

$$d_1 = \frac{\ln(60/65) + (0.05 + 0.30^2/2)(1)}{0.30\sqrt{1}} = 0.04986,$$

$$\Delta_P^{65\text{-strike}} = -N(-d_1) = -0.48012,$$

$$\Gamma_P^{65\text{-strike}} = \frac{1}{\sqrt{2\pi}} e^{-d_1^2/2} \times \frac{1}{S\sigma\sqrt{T}}$$

$$= \frac{1}{\sqrt{2\pi}} e^{-0.04986^2/2} \times \frac{1}{60(0.30)\sqrt{1}}$$

$$= 0.02214.$$

After buying 200 units of the 60-strike call, our gamma is $200\Gamma_C^{60\text{-strike}} = 4.216$. To gamma-hedge our position, we should $\boxed{\text{sell}}$ $4.216/0.02214 = \boxed{190.42}$ 65-strike puts. The overall delta at this point is

$$200\Delta_C^{60\text{-strike}} - 190.42\Delta_P^{65\text{-strike}} = 216.52,$$

so we should also $\boxed{\text{sell } 216.52}$ shares of the stock to make our position both delta-neutral and gamma-neutral.   □

Remark. There is no need to compute the two option prices.

9. *Solution.* (a) For the 50-strike call,

$$d_1 = \frac{\ln(50/50) + (0.05 + 0.25^2/2)(1)}{0.25\sqrt{1}} = 0.325,$$

$$\Delta^{50\text{-strike}} = N(d_1) = 0.62741,$$

$$\Gamma^{50\text{-strike}} = \frac{1}{\sqrt{2\pi}} e^{-d_1^2/2} \times \frac{1}{S\sigma\sqrt{T}}$$

$$= \frac{1}{\sqrt{2\pi}} e^{-0.325^2/2} \times \frac{1}{50(0.25)\sqrt{1}}$$

$$= 0.030274.$$

For the 60-strike call,

$$d_1 = \frac{\ln(50/60) + (0.05 + 0.25^2/2)(1)}{0.25\sqrt{1}} = -0.40429,$$

$$\Delta^{60\text{-strike}} = N(d_1) = 0.34300,$$

$$\Gamma^{60\text{-strike}} = \frac{1}{\sqrt{2\pi}}e^{-d_1^2/2} \times \frac{1}{S\sigma\sqrt{T}}$$

$$= \frac{1}{\sqrt{2\pi}}e^{-(-0.40429)^2/2} \times \frac{1}{50(0.25)\sqrt{1}}$$

$$= 0.02941.$$

After selling 1,000 50-strike calls, our gamma is $-1,000\Gamma^{50\text{-strike}} = -30.274$. To gamma-hedge our position, we should $\boxed{\text{buy}}$ $30.274/0.02941 = \boxed{1,029.38}$ 60-strike calls. The overall delta at this point is

$$-1,000\Delta^{50\text{-strike}} + 1,029.38\Delta^{60\text{-strike}} = -274.33,$$

so should also $\boxed{\text{buy } 274.33}$ shares for delta-hedging.

(b) Our initial investment as a result of the delta-gamma-hedged call position is

$$274.33(50) + 1,029.38(2.5127) - 1,000(6.1680) = 10,135.02313.$$

The payoff in one month is

$$274.33(50) + 1,029.38(2.2591) - 1,000(5.8611) = 10,180.87236.$$

Thus the one-month holding profit is

$$10,180.87236 - 10,135.02313e^{0.05/12} = \boxed{3.53187}.$$

$\square$

11. *Solution.* With $\epsilon = 10.50 - 10 = 0.5$, the estimated new call price is

$$C + \Delta\epsilon + \frac{1}{2}\Gamma\epsilon^2 = 2 + (0.6)(0.5) + \frac{1}{2}(0.2)(0.5)^2$$

$$= \boxed{2.325}.$$

$\square$

13. *Solution.* In this problem, we deduce from the two approximations the values of the delta and gamma of the futures put, which in turn allow us to infer the values of its initial price and volatility. Exact Black-Scholes calculations can then be performed.

Using the delta approximation, we have $4.148 = \Delta_P\epsilon = -10\Delta_P$, or $\Delta_P = -0.4148$. Because $\Delta_P = -e^{-\delta T}N(-d_1) = -e^{-0.08(0.75)}N(-d_1)$, we get $N(-d_1) = 0.4404$. Then $d_1 = -N^{-1}(0.4404) = 0.15$. As the futures put is at-the-money,

$$d_1 = \frac{\ln(1) + (\sigma^2/2)T}{\sigma\sqrt{T}} = \frac{1}{2}\sigma\sqrt{0.75} = 0.15,$$

leading to $\sigma = 0.346410$.

Turning to the delta-gamma approximation, we now have

$$4.231 = \Delta_P \epsilon + \frac{1}{2}\Gamma_P \epsilon^2 = 4.148 + \frac{1}{2}\Gamma_P(-10)^2,$$

which gives $\Gamma_P = 0.00166$. Using the gamma formula,

$$\Gamma_P = e^{-0.08(0.75)} \times \frac{1}{\sqrt{2\pi}}e^{-0.15^2/2} \times \frac{1}{F_0(0.346410)\sqrt{0.75}} = 0.00166,$$

so the *initial* futures price is $F = 745.9976$.

We are now ready for the exact calculation of the new price of the futures put. With

$$d_1 = \frac{\ln(\overbrace{735.9976}^{\text{drops by 10}}/745.9976) + (0.346410^2/2)(0.75)}{0.346410\sqrt{0.75}} = 0.10501,$$

$$d_2 = d_1 - 0.346410\sqrt{0.75} = -0.19499,$$

$$N(-d_1) = 0.45818,$$

$$N(-d_2) = 0.57730,$$

the exact put price when the futures price decreases to 735.9976 is

$$
\begin{aligned}
P &= e^{-rT}[KN(-d_2) - F_0^{\text{new}}N(-d_1)] \\
&= e^{-0.08(0.75)}[\ \underbrace{745.9976}_{\text{initial futures price}}(0.57730) - \underbrace{735.9976}_{\text{new futures price}}(0.45818)] \\
&= \boxed{88.00}.
\end{aligned}
$$

□

15. *Solution.* We need the current delta of Put A. By the Black-Scholes equation,

$$0.05(70)(-0.2867) + \frac{1}{2}(0.25)^2(70)^2(0.0112) + \theta = 0.05(6.9389),$$

which gives $\theta = -0.364605$.

Let x and y be, respectively, the number of units of Put B and the stock to *buy*. To maintain delta-neutrality and theta-neutrality, we are prompted to solve the following system of two equations:

$$
\begin{cases}
-1000(-0.2867) - 0.3433x + y = 0 & \text{(delta-neutrality)} \\
-1000(-0.364605) - 0.2060x = 0 & \text{(theta-neutrality)}
\end{cases}.
$$

The second equation implies that $x = 1,769.9272$. Then it follows from the first equation that $y = 320.9077$. In other words, 1769.9029 units of Put B and 320.9160 shares of stock should be bought.

The net investment is

$$1,769.9272\ \underbrace{(9.0062)}_{\text{Put B's price}} + 320.9160\ \underbrace{(70)}_{\text{stock price}} - 1000\ \underbrace{(6.9389)}_{\text{Put A's price}} = \boxed{31,466}.$$

□

17. *Solution.* From the current elasticity, we have

$$5.4417 = \Omega_C(1) = \frac{S(1)\Delta_C(1)}{C(1)} = \frac{S(0.7296)}{5.7653} \quad \Rightarrow \quad S(1) = 43.$$

Furthermore, the Black-Scholes equation applied to the "Now" column says that

$$5.7653r = -2.1911 + r(43)(0.7296) + \frac{1}{2}(0.2)^2(43)^2(0.0385),$$

which gives $r = 0.03$,

Back to one year ago, another application of the Black-Scholes equation yields

$$5.6295(0.03) = -1.6569 + 0.03(40)\Delta + \frac{1}{2}(0.2)^2(40)^2(0.0331),$$

resulting in $\Delta = 0.6388$. By buying 1,000 units of the call, you incurred a delta of $1,000(0.6388) = 638.8$. To neutralize delta, we should have sold 638.8 shares of the stock one year ago.

To conclude, the 1-year holding profit is

$$[-638.8(43) + 1,000(5.7653)] - e^{0.03}[-638.8(40) + 1,000(5.6295)]$$
$$= -21,703.1 - e^{0.03}(-19,922.5) = \boxed{-1,174}.$$

□

19. *Solution.* Because the delta three months ago is positive while the current delta is negative, the "Three months ago" column is for the otherwise identical call and the "Now" column is for the put. An application of the Black-Scholes equation to the values three months ago yields

$$5.2121r = -4.2164 + r(50)(0.5129) + \frac{1}{2}(0.3)^2(50)^2(0.0266),$$

which gives $r = 0.06$. By another application of the Black-Scholes equation to the current values, we have

$$0.06P = -2.1069 + 0.06(55)(-0.3809) + \frac{1}{2}(0.3)^2(55)^2(0.0267),$$

leading to $P = 4.511125$. Furthermore, by put-call parity, the price and delta of the put three months ago are respectively

$$P = C - S + Ke^{-rT} = 5.2121 - 50 + 55e^{-0.06} = 7.009149$$

and

$$\Delta_P = \Delta_C - 1 = 0.5129 - 1 = -0.4871.$$

By selling 1,000 units of the put, you incurred a delta of $-1,000(-0.4871) = 487.1$. To neutralize delta, we should have sold 487.1 shares of the stock three months ago.

The 3-month holding profit is

$$[-487.1(55) - 1,000(4.511125)] - e^{0.06(0.25)}[-487.1(50) - 1,000(7.009149)]$$
$$= -31,301.625 - e^{0.06(0.25)}(-31,364.149) = \boxed{-563.53}.$$

□

B.8   Chapter 8

1. *Solution.* Note that $\sigma = \sqrt{0.09} = 0.3$ (not 0.09). By (8.1.5),

$$d_1 = \frac{\ln[S(0)/1.1S(0)] + (0.06 - 0.02 + 0.09/2)(4)}{0.3\sqrt{4}} = 0.40782,$$

and

$$\pi = S(0)e^{-\delta T}N(d_1) = S(0)e^{-0.02(4)}N(0.40782),$$

so $\pi/S(0) = e^{-0.08}\underbrace{N(0.40782)}_{=0.65830} = \boxed{0.60769}$.   □

3. *Solution.* By risk-neutral pricing, the time-1 price of the derivative is given by

$$
\begin{aligned}
V(1) \quad &= \quad \overbrace{2S(1)e^{-0.02(2)}N(d_1)}^{\text{2 units of } 1.5S(1)\text{-strike A/N call}} \quad 1_{\{S(1)>120\}} \\
&\quad + \quad \underbrace{S(1)e^{-0.02(2)}}_{\text{1 unit of } 1.5S(1)\text{-strike A/N call}} \quad N(d_1)1_{\{S(1)<120\}} \\
&= \quad 2S(1)e^{-0.04}N(d_1)1_{\{S(1)>120\}} + S(1)e^{-0.04}N(d_1)1_{\{S(1)<120\}},
\end{aligned}
$$

where

$$d_1 = \frac{\ln[S(1)/1.5S(1)] + (0.05 - 0.02 + 1/2 \times 0.25^2)(2)}{0.25\sqrt{2}} = -0.80035 \approx -0.80,$$

which is free of $S(1)$. Thus, the special 3-year partial asset-or-nothing option is equivalent to $2e^{-0.04}N(d_1)$ units of a 1-year 120-strike asset-or-nothing *call*, plus $e^{-0.04}N(d_1)$ units of a 1-year 120-strike asset-or-nothing *put*. With

$$d_1' = \frac{\ln(100/120) + (0.05 - 0.02 + 1/2 \times 0.25^2)(1)}{0.25\sqrt{1}} = -0.48429 \approx -0.48,$$

the time-0 price of the derivative, by risk-neutral pricing again, is

$$
\begin{aligned}
V(0) \quad &= \quad [2e^{-0.04}N(d_1)][S(0)e^{-0.02}N(d_1')] + [e^{-0.04}N(d_1)][S(0)e^{-0.02}]N(-d_1')] \\
&= \quad 2e^{-0.06}(100)\underbrace{N(-0.80)}_{0.2119}\underbrace{N(-0.48)}_{0.3156} + e^{-0.06}(100)\underbrace{N(-0.80)}_{0.2119}\underbrace{N(0.48)}_{0.6844} \\
&= \quad \boxed{26.2541}.
\end{aligned}
$$

                                   □

5. *Solution.* With $K_1 = 100$, $K_2 = 110$,

$$d_1 = \frac{\ln(95/110) + (0.05 - 0.1 + 0.1^2/2)(0.5)}{0.1\sqrt{0.5}} = -2.39148,$$

$$d_2 = d_1 - 0.1\sqrt{0.5} = -2.46220,$$

the price of the gap call option is

$$C^{\text{gap}} = 95e^{-0.1(0.5)}\underbrace{N(-2.39148)}_{0.00839} - 100e^{-0.05(0.5)}\underbrace{N(-2.46220)}_{0.00690} = \boxed{0.0852}.$$

                                   □

7. *Solution.* Let $X(t)$ be the time-t exchange rate of US dollars per British pound. In three months, Michael will need to pay 200,000 pounds, or $\$200,000X(0.25)$, but he will receive $\$320,000$. The shortfall equals $(200,000X(0.25) - 320,000)_+ = 200,000(X(0.25) - 1.6)_+$, which can be compensated by 200,000 units of a 3-month $\$1.6$-strike European call option on pounds. To lower his cost, Michael can replace each plain vanilla call option by a 3-month $\$1.6$-strike European gap call option on pounds with a payment trigger of $\$1.5$, and with payoff equal to $[X(0.25) - 1.6]1_{\{X(0.25)>1.5\}}$. If the 3-month exchange rate is between 1.5 and 1.6, the payoff of such a gap option will be negative.

As

$$d_1 = \frac{\ln(1.6/\boxed{1.5}) + (0.01 - 0.02 + 0.2^2/2)(0.25)}{0.2\sqrt{0.25}} = 0.67039,$$

$$d_2 = d_1 - 0.2\sqrt{0.25} = 0.57039,$$

$$N(d_1) = 0.74870,$$

$$N(d_2) = 0.71579,$$

the price of each currency gap call option is

$$C^{\text{gap}} = 1.6e^{-0.02(0.25)}(0.74870) - \boxed{1.6}e^{-0.01(0.25)}(0.71579) = 0.04954.$$

The total cost of the currency gap options is $200,000(0.04954) = \boxed{9,908}$. □

Remark. Notice that the total cost decreases quite remarkably from 12,316.90 in the plain vanilla case (see Problem 6.4.11) to 9,908 in the current gap option case.

9. *Solution.* (A) Yes

(B) Yes

(C) No

(D) Yes

(E) No

We start with the most obvious representation *(is it obvious?)*

$$\begin{aligned} V(0) &= P(70,60) + C(70,80) + PV_{0,1}(10) \\ &= P(70,60) + C(70,80) + 10e^{-0.06}. \quad \textbf{(A)} \ \boxtimes \end{aligned} \quad (\text{B.8.1})$$

By put-call parity for gap options,

$$\begin{aligned} C(70,60) - P(70,60) &= C(70,80) - P(70,80) \\ &= F_{0,1}(S) - PV_{0,1}(70) \\ &= 70e^{-0.03} - \boxed{70}e^{-0.06}. \quad (\text{B.8.2}) \end{aligned}$$

Plugging (B.8.2) into (B.8.1) leads to

$$\begin{aligned} V(0) &= [C(70,60) - 70e^{-0.03} + \boxed{70}e^{-0.06}] + C(70,80) + 10e^{-0.06} \\ &= C(70,60) + C(70,80) - 70e^{-0.03} + 80e^{-0.06} \quad \textbf{(B)} \ \boxtimes \end{aligned}$$

and

$$\begin{aligned} V(0) &= P(70,60) + [P(70,80) + 70e^{-0.03} - \boxed{70}e^{-0.06}] + 10e^{-0.06} \\ &= P(70,60) + P(70,80) + 70e^{-0.03} - 60e^{-0.06} \quad \textbf{(D)} \ \boxtimes \end{aligned}$$

□

Remark. Answers (C) and (E) would be obtained if you erroneously replaced the strike price 70 by the payment trigger (60 or 80) when writing the put-call parity for gap options.

11. *Solution.* The 90% lognormal prediction interval for $S(0.5)$ is given by

$$\left(S(0)e^{(\alpha-\delta-\sigma^2/2)(0.5)-\sigma\sqrt{0.5}(1.645)}, S(0)e^{(\alpha-\delta-\sigma^2/2)(0.5)+\sigma\sqrt{0.5}(1.645)} \right).$$

The ratio of the upper end-point to the lower end-point of the prediction interval gives

$$e^{2\sigma\sqrt{0.5}(1.645)} = \frac{92.0014}{57.7734},$$

which in turn implies that $\sigma = 0.2$. Moreover, the geometric average of the two endpoints of the prediction interval for $S(0.5)$ gives

$$S(0)e^{(\alpha-\delta-\sigma^2/2)(0.5)} = \sqrt{57.7734(92.0014)} \quad \Rightarrow \quad S(0)e^{0.5(\alpha-\delta)} = 73.638363. \quad \text{(B.8.3)}$$

Now, we are also given in (i) that $\mathbb{P}(S(0.25) < 70) = N(-d_2^{@K=70}) = 0.3669$ (not using $K = 75$, why?), or

$$\hat{d}_2 = \frac{\ln[S(0)/70] + (\alpha - \delta - 0.2^2/2)(0.25)}{0.2\sqrt{0.25}} = -N^{-1}(0.3669) = 0.34008,$$

which, together with (B.8.3), results in

$$\frac{\ln(73.638363e^{-0.5(\alpha-\delta)}/70) + (\alpha - \delta - 0.2^2/2)(0.25)}{0.2\sqrt{0.25}} = 0.34.$$

This gives $\alpha - \delta = 0.046684$ and, by (B.8.3) again, $S(0) = 71.9394$. Finally, the first and second moments of $S(0.75)$, which is lognormal with parameters

$$\begin{cases} m = \ln S(0) + (\alpha - \delta - \sigma^2/2)(t) = \ln 71.9394 + (0.046684 - 0.2^2/2)(0.75) = 4.295837 \\ v^2 = \sigma^2 t = 0.2^2(0.75) = 0.03, \end{cases}$$

are, respectively,

$$\mathbb{E}[S(0.75)] = e^{m+v^2/2} = e^{4.295837+0.03/2} = 74.5028$$
$$\mathbb{E}[S(0.75)^2] = e^{2(m+v^2)} = e^{2(4.295837+0.03)} = 5,719.7135,$$

so $\mathrm{Var}(S(0.75)) = \mathbb{E}[S(0.75)^2] - \mathbb{E}[S(0.75)]^2 = \boxed{169.05}.$ ☐

Remark. You can also take the first and second moments directly from the stock price equation

$$S(0.75) = S(0)e^{(\alpha-\delta-\sigma^2/2)(0.75)+\sigma\sqrt{0.75}Z}$$

using the moment-generating function formula of a standard normal random variable.

13. *Solution.* Since

$$\begin{aligned} \text{Payoff} &= [S(1) - 60]1_{\{60 \le S(1) \le 80\}} \\ &= [S(1) - 60][1_{\{S(1) \ge 60\}} - 1_{\{S(1) > 80\}}] \\ &= [S(1) - 60]_+ - [S(1) - 60]1_{\{S(1) > 80\}}, \end{aligned}$$

the given "truncated" call option is equivalent to a long 60-strike plain vanilla call option plus a short 60-strike 80-trigger gap call option. Let $d_i^{@K}$ be the value of d_i evaluated at K for $i = 1, 2$. With

$$d_1^{@60} = \frac{\ln(65/60) + (0.06 - 0.03 + 0.25^2/2)(1)}{0.25\sqrt{1}} = 0.56517,$$

$$N(d_1^{@60}) = 0.71402,$$

$$d_1^{@80} = \frac{\ln(65/80) + (0.06 - 0.03 + 0.25^2/2)(1)}{0.25\sqrt{1}} = -0.58556,$$

$$d_2^{@80} = d_1^{@60} - 0.25\sqrt{1} = -0.83556,$$

$$N(d_1^{@80}) = 0.27909,$$

$$N(d_2^{@80}) = 0.20170,$$

the current delta of the "truncated" call option is

$$\Delta = \underbrace{e^{-0.03}(0.7157)}_{\text{delta of 60-strike plain vanilla call}}$$

$$- \left[\underbrace{e^{-0.03}(0.27909)}_{\text{delta of 80-strike plain vanilla call}} \underbrace{+ 20e^{-0.06}N'(-0.83556) \times \frac{1}{65(0.25)\sqrt{1}}}_{\text{delta of 80-strike C/N call of \$20}} \right]$$

$$= 0.69292 - (0.27084 + 0.32616)$$

$$= \boxed{0.0959}.$$

□

15. *Solution.* By put-call parity for European gap options,

$$7.3528 - (-3.4343) = 100e^{-0.02(0.5)} - \underbrace{90}_{\text{not 110}} e^{-0.5r},$$

which gives $r = 4\%$. As the 90-strike and 110-payment-trigger gap call is equivalent to a 110-strike plain vanilla call plus a 110-strike cash-or-nothing call of 20, and

$$d_1 = \frac{\ln(100/110) + (0.04 - 0.02 + 0.2^2/2)(0.5)}{0.2\sqrt{0.5}} = -0.53252,$$

$$N(d_1) = 0.29718,$$

$$d_2 = d_1 - 0.2\sqrt{0.5} = -0.67394,$$

the delta of the gap call is

$$e^{-0.02(0.5)}(0.29718) + 20e^{-0.04(0.5)}\frac{1}{\sqrt{2\pi}}e^{-(-0.67394)^2/2} \times \frac{1}{100(0.2)\sqrt{0.5}} = 0.73489.$$

To delta-hedge his position, Ryan buys $10(0.73489) = 7.3489$ shares of the stock at time 0. After one month, his holding profit is

$$\underbrace{[7.3489e^{0.02/12}(95) - 10(4.2437)]}_{656.87305} - \underbrace{[7.3489(100) - 10(7.3528)]e^{0.04/12}}_{663.57027} = \boxed{-6.70}.$$

□

17. *Solution.* (a) The eight blanks should be filled in as follows:

- $\boxed{1} = 0$
- $\boxed{2} = F^P_{0,T}(K)$ or Ke^{-rT}
- $\boxed{3} = F^P_{0,T}(S)$
- $\boxed{4} = 0$
- $\boxed{5} = F_{0,T_f}$
- $\boxed{6} = T$ (not T_f)
- $\boxed{7} = F^P_{0,T}(S_2)$
- $\boxed{8} = 2F^P_{0,T}(S_1)$

  (b)   • For option B, σ is the volatility of the prepaid forward on the stock.
  - For option D, σ is the volatility of the ratio of the prepaid forward on Stock 1 to the prepaid forward on Stock 2.

                                                                                         □

19. *Solution.* The call price computed by Actuary A turns out to be identical to the put price computed by Actuary B.

To see this, note that the two options are related by swapping the price of the underlying stock and the strike price, and swapping the value of the dividend yield and value of the risk-free interest rate. From the perspective of an exchange option, the two options are the same in terms of the blended volatility ($\sigma = 0.3$), the current price and "dividend yield" of the asset acquired (190 and 5%, respectively), and the current price and "dividend yield" of the asset given up (200 and 3%, respectively). As a result, the two options share the same current price, which, by virtue of (8.2.2), is given by

$$BS(190, 0.05; 200, 0.03; 0.3, 1) = 190e^{-0.05(1)}N(d_1) - 200e^{-0.03(1)}N(d_2),$$

with

$$d_1 = \frac{\ln(190/200) + (0.03 - 0.05 + 0.3^2/2)(1)}{0.3\sqrt{1}} \quad \text{and} \quad d_2 = d_1 - 0.3\sqrt{1}.$$

                                                                                         □

21. *Solution.* (A) Decrease.

    (B) Increase.

    (C) Decrease.

    (D) Decrease.

    (E) No change. When the underlying and strike assets both pay continuous proportional dividends, the exchange option pricing formula is free of the continuously compounded risk-free interest rate.

                                                                                         □

    Remark. The conclusion for (E) may not be true for assets paying discrete, non-random dividends. The risk-free rate is used in determining the prepaid forward prices.

23. *Solution.* Note that the three options all share the same current price and "dividend yield" of the asset acquired, namely Stock 2 (200 and 5%, respectively), and the same current price and "dividend yield" of the asset given up (200 and 8%, respectively). Only their blended volatilities differ.

I. The volatility is simply the volatility of Stock 2, or σ_2.

II. The blended volatility equals

$$\sigma = \sqrt{\sigma_1^2 + \sigma_2^2 - 2\rho\sigma_1\sigma_2} \stackrel{(\sigma_1=\sigma_2)}{=} \sqrt{2\sigma_2^2 - 2(0.7)\sigma_2^2} = \sqrt{0.6}\sigma_2,$$

which is less than σ_2.

III. The blended volatility equals

$$\sigma = \sqrt{2\sigma_2^2 - 2(-0.25)\sigma_2^2} = \sqrt{2.5}\sigma_2,$$

which is higher than σ_2.

Recall that the price of a plain vanilla option increases with the volatility (equivalently, vega is positive; see Problem 6.4.18) and that the Black-Scholes price of an exchange option shares the same structure as that of a plain vanilla option. Because the blended volatilities of the three options are ordered as II $<$ I $<$ III, their prices are ordered as $\boxed{\text{II} < \text{I} < \text{III}}$ as well. □

25. *Solution.* (a) Since $\max(S_1(T), S_2(T)) = (S_1(T) - S_2(T))_+ + S_2(T)$, the time-0 price of the maximum option is

$$
\begin{aligned}
V^{\max} &= [F_{0,T}^P(S_1)N(d_1) - F_{0,T}^P(S_2)N(d_2)] + F_{0,P}^P(S_2) \\
&= F_{0,T}^P(S_1)N(d_1) + F_{0,T}^P(S_2)N(-d_2) \\
&= F_{0,T}^P(S_1)N\left(\frac{\ln[F_{0,T}^P(S_1)/F_{0,T}^P(S_2)] + (\sigma^2/2)T}{\sigma\sqrt{T}}\right) \\
&\quad + F_{0,T}^P(S_2)N\left(\frac{\ln[F_{0,T}^P(S_2)/F_{0,T}^P(S_1)] + (\sigma^2/2)T}{\sigma\sqrt{T}}\right),
\end{aligned}
$$

where the first equality follows from $N(x) + N(-x) = 1$ for any $x \in \mathbb{R}$. Observe that the two terms

$$F_{0,T}^P(S_1)N\left(\frac{\ln[F_{0,T}^P(S_1)/F_{0,T}^P(S_2)] + (\sigma^2/2)T}{\sigma\sqrt{T}}\right)$$

and

$$F_{0,T}^P(S_2)N\left(\frac{\ln[F_{0,T}^P(S_2)/F_{0,T}^P(S_1)] + (\sigma^2/2)T}{\sigma\sqrt{T}}\right)$$

are symmetric with each other.

Remark. Note that

$$\max(S_1(T), S_2(T)) = S_1(T)1_{\{S_1(T)>S_2(T)\}} + S_2(T)1_{\{S_2(T)\geq S_1(T)\}}.$$

The two terms above correspond to the time-0 prices of the two asset-or-nothing options with time-T payoffs $S_1(T)1_{\{S_1(T)>S_2(T)\}}$ and $S_2(T)1_{\{S_2(T)\geq S_1(T)\}}$.

(b) As $\min(S_1(T), S_2(T)) = S_1(T) + S_2(T) - \max(S_1(T), S_2(T))$, the time-0 price of the minimum option is

$$
\begin{aligned}
V^{\min} &= F_{0,T}^P(S_1) + F_{0,T}^P(S_2) - [F_{0,T}^P(S_1)N(d_1) + F_{0,T}^P(S_2)N(-d_2)] \\
&= F_{0,T}^P(S_1)N(-d_1) + F_{0,T}^P(S_2)N(d_2) \\
&= F_{0,T}^P(S_1)N\left(\frac{\ln[F_{0,T}^P(S_2)/F_{0,T}^P(S_1)] - (\sigma^2/2)T}{\sigma\sqrt{T}}\right) \\
&\quad + F_{0,T}^P(S_2)N\left(\frac{\ln[F_{0,T}^P(S_1)/F_{0,T}^P(S_2)] - (\sigma^2/2)T}{\sigma\sqrt{T}}\right),
\end{aligned}
$$

where the first line follows from the result of part (a). Observe again that the last two terms are symmetric with each other.

Remark. The two terms above correspond to the time-0 prices of the two asset-or-nothing options with time-T payoffs $S_2(T)1_{\{S_1(T)>S_2(T)\}}$ and $S_1(T)1_{\{S_2(T)\geq S_1(T)\}}$.

□

27. *Solution.* Rewrite the payoff of the contingent claim as

$$
\begin{aligned}
\min(S_1(4), S_2(4)) &= S_1(4) + \min(S_2(4) - S_1(4), 0) \\
&= S_1(4) - (S_1(4) - S_2(4))_+ .
\end{aligned}
$$

As

$$
\begin{aligned}
\sigma &= \sqrt{0.12^2 + 0.15^2 - 2(0.12)(0.15)(-0.25)} = 0.214243, \\
F_{0,4}^P(S_1) &= 100e^{-0.03(4)} = 88.692044, \\
F_{0,4}^P(S_2) &= 120 - 5e^{-0.04(2)} = 115.384418, \\
d_1 &= \frac{\ln(88.692044/115.384418) + (0.214243^2/2)(4)}{0.214243\sqrt{4}} = -0.39978, \\
d_2 &= d_1 - 0.214243\sqrt{4} = -0.82826, \\
N(d_1) &= 0.34466, \\
N(d_2) &= 0.20376,
\end{aligned}
$$

the price of the exchange option giving up one unit of Stock 2 in return for one unit of Stock 1 is

$$
V = 88.692044(0.34466) - 115.384418(0.20376) = 7.05787.
$$

Therefore, the price of the minimum option is

$$
V^{\min} = 88.692044 - 7.05787 = \boxed{81.63417}.
$$

□

29. *Solution.* Note that Derivatives A and B form a call-put pair with the underlying asset being the European call which matures at time 1. By compound option parity,

$$
V_0^A - V_0^B = 6.4225 - 6 \times \underbrace{e^{-0.5r}}_{\text{discount for 6 months only}}.
$$

To find the 6-month discount factor, we apply put-call parity to the 6-month call-put pair and the 1-year call-put pair, yielding the following two equations:

$$
\begin{cases}
4.2581 - 2.7804 = S(0) - S(0)e^{-0.5r} \\
6.4225 - 3.5107 = S(0) - S(0)e^{-r}
\end{cases}
\Rightarrow
\begin{cases}
1.4777 = S(0)(1 - e^{-0.5r}) \\
2.9118 = S(0)(1 - e^{-r})
\end{cases}.
$$

Dividing the first equation from the second one (recall the identity $1-x^2 \equiv (1+x)(1-x)$) and noting that $e^{-0.5r} \neq 1$, we have

$$\frac{1-e^{-r}}{1-e^{-0.5r}} = 1 + e^{-0.5r} = \frac{2.9118}{1.4777} \quad \Rightarrow \quad e^{-0.5r} = 0.970495.$$

Hence the price of Derivative B is

$$V_0^B = V_0^A - 6.4225 + 6e^{-0.5r} = 2.5944 - 6.4225 + 6(0.970495) = \boxed{1.9949}.$$

□

31. *Solution.* The forward tree is built by setting

$$u = \exp[(r-\delta)h + \sigma\sqrt{h}] = \exp[(0.1-0.02)(0.5) + 0.3\sqrt{0.5}] = 1.286766,$$
$$d = \exp[(r-\delta)h - \sigma\sqrt{h}] = \exp[(0.1-0.02)(0.5) - 0.3\sqrt{0.5}] = 0.841868,$$

and the risk-neutral probability of an up move is

$$p^* = \frac{1}{1+e^{\sigma\sqrt{h}}} = \frac{1}{1+e^{0.3\sqrt{0.5}}} = 0.447165.$$

The resulting stock prices are depicted below:

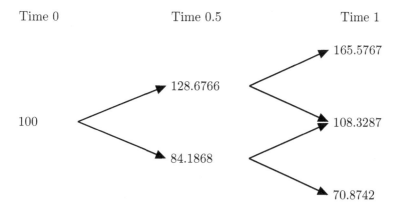

Time 0	Time 0.5	Time 1

(a) The arithmetic averages and the payoffs ($= (A - S(1))_+$) of the Asian put option corresponding to different stock price paths are:

Path	Arithmetic Average	Payoff
uu	$(128.6766 + 165.5767)/2 = 147.1266$	$(147.1266 - 165.5767)_+ = 0$
ud	$(128.6766 + 108.3287)/2 = 118.5026$	$(118.5026 - 108.3287)_+ = 10.1739$
du	$(84.1868 + 108.3287)/2 = 96.2578$	$(96.2578 - 108.3287)_+ = 0$
dd	$(84.1868 + 70.8742)/2 = 77.5305$	$(77.5305 - 70.8742)_+ = 6.6563$

Upon discounting the risk-neutral expected value of these payoffs back to time 0 at the risk-free interest rate, the current price of the arithmetic average strike Asian put option is

$$V_0 = e^{-rT}[p^*(1-p^*)V_{ud} + (1-p^*)^2 V_{dd}]$$
$$= e^{-0.1(1)}[(0.447165)(1-0.447165)(10.1739) + (1-0.447165)^2(6.6563)]$$
$$= \boxed{4.1165}.$$

(b) The geometric averages and the payoffs $(= (G - S(1))_+)$ of the Asian put option corresponding to different stock price paths are:

Path	Geometric Average	Payoff
uu	$\sqrt{128.6766(165.5767)} = 145.9652$	$(145.9652 - 165.5767)_+ = 0$
ud	$\sqrt{128.6766(108.3287)} = 118.0651$	$(118.0651 - 108.3287)_+ = 9.7364$
du	$\sqrt{84.1868(108.3287)} = 95.4979$	$(95.4979 - 108.3287)_+ = 0$
dd	$\sqrt{84.1868(70.8742)} = 77.2442$	$(77.2442 - 70.8742)_+ = 6.3700$

By risk-neutral pricing again, the current price of the geometric average strike Asian put option is

$$
\begin{aligned}
V_0 &= e^{-rT}[p^*(1-p^*)V_{ud} + (1-p^*)^2 V_{dd}] \\
&= e^{-0.1(1)}[(0.447165)(1 - 0.447165)(9.7364) + (1 - 0.447165)^2 (6.3700)] \\
&= \boxed{3.9394}.
\end{aligned}
$$

□

33. *Solution.* Put-call parity for otherwise identical European Asian options asserts that

$$
\begin{aligned}
C^{\text{Asian}} - P^{\text{Asian}} &= \text{Time-0 price of } \frac{S(1) + S(2) + S(3)}{3} \text{ payable at time 3} \\
&\quad - 15e^{-rT}.
\end{aligned}
$$

To evaluate the time-0 price of the arithmetic average, we reason as follows:

- To obtain $S(1)$ payable at $t = 3$ (not $t = 1!$), we need $e^{-2r}S(1)$ at $t = 1$ (this known amount will grow at the risk-free interest to exactly $S(1)$ at $t = 3$), or equivalently, $e^{-2r}F_{0,1}^P(S) = e^{-2r} \times S(0)e^{-\delta} = e^{-2(0.06)-0.03}(15) = 15e^{-0.15}$ at $t = 0$.

- Similarly, the time-0 price of $S(2)$ payable at $t = 3$ is

$$
e^{-r}F_{0,2}^P(S) = e^{-0.06-2(0.03)}(15) = 15e^{-0.12}.
$$

- The time-0 price of $S(3)$ payable at $t = 3$ is simply

$$
F_{0,3}^P(S) = S(0)e^{-3\delta} = 15e^{-3(0.03)} = 15e^{-0.09}.
$$

Consequently, the time-0 price of the arithmetic average payable at $t = 3$ equals

$$
\frac{15e^{-0.15} + 15e^{-0.12} + 15e^{-0.09}}{3} = 13.3078,
$$

and the required Asian put price is $1.5 - 13.3078 + 15e^{-0.06(3)} = \boxed{0.7213}$.

□

Remark. Note that the time-0 price of the arithmetic average is *not*

$$
\frac{F_{0,1}^P(S) + F_{0,2}^P(S) + F_{0,3}^P(S)}{3}.
$$

35. *Solution.* The forward tree is built by setting

$$
u = \exp[(r_\$ - r_{\euro})h + \sigma\sqrt{h}] = \exp[(0.04 - 0.05)(0.25) + 0.2\sqrt{0.25}] = 1.102411,
$$
$$
d = \exp[(r_\$ - r_{\euro})h - \sigma\sqrt{h}] = \exp[(0.04 - 0.05)(0.25) - 0.2\sqrt{0.25}] = 0.902578,
$$

and the risk-neutral probability of an up move is

$$
p^* = \frac{1}{1 + e^{\sigma\sqrt{h}}} = \frac{1}{1 + e^{0.2\sqrt{0.25}}} = 0.475021.
$$

The resulting dollar/euro exchange rates are depicted below:

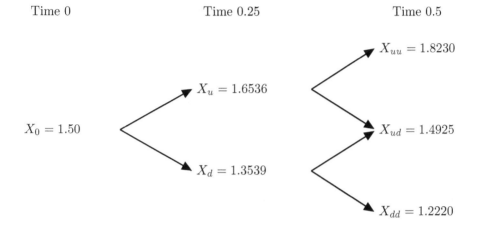

From the binomial tree, we can calculate the 6-month payoffs of the standard American lookback call corresponding to different exchange rate paths:

Path	Terminal Exchange Rate (A)	Min. Exchange Rate (B)	Terminal Payoff = (A) − (B)
uu	1.8230	1.50	0.323
ud	1.4925	1.4925	0
du	1.4925	1.3539	0.1386
dd	1.2220	1.2220	0

Moving back to the end of three months, we have:

Path	Current Exchange Rate (A)	Min. Exchange Rate (B)	Holding Value	Exercise Value = (A) − (B)
u	1.6536	1.50	$e^{-0.04/4}[p^*(0.323)] = 0.1519$	0.1536
d	1.3539	1.3539	$e^{-0.04/4}[p^*(0.1386)] = 0.0652$	0

Early exercise is optimal at the u node, with $V_u = 0.1536$ and $V_d = 0.0652$. The replicating portfolio of the lookback call at the initial node is set up by

$$
\Delta = e^{-r_{\euro}h}\left(\frac{V_u - V_d}{X_u - X_d}\right) = e^{-0.05(0.25)}\left(\frac{0.1536 - 0.0652}{1.6536 - 1.3539}\right) = 0.2913
$$

and

$$B = e^{-r_\$ h}\left(\frac{uV_d - dV_u}{u - d}\right)$$

$$= e^{-0.04(0.25)}\left[\frac{1.102411(0.0652) - 0.902578(0.1536)}{1.102411 - 0.902578}\right]$$

$$= -0.3307.$$

It follows that the fair price of the lookback call is

$$V_0 = \Delta X + B = 0.2913(1.5) + (-0.3307) = 0.1062,$$

which is higher than the observed price. To pursue an arbitrage strategy, at the initial exchange rate node one should:

- Buy the lookback call from the market.
- Sell the replicating portfolio by selling 0.2913 units of euro and buying 0.3307 units of dollar.

$\square$

37. *Solution.* In terms of payoff, the four options are ordered as

$$\text{II} \le \text{I} \le \text{IV} \le \text{III}.$$

It follows that their (fair) prices are ordered in the same way. $\square$

39. *Solution.* This problem uses an indirect way to request the value of the right to shout over the 1-year life of the shout option. Given that $u = 1.2212$ and $d = 0.8637$, we first construct the binomial tree below:

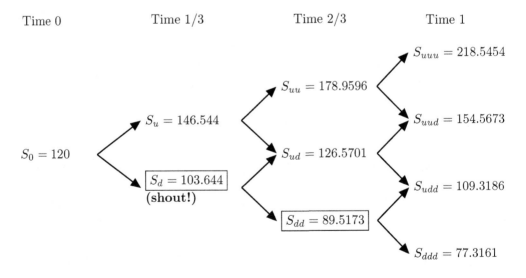

The risk-neutral probability of an up move is

$$p^* = \frac{e^{(r-\delta)h} - d}{u - d} = \frac{e^{(0.08-0)/3} - 0.8637}{1.2212 - 0.8637} = 0.456854.$$

Note that if we never shout during the life of the shout put, then its possible terminal payoffs are

$$P_{uuu} = P_{uud} = 0, \quad P_{udd} = 10.6814, \quad P_{ddd} = 42.6839.$$

We now investigate the optimal time to shout. Prior to maturity, the put option is in-the-money only at the d node and the dd node, so it suffices to focus on these two nodes.

- dd node: If we shout at the dd node, we can guarantee a minimum payoff of $120 - 89.5173 = 30.4827$ and the two possible terminal payoffs become

$$\begin{aligned} V_{udd} &= \max(10.6814, 30.4827) = 30.4827, \\ V_{ddd} &= \max(42.6839, 30.4827) = 42.6839. \end{aligned}$$

By risk-neutral pricing, the value of the option at the dd node is

$$V_{dd} = e^{-0.08/3}[0.456854(30.4827) + (1 - 0.456854)(42.6839)] = 36.1332.$$

- d node: We now further roll back one period to the d node. To determine whether to shout at the d node or to wait until the dd node (if the stock price indeed drops after 4 months), we have to evaluate the value of the shout put under these two mutually exclusive actions.

  Case 1. If we shout at the d node, then we will receive a payoff of at least $120 - 103.644 = 16.356$ at maturity, or more if the shout put expires deeper in the money. From the perspective of the d node, only the uud, udd, and ddd nodes are possible outcomes, with respective terminal payoffs being

$$\begin{aligned} V_{uud} &= \max(0, 16.356) = 16.356, \\ V_{udd} &= \max(10.6814, 16.356) = 16.356, \\ V_{ddd} &= \max(42.6839, 16.356) = 42.6839. \end{aligned}$$

  By risk-neutral pricing, the value of the shout put is

$$\begin{aligned} V_d &= e^{-0.08(2/3)}\{16.356 + (1 - 0.456854)^2(42.6839 - 16.356)]\} \\ &= 22.8701. \end{aligned}$$

  Case 2. If we do not shout at the d node, but shout at the dd node, if we happen to reach that node, then the shout put will be worth $V_{ud} = e^{-0.08/3}(1 - 0.456854)(10.6814) = 5.6489$ at the ud node or $V_{dd} = 36.1332$ at the dd node (where we shout). By risk-neutral pricing again, the value of the shout call is

$$V_d = e^{-0.08/3}[0.456854(5.6489) + (1 - 0.456854)(36.1332)] = 21.6220.$$

Since the shout put is worth more in Case 1, it is advisable to shout at the d node, clinching the higher value of $V_d = 22.8701$.

For the corresponding plain vanilla put, its value at the d node is

$$\begin{aligned} P_d &= e^{-0.08(2/3)}[2(0.456854)(1 - 0.456854)(10.6814) + (1 - 0.456854)^2(42.6839)] \\ &= 16.9637. \end{aligned}$$

The value of the right to shout is

$$P_{II} - P_I = e^{-0.08/3}(1 - 0.456854)\underbrace{(22.8701 - 16.9637)}_{\text{change in } V_d} = \boxed{3.1236}.$$

□

Remark. It can be shown that $P_I = 10.3002$ and $P_{II} = 13.4238$, hence $P_{II} - P_I = 3.1236$, agreeing with the answer above.

41. *Solution.* For options I and II, $S(0) = 50 < B \leq K = 60$, so their payoff formula can be simplified into

$$(S(T) - K)_+ \times 1_{\{M(T) > B\}} = (S(T) - K) \times 1_{\{S(T) > K, M(T) > B\}} = (S(T) - K) \times 1_{\{S(T) > K\}},$$

which is that of a 60-strike plain vanilla call. Hence options I and II have the same price. When $B > K = 60$, the payoff formula is

$$(S(T) - K)_+ \times 1_{\{M(T) > B\}},$$

which is non-increasing in B. Thus the prices of options III and IV are ordered as (III) $\geq$ (IV).

In conclusion, $\boxed{(I) = (II) \geq (III) \geq (IV)}$. (**Answer: (C)**)   □

43. *Solution.* Given that $u = 1.15$ and $d = 0.95$, the three-period binomial stock price tree is exhibited overleaf:

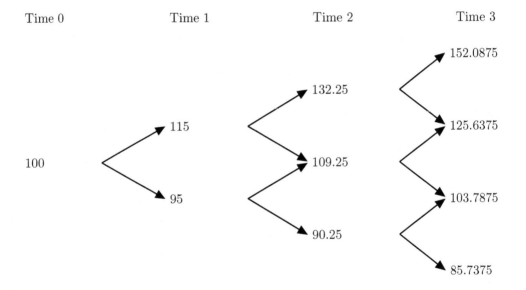

| Time 0 | Time 1 | Time 2 | Time 3 |

The risk-neutral probability of an up move is

$$p^* = \frac{e^{0.1(1)} - 0.95}{1.15 - 0.95} = 0.775855.$$

We tabulate below the possible payoffs of the up-and-in call along with the associated risk-neutral probabilities:

Node	Payoff	Number of Paths	Risk-neutral Probability
uuu	52.0875	1	$p^3 = 0.467027$
uud	25.6375	3	$3(p^*)^2(1 - p^*) = 0.404773$
udd	3.7875	1	$p^*(1 - p^*)^2 = 0.038980$

Note that the up-and-in call pays off at the *udd* node only via the *udd* path; both the *dud* and *ddu* paths do not cross the barrier of 110. By risk-neutral pricing, the price of the up-and-in call is

$$V = e^{-0.1(3)}[0.467027(52.0875) + 0.404773(25.6375) + 0.038980(3.7875)]$$
$$= \boxed{25.82}.$$

□

45. *Solution.* The forward tree parameters are

$$u = \exp[(r-\delta)h + \sigma\sqrt{h}] = \exp[(0.05-0.03)(1) + 0.3\sqrt{1}] = 1.377128,$$
$$d = \exp[(r-\delta)h - \sigma\sqrt{h}] = \exp[(0.05-0.03)(1) - 0.3\sqrt{1}] = 0.755784.$$

The risk-neutral probability of an up move is

$$p^* = \frac{e^{(r-\delta)h} - d}{u - d} = \frac{e^{(0.05-0.03)(1)} - 0.755784}{1.377128 - 0.755784} = 0.425557,$$

or

$$p^* = \frac{1}{1 + e^{\sigma\sqrt{h}}} = \frac{1}{1 + e^{0.3\sqrt{1}}} = 0.425557.$$

The stock price tree is depicted below:

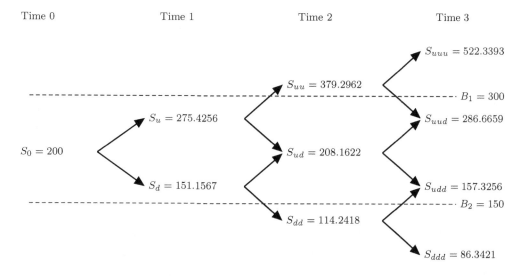

Observe that the special collar may pay off only in the *uud* node attainable by 2 paths (the *udu* path and the *duu* path; the *uud* path crosses the upper barrier of 300), and the *udd* node attainable by 2 paths (the *udd* path and the *dud* path; the *ddu* path crosses the lower barrier of 150). The payoffs are summarized as follows:

Path	Geometric Average (G)	Payoff of Collar $= (180 - G)_+ - (G - 250)_+$
udu	$(275.4256 \times 208.1622 \times 286.6659)^{1/3} = 254.2499$	-4.2499
duu	$(151.1567 \times 208.1622 \times 286.6659)^{1/3} = 208.1622$	0
udd	$(275.4256 \times 208.1622 \times 157.3256)^{1/3} = 208.1622$	0
dud	$(151.1567 \times 208.1622 \times 157.3256)^{1/3} = 170.4288$	9.5712

By risk-neutral pricing, the price of the collar is

$$V_0 = e^{-0.05(3)}[(p^*)^2(1 - p^*)(-4.2499) + p^*(1 - p^*)^2(9.5712)] = \boxed{0.7763}.$$

☐

Remark. The presence of the double barriers actually simplifies your computations—you need not deal with $2^3 = 8$ stock price paths. Only four matter.

47. *Solution.* (a) The payoff at time $t = T = 6$ of the chooser option is

$$\max\{(S(6) - 30)_+, (30 - S(6))_+\} = \begin{cases} 30 - S(6), & \text{if } S(6) < 30 \\ S(6) - 30, & \text{if } S(6) \geq 30 \end{cases}$$

$$= |S(6) - 30|,$$

which is the payoff of a $\boxed{\text{long 30-strike (6-year) straddle}}$. The price of the chooser option is therefore the sum of the price of a 30-strike 6-year call and that of a 30-strike 6-year put. To obtain the put price, we use put-call parity:

$$\begin{aligned} P(32, 30, 6) &= C(32, 30, 6) - S(0) + \text{PV}_{0,6}(\text{Div}) + Ke^{-rT} \\ &= 9.48 - 32 + 1.5(e^{-0.05} + e^{-0.05(3)} + e^{-0.05(5)}) + 30e^{-0.05(6)} \\ &= 3.590654. \end{aligned}$$

The price of the chooser option is then $C(32, 30, 6) + P(32, 30, 6) = 9.48 + 3.590654 = \boxed{13.07}$.

(b) The time-2 payoff of the chooser option is

$$P(S(2), 30, 4) + (C(S(2), 30, 4) - P(S(2), 30, 4))_+,$$

which, by put-call parity, equals

$$P(S(2), 30, 4) + \left(S(2) - \text{PV}_{2,6}(\text{Div}) - 30e^{-0.05(4)}\right)_+$$

$$= P(S(2), 30, 4) + \left(S(2) - 1.5(e^{-0.05} + e^{-0.05(3)}) - 30e^{-0.05(4)}\right)_+$$

$$= P(S(2), 30, 4) + (S(2) - 27.279829)_+.$$

It follows that the time-0 price of the chooser option is the sum of the price of a 6-year 30-strike put option and the price of a 2-year 27.28-strike call option, i.e.

$$P(32, 30, 6) + C(32, 27.28, 2) = \underset{\text{from part (a)}}{\underbrace{3.590654}} + 7.60 = \boxed{11.19}.$$

☐

Remark. The performance in this question was catastrophic when it was offered in a real exam. As you can see from the above solutions, the question is not ridiculously hard. You can do it as long as you know how to manipulate put-call parity. Incidentally, this question also illustrates the futility of slavishly memorizing pricing formulas without understanding the assumptions governing their validity and the importance of being able to work out problems from first principles.

49. *Notation:* Let $V^{2B \to A}(t, T)$ (resp. $V^{A \to 2B}(t, T)$) be the time-t price of the European option maturing at time T and giving you the right to exchange two units of Stock B for one unit of Stock A (resp. exchange one unit of Stock A for two units of Stock B).

Solution 1. The time-1 payoff of the chooser option is

$$\max\{V^{2B \to A}(1, 3), V^{A \to 2B}(1, 3)\}$$

$$= \qquad V^{A \to 2B}(1, 3) + \left(V^{2B \to A}(1, 3) - V^{A \to 2B}(1, 3)\right)_+$$

$$\overset{\text{(exchange option parity)}}{=} \quad V^{A \to 2B}(1, 3) + \left(F_{1,3}^P(A) - 2F_{1,3}^P(B)\right)_+$$

$$= \qquad V^{A \to 2B}(1, 3) + \left(A(1)e^{-0.02(2)} - 2B(1)e^{-0.02(2)}\right)_+$$

$$= \qquad \underbrace{V^{A \to 2B}(1, 3)}_{\substack{\text{time-1 payoff of option to} \\ \text{exchange A for 2 units of B at time 3}}}$$

$$+ e^{-0.02(2)} \underbrace{V^{2B \to A}(1, 1)}_{\text{exchange 2 units of B for A at time 1}} \quad .$$

Thus the time-0 price of the chooser is

$$V^{A \to 2B}(0, 3) + e^{-0.04} \underbrace{V^{2B \to A}(0, 1)}_{=5 \text{ (from (iv))}} = V^{A \to 2B}(0, 3) + 5e^{-0.04}.$$

By the exchange option parity at time 0, we have

$$V^{A \to 2B}(0, 3) - \underbrace{V^{2B \to A}(0, 3)}_{=8 \text{ from (v)}} = 2F_{0,3}^P(B) - F_{0,3}^P(A) = e^{-0.02(3)}[2(100) - 205],$$

so $V^{A \to 2B}(0, 3) = 3.291177$. Hence the answer is $3.291177 + 5e^{-0.04} = \boxed{8.0951}$. □

Solution 2. The time-1 payoff of the chooser option is

$$\max\{V^{2B \to A}(1, 3), V^{A \to 2B}(1, 3)\}$$

$$= \qquad V^{2B \to A}(1, 3) + \left(V^{A \to 2B}(1, 3) - V^{2B \to A}(1, 3)\right)_+$$

$$\overset{\text{(exchange option parity)}}{=} \quad V^{2B \to A}(1, 3) + \left(2F_{1,3}^P(B) - F_{1,3}^P(A)\right)_+$$

$$= \qquad V^{2B \to A}(1, 3) + \left(2B(1)e^{-0.02(2)} - A(1)e^{-0.02(2)}\right)_+$$

$$= \qquad \underbrace{V^{2B \to A}(1, 3)}_{\substack{\text{time-1 payoff of option to} \\ \text{exchange 2 units of B for A at time 3}}}$$

$$+ e^{-0.02(2)} \underbrace{V^{A \to 2B}(1, 1)}_{\text{exchange A for 2 units of B at time 1}} \quad .$$

Thus the time-0 price of the chooser is

$$\underbrace{V^{2B \to A}(0, 3)}_{=8 \text{ (from (v))}} + e^{-0.04} V^{A \to 2B}(0, 1) = 8 + e^{-0.04} V^{A \to 2B}(0, 1).$$

By the exchange option parity at time 0, we have

$$V^{A \to 2B}(0, 1) - \underbrace{V^{2B \to A}(0, 1)}_{=5 \text{ (from (iv))}} = 2F_{0,1}^P(B) - F_{0,1}^P(A) = e^{-0.02(1)}[2(100) - 205],$$

so $V^{2B \to A}(0, 1) = 0.099007$. Hence the answer is $8 + e^{-0.04}(0.099007) = \boxed{8.0951}$.   □

51. *Solution.* (a) From the perspective of time 1, the contingent claim maturing at time 3 is a 2-year $S(1)$-strike European cash-or-nothing call of $\$S(1)$ payable at time 3. Since

$$d_1 = \frac{\ln[S(1)/S(1)] + (0.08 - 0.02 + 0.3^2/2)(2)}{0.3\sqrt{2}} = 0.49497,$$

$$d_2 = d_1 - 0.3\sqrt{2} = 0.07071,$$

$$N(d_2) = 0.52819,$$

the time-1 price of the contingent claim is

$$S(1)e^{-2r}N(d_2) = S(1)e^{-0.08(2)}(0.52819).$$

Hence the time-0 price is

$$F_{0,1}^P(S)e^{-0.08(2)}(0.52819) = 100e^{-0.02-0.08(2)}(0.52819) = \boxed{44.12}.$$

(b) From the point of view of time 1, the contingent claim is a 2-year $S(1)$-strike European asset-or-nothing call. With $N(d_1) = 0.68969$, the time-1 price of the contingent claim is

$$S(1)e^{-2\delta}N(d_1) = S(1)e^{-0.02(2)}(0.68969).$$

Thus the time-0 price is

$$F_{0,1}^P(S)e^{-0.02(2)}(0.68969) = 100e^{-0.02(3)}(0.68969) = \boxed{64.95}.$$

   □

53. *Solution.* The time-3 payoff of the forward start gap call option is

$$V(3) = [S(3) - 100]1_{\{S(3) > S(2)\}}.$$

At time 2, $S(2)$ is known and the forward start, by the Black-Scholes formula for gap options, is worth

$$V(2) = S(2)e^{-\delta}N(d_1) - 100e^{-r}N(d_2),$$

where

$$d_1 = \frac{\ln[S(2)/S(2)] + (0.06 - 0.025 + 0.24^2/2)(1)}{0.24\sqrt{1}} = 0.26583,$$

$$d_2 = d_1 - 0.24\sqrt{1} = 0.02583,$$

both of which do not depend on $S(2)$. It follows that the forward start is equivalent to $e^{-\delta}N(d_1)$ units of a 2-year prepaid forward on the stock and a short 2-year zero-coupon bond of face value $100e^{-r}N(d_2)$. As $N(d_1) = 0.6064$ (0.60481) and $N(d_2) = 0.5120$ (0.51030), the time-0 price of the forward start is

$$
\begin{aligned}
V &= F_{0,2}^P(S)e^{-\delta}N(d_1) - 100e^{-3r}N(d_2) \\
&= S(0)e^{-3\delta}N(d_1) - 100e^{-3r}N(d_2) \\
&= 100e^{-3(0.025)}(0.60481) - 100e^{-3(0.06)}(0.51030) \\
&= \boxed{13.48701}.
\end{aligned}
$$

□

Remark. If it is the strike price (instead of the payment trigger) of the gap call that equals $S(2)$, then d_1 and d_2 will depend on $S(2)$ and the current price of the forward start cannot be easily found.

55. *Solution.* Let's first derive a general expression for the current price of the T-year contingent claim which pays at time T $\min[S(t), S(T)]$ for some $t \le T$, then evaluate the price at the numerical values given in the problem.

- *Method 1:* To begin with, we rewrite the time-T payoff of the contingent claim as

$$
\min[S(t), S(T)] = S(T) + \min[S(t) - S(T), 0] = S(T) - (S(T) - S(t))_+,
$$

which shows that the contingent claim is equivalent to a T-year prepaid forward on the stock plus a short T-year forward start option whose strike is determined by the time-t stock price. It follows from the results in Subsection 8.8.2 that the time-0 price of the contingent claim is

$$
\begin{aligned}
V(0) &= S(0)e^{-\delta T} - S(0)[e^{-\delta T}N(d_1) - e^{-\delta t - r(T-t)}N(d_2)] \\
&= S(0)e^{-\delta T}N(-d_1) + S(0)e^{-\delta t - r(T-t)}N(d_2),
\end{aligned}
$$

where

$$
\begin{aligned}
d_1 &= \frac{\ln[S(t)/S(t)] + (r - \delta + \sigma^2/2)(T - t)}{\sigma\sqrt{T - t}} = \frac{(r - \delta + \sigma^2/2)(T - t)}{\sigma\sqrt{T - t}}, \\
d_2 &= d_1 - \sigma\sqrt{T - t},
\end{aligned}
$$

both of which are free of $S(t)$.

- *Method 2:* The time-T payoff of the contingent claim is

$$
\begin{aligned}
\text{Payoff} &= \begin{cases} S(t), & \text{if } S(t) < S(T) \\ S(T), & \text{if } S(T) \le S(t) \end{cases} \\
&= S(t)\mathbf{1}_{\{S(T)>S(t)\}} + S(T)\mathbf{1}_{\{S(T)\le S(t)\}}.
\end{aligned}
$$

From the perspective of time t, the contingent claim is equivalent to a $(T - t)$-year $S(t)$-strike *cash-or-nothing call* of $\$S(t)$, plus a $(T - t)$-year $S(t)$-strike *asset-or-nothing put*. Thus the time-t price is

$$
\begin{aligned}
V(t) &= S(t)e^{-r(T-t)}N(d_2) + S(t)e^{-\delta(T-t)}N(-d_1) \\
&= S(t)[e^{-r(T-t)}N(d_2) + e^{-\delta(T-t)}N(-d_1)],
\end{aligned}
$$

where

$$d_1 = \frac{\ln[S(t)/S(t)] + (r - \delta + \sigma^2/2)(T - t)}{\sigma\sqrt{T - t}} = \frac{(r - \delta + \sigma^2/2)(T - t)}{\sigma\sqrt{T - t}},$$

$$d_2 = d_1 - \sigma\sqrt{T - t},$$

both of which are free of $S(t)$. It follows that the time-0 price is

$$\begin{aligned} V(0) &= S(0)e^{-\delta t}[e^{-r(T-t)}N(d_2) + e^{-\delta(T-t)}N(-d_1)] \\ &= S(0)[e^{-\delta t - r(T-t)}N(d_2) + e^{-\delta T}N(-d_1)]. \end{aligned}$$

With $S(0) = 20, r = 0.04, \delta = 0.015, \sigma = 0.25, t = 2, T = 3$, we have

$$d_1 = \frac{(0.04 - 0.015 + 0.25^2/2)(3 - 2)}{0.25\sqrt{3 - 2}} = 0.225,$$

$$d_2 = d_1 - 0.25\sqrt{1} = -0.025,$$

$$N(-d_1) = 0.41099,$$

$$N(d_2) = 0.49003,$$

so the time-0 price of the contingent claim is

$$\begin{aligned} V &= 20e^{-0.015(3)}(0.41099) + 20e^{-0.015(2)-0.04(1)}(0.49003) \\ &= \boxed{16.9961}. \end{aligned}$$

□

B.9   Chapter 9

1. *Solution.* By put-call parity for nondividend-paying stocks, we have

$$0.15 = C - P = S(0) - 70e^{-rT} = 60 - 70e^{-4r},$$

which yields $r = \boxed{0.039}$. (**Answer: (A)**)   □

3. *Solution.* By currency option put-call duality, the prices of 3-year dollar-denominated put options on euros with strike prices of \$1.2 and \$1.5 are, respectively,

$$P_\${}(1.25, 1.2) = 1.25 \times 1.2 \times \underbrace{C_{€}(1/1.25, 1/1.2)}_{0.03315 \text{ from (v)}} = 0.049725$$

and

$$P_\${}(1.25, 1.5) = 1.25 \times 1.5 \times \underbrace{C_{€}(1/1.25, 1/1.5)}_{0.1641 \text{ from (iv)}} = 0.3076875.$$

Two applications of currency put-call parity, one at the strike of \$1.2, and one at the strike of \$1.5, yield

$$\begin{cases} C_\${}(1.25, 1.2) - P_\${}(1.25, 1.2) = X(0)e^{-r_{€}T} - K_1 e^{-r_\${}T} \\ C_\${}(1.25, 1.5) - P_\${}(1.25, 1.5) = X(0)e^{-r_{€}T} - K_2 e^{-r_\${}T} \end{cases}$$

or

$$\begin{cases} 0.015725 = 1.25e^{-3r_{\text{€}}} - 1.2e^{-3r_{\text{\$}}} \\ -0.2492675 = 1.25e^{-3r_{\text{€}}} - 1.5e^{-3r_{\text{\$}}} \end{cases}$$

Solving this system of equations leads to $r_{\text{\$}} = \boxed{4.14\%}$ (and $r_{\text{€}} = 5.01\%$). □

5. *Proof.* We justify $P^E \leq F_{0,T}^P(K)$ in three ways:

(a) With a put option, the most you can receive at expiration is the strike price K. One will not pay more than the present value of the strike price at time 0 just to acquire a right to receive the strike price at time T.

(b) Assume on the contrary that $P^E > F_{0,T}^P(K)$. Then we lend $F_{0,T}^P(K)$ and sell the European put, for a strictly positive net cash inflow of $P^E - F_{0,T}^P(K)$ at time 0. At expiration, our payoff is

$$-(K - S(T))_+ + K = \begin{cases} S(T), & \text{if } S(T) \leq K, \\ K, & \text{if } S(T) > K, \end{cases}$$

which must be non-negative. We have thus set up an arbitrage strategy. To preclude such an arbitrage opportunity, we must have $P^E \leq F_{0,T}^P(K)$.

(c) Since $C^E \leq F_{0,T}^P(S)$, it follows from put-call parity that

$$P^E = C^E - F_{0,T}^P(S) + F_{0,T}^P(K) \leq [F_{0,T}^P(S) - F_{0,T}^P(S)] + F_{0,T}^P(K) = F_{0,T}^P(K).$$

□

7. *Solution.* Because one year from now Jeff will receive a fixed sum in pounds, which he plans to exchange to dollars to pay for his debt in US dollars, he will be worried that the dollar/pound exchange rate will go down in one year. Therefore, he should need European/American dollar-denominated \$1.6-strike *put* options on pounds.

Note that

$$F_{0,1}^P(X) = X(0)e^{-r_{\text{£}}(1)} = X(0)e^{-0.08} = 0.9231X(0)$$

and

$$F_{0,1}^P(K) = Ke^{-r_{\text{\$}}(1)} = 1.6e^{-0.06} = 1.5068$$

The European put option price is bounded from above and below as follows:

$$(1.5068 - 0.9231X(0))_+ = (F_{0,1}^P(K) - F_{0,1}^P(X))_+ \leq P^E \leq F_{0,1}^P(K) = 1.5068.$$

The American put option price is bounded from above and below as follows:

$$\max\{(K - X(0))_+, (F_{0,1}^P(K) - F_{0,1}^P(X))_+\} \leq P^A \leq K,$$

or

$$\max\{(1.6 - X(0))_+, (1.5068 - 0.9231X(0))_+\} \leq P^A \leq 1.6.$$

The possible American and European put prices are sketched in Figure B.9.1. □

9. *Solution.* (a) This is just the ordinary convexity condition. The answer is $\boxed{0 \leq \lambda \leq 1}$.

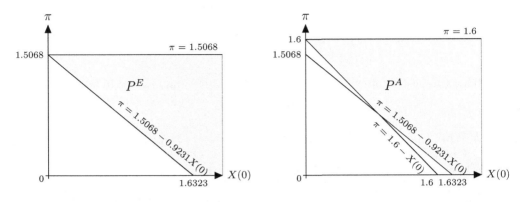

FIGURE B.9.1
The sets of allowable values for the European put price (left) and American put price (right).

(b) If $\lambda \in [0,1]$, then $K_\lambda \in [K_1, K_2]$, so that convexity implies the reverse inequality, which we do not desire. In order that the stated inequality holds, we must have $\lambda \leq 0$ or $\lambda \geq 1$. We explore these two cases separately:

Case 1. $\lambda \leq 0$.
In this case, $K_1 < K_2 < K_\lambda$ with $K_2 = \frac{1}{1-\lambda}K_\lambda - \frac{\lambda}{1-\lambda}K_1$. Note that $\frac{1}{1-\lambda} \in (0,1]$, $-\frac{\lambda}{1-\lambda} \in [0,1)$ and they sum to 1, so they can be regarded as weights. Convexity results in

$$C(K_2) \leq \frac{1}{1-\lambda}C(K_\lambda) - \frac{\lambda}{1-\lambda}C(K_1),$$

or (multiplying both sides by $1 - \lambda > 0$)

$$C(K_\lambda) \geq \lambda C(K_1) + (1-\lambda)C(K_2).$$

Case 2. $\lambda \geq 1$.
In this case, $K_\lambda < K_1 < K_2$ with $K_1 = \frac{1}{\lambda}K_\lambda + \frac{\lambda-1}{\lambda}K_2$, $0 < \frac{1}{\lambda} \leq 1$ and $0 < \frac{\lambda-1}{\lambda} < 1$. Convexity leads to

$$C(K_1) \leq \frac{1}{\lambda}C(K_\lambda) + \frac{\lambda-1}{\lambda}C(K_2),$$

or (multiplying both sides by $\lambda > 0$)

$$C(K_\lambda) \geq \lambda C(K_1) + (1-\lambda)C(K_2).$$

Since strike prices should not be negative, we also have

$$K_\lambda = \lambda(K_1 - K_2) + K_2 \geq 0 \overset{(K_1 \leq K_2)}{\Rightarrow} \lambda \leq \frac{K_2}{K_2 - K_1}.$$

In conclusion, the range of values of λ is $\boxed{\lambda \leq 0}$ or $\boxed{1 \leq \lambda \leq K_2/(K_2 - K_1)}$. $\square$

11. *Solution.* Here are two sets of arbitrage strategies based on different principles:

- *Strategy 1 (Convexity):* Note that the put prices violate convexity, because

$$P(25) = 2.64 > \frac{2}{3}P(20) + \frac{1}{3}P(35) = \frac{2}{3}(1.35) + \frac{1}{3}(4.36) = 2.3533.$$

To pursue an arbitrage strategy, we buy two 20-strike puts, sell three 25-strike puts, and buy one 35-strike put. At time 0, we receive $3(2.64 - 2.3533) = 0.86$. At expiration, our payoff from the asymmetric put butterfly spread is always non-negative.

- *Strategy 2 (Call butterfly spread and put butterfly spread price equality / Put-call parity):* Note that the (asymmetric) call 20-25-35 butterfly spread costs $2C(20) - 3C(25) + C(35) = 2(5.16) - 3(3.59) + 2.45 = 2$ whereas the (asymmetric) put 20-25-35 butterfly spread costs -0.86. To exploit the mispricing, we buy the put 20-25-35 butterfly spread and sell the call 20-25-35 butterfly spread. At time 0, we receive $2 - (-0.86) = 2.86$. At expiration, our payoff is exactly zero.

□

Remark. (i) Note that if put-call parity holds, then the European call prices satisfy the convexity inequality if and only if the corresponding European put prices satisfy the convexity inequality. In this problem, the put prices violate convexity, but the call prices do not. In other words, the three pairs of call-put prices do not conform to put-call parity.

(ii) That the call butterfly spread must cost the same as an otherwise identical put butterfly spread is a consequence of put-call parity. Strategy 2, in essence, is based on put-call parity.

Bibliography

Back, K. (2005), *A Course in Derivative Securities*, Springer

Bean, M. A. (2018), Actuarial applications of options and other financial derivatives, Investment and Financial Markets Exam Study Note IFM-22-18, Society of Actuaries.

Black, F. and Scholes, M. (1973), 'The pricing of options and corporate liabilities', *The Journal of Political Economy* **81**(3), 637–654.

Garman, M. B. and Kohlhagen, S. W. (1983), 'Foreign currency option values', *Journal of International Money and Finance* **2**(3), 231–237.

Gerber, H. U. and Shiu, E. (1994), 'Option pricing by Esscher transform', *Transactions of the Society of Actuaries* **46**, 99–191.

Geske, R. (1979), 'The valuation of compound options', *Journal of Financial Economics* **7**, 63–81.

Goldman, M. B., Sosin, H. B. and Gatto, M. A. (1979), 'Path dependent options: "buy at the low, sell at the high"', *Journal of Finance* **34**(5), 1111–1127.

Hogg, R. V., McKean, J. W. and Craig, A. T. (2013), *Introduction to Mathematical Statistics*, Seventh edn, Pearson.

Hogg, R. V., Tanis, E. and Zimmerman, D. (2014), *Probability and Statistical Inference*, Ninth edn, Pearson.

Hull, J. C. (2015), *Options, Futures and Other Derivatives*, Ninth edn, Pearson.

Margrabe, W. (1978), 'The value of an option to exchange one asset for another', *The Journal of Finance* **33**(1), 177–186.

McDonald, R. L. (2013), *Derivatives Markets*, Third edn, Pearson.

Musiela, M. and Rutkowski, M. (2005), *Martingale Methods in Financial Modelling*, Stochastic Modelling and Applied Probability, Second edn, Springer.

Panjer, H. H., ed. (1998), *Financial Economics: With Applications to Investments, Insurance and Pensions*, The Actuarial Foundation.

Shreve, S. E. (2004), *Stochastic Calculus for Finance II*, Springer.

Sundaram, R. K. and Das, S. R. (2016), *Derivatives: Principles and Practice*, Second edn, McGraw Hill Education, New York.

Index

Printed in the United States
by Baker & Taylor Publisher Services